Electricity and Electronics

by

Howard H. Gerrish
Professor Emeritis
Humboldt State University
Humboldt, California

William E. Dugger, Jr.
Program Area Leader
Industrial Arts Education
Virginia Polytechnic Institute
and State University
Blacksburg, Virginia

South Holland, Illinois
THE GOODHEART-WILLCOX COMPANY, INC.
Publishers

Library of Congress Cataloging in Publication Data

Gerrish, Howard H.
  Electricity and electronics

  Includes index.
  1. Electric engineering.   2. Electronics.
I. Dugger, William, joint author.   II. Title.
TK146.G44   1977   621.3   77-7937
ISBN 0—87006—232—8

INTRODUCTION

Electricity and Electronics provides an easy-to-understand source of information on basic fundamentals for the student who is interested in the rapidly progressing field of Electricity and Electronics. These fundamentals are applicable regardless of changes in circuit concepts.

ELECTRICITY AND ELECTRONICS employs a dual approach to learning:

1. Experimentation and demonstration to assure thorough understanding of principles. The instructional material is arranged to provide continuity in learning, and learning situations which are challenging to the student.

2. Practical application of the principles and skills developed by constructing inexpensive, professional appearing take-home projects. The projects have proven student appeal and have been thoroughly shop and laboratory tested.

Every effort has been made to relate the principles and projects to actual applications in home and industry.

Upon completion of this course, you should have a thorough understanding of the modern concepts of Electricity and Electronics.

Today, broad horizons in Electronics offer many opportunities for a specialized education, and a lifetime of challenging and personally rewarding employment.

Howard H. Gerrish
William E. Dugger, Jr.

CONTENTS

1 SCIENCE OF ELECTRONICS 7

2 SOURCES OF ELECTRICITY 15

3 CIRCUITS AND POWER 28

4 MAGNETISM 49

5 GENERATORS 64

6 INDUCTANCE AND RL CIRCUITS 78

7 CAPACITANCE IN ELECTRICAL CIRCUITS 96

8 TUNED CIRCUITS: RCL NETWORKS 111

9 ELECTRIC MOTORS 127

10 INSTRUMENTS AND MEASUREMENTS 143

11 VACUUM TUBES AND SEMICONDUCTORS 157

12 POWER SUPPLIES 165

13 ELECTRON AMPLIFIERS 179

14 ELECTRONIC OSCILLATORS 209

15 RADIO TRANSMITTERS 218

16 RADIO RECEIVERS 235

17 TELEVISION 256

18 ELECTRONICS IN INDUSTRY 271

19 TEST INSTRUMENTS 282

20 INTEGRATED CIRCUITS, DIGITAL AND LINEAR 292

21 CAREER OPPORTUNITIES 300

REFERENCE SECTION 305

DICTIONARY OF TERMS 317

INDEX 333

SAFETY PRECAUTIONS FOR THE
ELECTRICITY-ELECTRONICS SHOP

There is always an element of danger when working with electricity. Observe all safety rules that concern each project and be particularly careful not to contact any live wire or terminal regardless of whether it is connected to either a low voltage or a high voltage. Projects do not specify dangerous voltage levels. However, keep in mind at all times that it is possible to experience a surprising electric shock under certain circumstances. Even a normal healthy person can be injured or seriously hurt by the shock or what happens as a result of it. Do not fool around. The lab is no place for horseplay.

PROJECTS AND ACTIVITIES

1 STATIC ELECTRICITY
   EXPERIMENTS 9, 10, 11, 12
2 ELECTRICITY FROM
   A GRAPEFRUIT 15, 16
3 ELECTRICITY FROM
   NICKELS AND PENNIES 15, 16
4 VOLTAIC CELL 16
5 THERMOCOUPLE 25, 26
6 ELECTRIC STOVES 45, 46
7 SERIES AND PARALLEL CIRCUITS . . 46, 47
8 ELECTRODEMONSTRATOR 46, 47
9 LAWS OF MAGNETISM 50, 51
10 MAGNETIC LINES OF FORCE 51
11 SOLENOID SUCKING COIL 55
12 ELECTROMAGNET 54, 55
13 CIRCUIT BREAKER 58
14 DOOR BELL 58
15 BUZZER 58
16 MAGNETIC PUTTING GREEN 60, 61, 62
17 TACHOMETER 76
18 RPM METER 76
18 SELF-INDUCTION OF COIL 80
19 TRANSFORMER ACTION 87
20 INDUCTION COIL AND SHOCKER 87
21 AUTOMATIC GARAGE LIGHT
   CONTROL 107, 108, 109

22 WISE OWL PROJECT 102, 103
23 NOMOGRAPH 124, 125
24 SERIES MOTOR 130
25 SHUNT MOTOR 130
26 AC MOTOR 138
27 APPLAUSE METER 154, 155
28 TRICKLE CHARGER FOR
   AUTOMOBILE BATTERY 176, 177
29 POWER SUPPLY 177, 178
30 TELEPHONE AMPLIFIER 205, 206
31 PAGING AMPLIFIER 207, 208
32 CODE PRACTICE OSCILLATOR . . . 214, 215
33 SQUAWKER HORN 215, 216
34 ELECTROLUMINESCENT
   NIGHT LIGHT 216, 217
35 AM TRANSMITTER 232, 233, 234
36 CRYSTAL RADIO 251, 252
37 FOUR TRANSISTOR RADIO 252, 253
38 SUPERHETERODYNE RADIO 253, 254
39 TIME CONTROL CIRCUIT 273
40 RADIO CONTROL TRANSMITTER 276
41 OSCILLOSCOPE FAMILIARIZATION . . . 290
42 SIGNAL GENERATOR PRACTICE 290
43 AUDIO GENERATOR 290
44 RF GENERATOR 290
45 ELECTRODEMONSTRATOR . . 314, 315, 316

FOR YOUR INFORMATION

A commercial version of the electrodemonstrator as described in Appendix 6 of this text is available from Lab-Volt Division, Buck Engineering Co., Inc., Box 686, Farmingdale, NJ 07727. Full information may be obtained without obligation.

Chapter 1

SCIENCE OF ELECTRONICS

You are fortunate to live in an age in which the opportunity exists to study the electron. New discoveries, developments, and applications in electronics almost daily open a promising vista of unlimited opportunities for the creative scientist, as well as for the skilled technician. We are living in a truly scientific age.

THE NATURE OF MATTER

The scientist tells us that everything is made up of matter. MATTER may be defined as anything which occupies space or has mass. Also some matter differs from other matter by frequent observable characteristics or properties. It may differ in color, taste or hardness. These characteristics permit us to identify the various forms of matter. Sometimes observable characteristics will not permit identification of a substance and it is necessary to make a chemical analysis to determine its nature and behavior. Most materials we use are made of a combination of various kinds of matter or mixtures of matter.

LESSON IN SAFETY: The experimenter, the technician, the scientist, the engineer, must respect the power of electricity. Although voltage used in projects and experiments described in this book are not dangerous to the normal, healthy individual, there is never an excuse to be careless.

In the basic study of the electron theory, it must be understood that some types of elementary matter exists which are mixtures and that further effort to break this matter into parts by chemical decomposition will produce no further change in the characteristics of the matter. These simple forms of material have been called ELEMENTS. Scientists have made further subdivisions of the elements by physical forces, such as those generated in the "atom smasher." Our concern is the nature of an element. There are many of them. Some familiar elements are iron, copper, gold, aluminum, carbon and oxygen. Chemists have isolated over ninety different kinds of elements existing in nature and several more found in their laboratories. A COMPOUND is a mixture of two or more elements.

MOLECULE AND THE ATOM

If you were to take a crystal of table salt and cut it in half, you would have two parts, but both would be common salt. Salt is a chemical compound of sodium (Na) and chlorine (Cl) or NaCl. In your imagination you might conceive the further division of the crystal of salt until it could not be divided again. This question has disturbed scientists and philosophers for hundreds of years. Can a particle of matter be divided to the point that if it is divided again it will disappear?

In science the smallest particle that can exist and still retain the properties of the original compound or element is called a MOLECULE. Referring again to the crystal of salt, even though the salt has been divided to the smallest particle that can exist by itself, it is still a compound of sodium and chlorine. So this tiny particle must consist of two parts; one of sodium and one of chlorine. The molecule of salt would no longer be salt if it were divided. However, if a molecule of an element were subdivided, the parts would be all alike. These tiny particles of an element are named ATOMS. The smallness of just one molecule will stagger your imagination. If you could start filling a small match box with molecules at the rate of ten million per second, it would take you over a billion years to fill the box!

The word atom is derived from the Greek word meaning "indivisible." It was not until recent years that the atom was divided into parts. It is not within the scope of this text to describe the chemical and physical reactions that take place when an atom is "split." However, it is important to learn some things about the structure of an atom.

ELECTRONS – PROTONS – NEUTRONS

Physicists have discovered that atoms are composed of minute particles of electricity. The center of each atom or NUCLEUS contains positively charged particles of electricity called PROTONS, and also neutral (neither positive or negative) particles. Most of the mass or weight of the atom is in its nucleus. Surrounding the nucleus, in rapidly revolving orbits, are negatively charged particles of electricity called ELECTRONS. The structure of the atom may be compared to the solar system, where the sun is at the center, or nucleus. The planets (Earth, Venus, Mars, etc.) revolving around the sun may be compared to the whirling electrons in orbit. The whirling electrons or planetary electrons are kept from falling into the center by centrifugal force. Normally for each proton in the nucleus there is one electron in orbit. The positive and negative charges cancel, leaving the atom electrically neutral. The atomic structure of each element may be described as having a fixed number of electrons in orbit. Examples of the atomic structure of two common elements are displayed in Fig. 1-1. All elements are arranged in order in the Periodic Table of Elements according to their basic ATOMIC NUMBER, which is the number of electrons in the orbits. Or they may be arranged by their ATOMIC WEIGHT, which is approximately the number of protons and neutrons in the nucleus. Referring again to Fig. 1-1, the atomic weight of hydrogen is one (scientifically 1.008), and its atomic number is one. The atomic weight of oxygen is sixteen; its atomic number is eight.

IONIZATION

Usually an atom remains in its normal state unless energy is added to the atom by some exterior force such as heat, friction or bom-

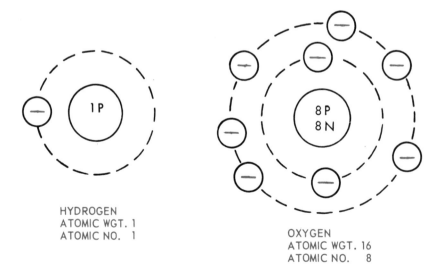

HYDROGEN
ATOMIC WGT. 1
ATOMIC NO. 1

OXYGEN
ATOMIC WGT. 16
ATOMIC NO. 8

Fig. 1-1. The atomic structure of hydrogen and oxygen.

bardment by other electrons. When energy is thus added to the atom, it will become excited and electrons which are loosely bound to the atom in the outer rings or orbits may leave the atom. If electrons leave the balanced or neutral atom, then there is a deficiency of electrons and the atom is no longer in electrical balance. The atom is then said to be IONIZED. If the atom has lost electrons, which are negative particles of electricity, the atom then would be a positively charged or positive ion. If the atom gained some electrons, then the atom would be negatively charged or a negative ion. This is an important lesson in electronic theory.

STATIC ELECTRICITY — LAW OF CHARGES

The word static means "at rest." Electricity can be at rest. The generation of static electricity may be demonstrated in many ways. By stroking the fur of a cat, you will notice that its fur will tend to be attracted to your hands as you bring your hand back over the cat. If this is done at night, you may see tiny sparks and hear a cracking sound which is caused by the discharge of static electricity. Actually the friction between your hand and the cat is exciting the atoms and they are becoming unbalanced. The sparks are created when the atoms attempt to neutralize themselves. You may have experienced the building-up of a static charge in your body while walking across a carpet or sliding across the seats of your car. Sometimes a spark discharge may result when you touch another person or the door handle of a car.

One of the fundamental laws in the study of electricity is the Law of Charges.

LIKE CHARGES REPEL EACH OTHER
UNLIKE CHARGES ATTRACT EACH OTHER

This law will be used many times in experiments described in this text. It is important to remember that <u>ONE ELECTRON WILL REPEL ANOTHER ELECTRON</u>. A negatively charged mass will be attracted by a positively charged mass. The Law of Charges may be demonstrated

in the electrical laboratory by means of an electroscope or pith balls, Fig. 1-2.

Materials needed:

1. Electroscope.
2. Vulcanite Rod and Cat Fur.
3. Glass Rod and Silk.
4. Single and Double Pith Balls on Stands.

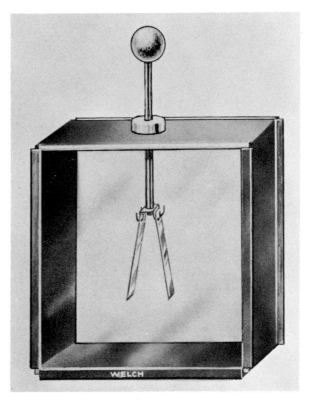

Fig. 1-2. An electroscope. The gold leaves in the glass enclosure will indicate an electric charge.
(Welch Scientific Co.)

EXPERIMENT I. Rub the vulcanite rod fast with the cat fur. You now have a negative charge on the rod as electrons were transferred to the rod from the fur by friction. Bring the charged rod close to the ball on top of the electroscope. You will notice that the gold leaves in the electroscope will expand. Remove the rod and the leaves will close. The electroscope is used in this case to show the presence of an electrical charge. The negatively charged rod repels the electrons on the ball of the electroscope and forces them down to the leaves. The leaves being both negatively charged will repel each other and expand, Fig. 1-3.

EXPERIMENT II. Once again rub the vulcanite rod in order to charge it. Now touch the ball on the electroscope with the rod. You will notice that the leaves expand and <u>remain in this manner after the rod is removed</u>. The rod has

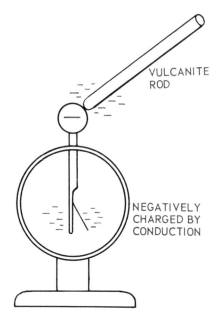

Fig. 1-4. The electroscope becomes charged when the rod touches the ball, and will remain charged.

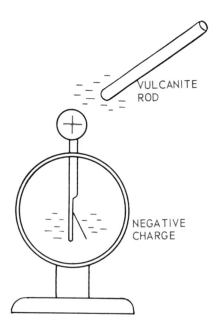

Fig. 1-3. The negatively charged rod forces electrons down to the leaves of the electroscope. The leaves expand because they are both negative and repel each other.

shared its charge with the electroscope and the electroscope remains charged, Fig. 1-4. Now touch the ball with your finger and the leaves will close, showing that the electroscope has been discharged through your body.

EXPERIMENT III. Repeat experiments I and II, using the glass rod and the silk. The glass rod will charge positively; however the results will be the same, with opposite polarity.

EXPERIMENT IV. Repeat experiment I by bringing the charged <u>rod close to, but not touching</u>, the electroscope. In this position, touch the electroscope with your other hand. Remove your hand and then the rod. You will observe that the electroscope will remain

charged. As no actual contact was made between the rod and the electroscope, the electroscope was charged by <u>induction.</u> <u>The nearness of the negative rod</u> forced the electrons on the electroscope to travel away, through your hand. Of course, when your hand was removed, there was no way for the electroscope to regain its lost electrons so it must remain positively charged as indicated by the expanded leaves.

EXPERIMENT V. Once again, rub the vulcanite rod and charge it negatively. Bring the rod close to a hanging pith ball, Fig. 1-5. The ball will at first be attracted to the rod due to unlike charges. When the pith ball touches the rod, it assumes the charge of the rod and is immediately repelled. Now both the ball and rod are negative. Try once again to touch the ball with the rod. Notice that the ball is always <u>repelled</u>.

EXPERIMENT VI. Using a second pith ball, charge it in the same manner as Experiment V. You will now have two pith balls charged negatively. Try to bring the balls close together and notice that they will not touch. There is a repulsive force existing between then, Fig. 1-6.

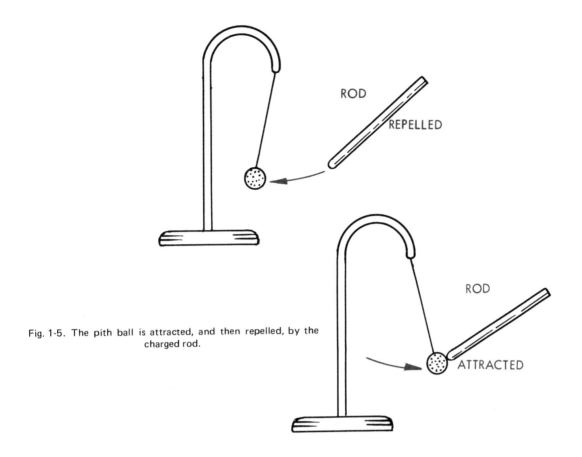

Fig. 1-5. The pith ball is attracted, and then repelled, by the charged rod.

EXPERIMENT VII. Leaving one pith ball negative, charge the second pith ball positively by using the glass rod and the silk. As you bring the two balls close to each other, you will notice that they are attracted to each other, Fig. 1-7.

These seven experiments prove the Law of Charges. You must know this law, or your lessons in electricity will be hard to understand.

THE COULOMB

In the earlier part of this lesson it was indicated that the atom and parts of the atom must be very, very small particles. It would not be practical to assume that the charge on a body is one or even a dozen more or less electrons. A larger quantity would be more convenient to use. A quantity of electricity is known as a COULOMB and represents

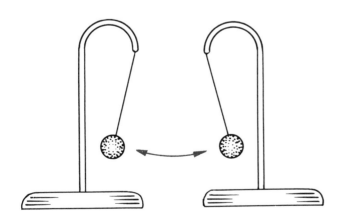

Fig. 1-6. The like charged pith balls repel each other.

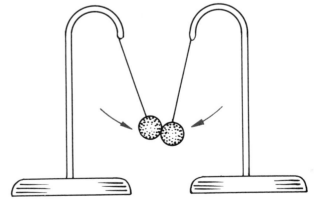

Fig. 1-7. Unlike charged pith balls attract each other.

approximately 6.24 x 10^{18} electrons or 6,240,000,000,000,000,000 electrons. The French scientist, Charles A. Coulomb, discovered many of the principles and laws of electrical charges. The Coulomb was named in his memory.

ELECTROSTATIC FIELDS

Once again perform Experiment IV and ask this question. When the vulcanite rod was brought close to, but not touching the electroscope, why did it affect the electroscope? This is so because of the invisible lines of force existing around a charged body. These lines of force are called the ELECTROSTATIC FIELD or DIELECTRIC FIELD. The field is strongest very close to the charged body. It diminishes at a distance inversely porportional to the square of the distance. Referring to Fig. 1-8, two charged balls are shown with lines representing the electrostatic fields of opposite polarity and the attractive force existing between them. In Fig. 1-9, two charged balls are shown with like

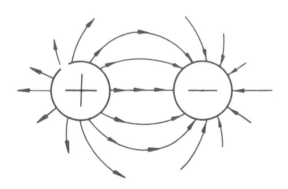

Fig. 1-8. The electrostatic fields of unlike charged bodies, showing attractive force.

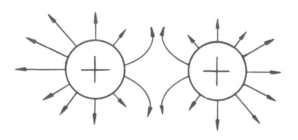

Fig. 1-9. The electrostatic fields of like charged bodies, showing repulsive force.

polarities and the repulsive force existing between the balls due to the electrostatic fields. Once again, the electrostatic fields add further proof to the Law of Charges. Like charges repel. Unlike charges attract.

LESSON IN SAFETY: Horseplay has no place in the electricity shop, nor in any shop where machinery is operating and tools are being used. Do not pay the penalty for carelessness or poor working habits. It may be painful and very expensive.

ELECTRIC CURRENT

At this point, you should have a thorough understanding of static electricity and the electrostatic fields existing around charged bodies. Consider the illustration in Fig. 1-10 which is very similar to previous Fig. 1-8,

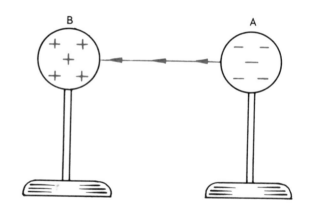

Fig. 1-10. Electrons flow from the negatively charged ball to the positive ball.

except that a short piece of copper wire has been connected between the balls, A and B. Again, you should realize what is meant by a negatively charged body such as ball A. It actually has an excess of negative charges of electrons. In the case of ball B, it will have a lack of electrons because of its positive charge. When the wire or CONDUCTOR is connected between the two balls, then the excess electrons on ball A will follow along the conductor to ball B. Electrons will continue to flow until the electron on <u>both balls are equal</u>. Then there is no difference in electrons between them. The force which caused these electrons to move is called <u>electromotive force</u> (emf) or <u>difference in potential</u>. It is known as VOLTAGE.

Voltage is the electrical pressure or force which causes the flow of electrons. It is measured in VOLTS.

This flow of electrons is called an ELECTRIC CURRENT. It is necessary for the technician to measure the flow of electrons or current in a circuit, so units of measurements have been standarized. Comparing the electrical circuit to a water pipe, one would measure the flow of water as so many gallons per minute. Electric current is measured by the number of electrons flowing per second past a given point in the conductor. The term coulomb is useful as a representation of a quantity of electrons. The standard unit for the measurement of electric current is the AMPERE. This represents the flow of one coulomb per second past a fixed point in the circuit. Mathematically,

$$I = \frac{Q}{t} \text{ or } Q = It$$

where Q = coulombs

I = current in amperes

t = time in seconds

Example: If 40 coulombs of electrons passed a given point in 10 seconds, then a current of 4 amperes would flow.

$$I = \frac{40 \text{ coulombs}}{10 \text{ seconds}} = 4 \text{ amperes}$$

DIRECTION OF CURRENT FLOW

It is apparent that the flow of electrons then makes an electric current. But <u>electrons are negative charges of electricity</u>, and therefore are attracted to the positively charged body. One may assume that electron flow is in the direction of negative to positive. This may cause some misunderstanding because many older texts describe electric current as flowing from positive to negative. This stems from an early belief that electricity was some kind of "liquid" or "juice" flowing in the wires. Frequently one will still hear an electrician say, "Turn on the juice." For the purposes of this text and this study of electronics, current <u>will always be considered as flowing from negative to positive</u> in external circuits. Any exceptions will be clearly stated.

CONDUCTORS AND INSULATORS

If a source of electrical voltage is connected to a device, such as a lamp, by means of copper wires, the electrons will flow or be conducted from the negative terminal of the source, through the lamp, and back to the positive terminal of the source. These copper wires are the path or road along which will flow the electric current. The wire is called a CONDUCTOR. Some materials are good conductors of electricity. Materials which have a large number of free electrons are considered good conductors. The actual conduction is accomplished by transferring electrons from one atom to the next in the conductor. One may visualize this flow by assuming that a piece of copper wire is neutral. If an electron is forced onto one

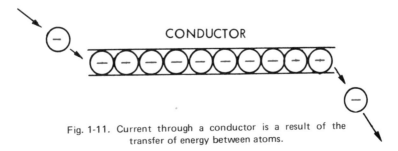

Fig. 1-11. Current through a conductor is a result of the transfer of energy between atoms.

end of the wire, an electron is forced off from the other end, as illustrated in Fig. 1-11. The original electron did not flow through the conductor, yet the energy was transferred by the interaction between the electrons in the conductor. The actual transfer of electrical energy has been accurately measured and approaches the speed of light or 186,000 miles per second. In terms of the metric system, this would be 300,000 kilometres per second or 300,000,000 metres per second.

Copper conductors are used in most electrical circuits. Copper is an excellent conductor. Silver is even better than copper, but it is too expensive for general use. Many high power lines use aluminum for added strength and light

weight. Brass, zinc and iron are also fair conductors of electricity.

Materials which have only a few free electrons do not readily conduct electrons and are called INSULATORS. Insulation is used on wires to protect personnel from accidental contact with circuits and to avoid electrical contact between wires and supporting devices. Common materials used for insulation are rubber, glass, air, bakelite, mica and asbestos. It is apparent then, that some materials resist the flow of electrons, while others have less resistance to the flow of electrons. This is true and resistance must always be considered as an important characteristic of electrical circuits. Resistance will be studied in Chapter 3.

FOR DISCUSSION

1. How do lightning rods protect a building during a storm?
2. Which great scientist discovered the relationship between lightning and electricity? Describe his experiments.
3. A police car seems to have some unusual radio interference while driving along the highway. This interference was not present when the car was not moving. Suggest a possible cause and explain.
4. Electrostatic fields are used in industry to reclaim by-products and prevent air pollution from smoke. Explain.

TEST YOUR KNOWLEDGE

Write your answers to these questions on a separate sheet of paper. Do not write in this book.

1. A negative charge of electricity is called an _____ ; a positive charge, a _____ a neutral particle, a _____ .
2. A unit of measurement for a quantity of electricity is a _____ and represents _____ x 10^{18} electrons.
3. What are the two laws of electrostatic charges?

4. A negative rod will charge an electroscope by contact. The leaves will both be _____ .
5. A positive rod will charge an electroscope by induction. The leaves will be _____ .
6. Electricity moves at the speed of light which is _____ miles per second or _____ metres per second.
7. Name four insulators. Why are insulators used?
8. When a charged rod is brought close to a pith ball, explain the action. Why?
9. Electromotive force or potential difference is measured in _____ .
10. Current is measured in _____ and represents the flow of _____ per _____ .
11. Check in a reference book in your library and discover the atomic weight and atomic number of:

	weight	number
Cu Copper	_____	_____
Al Aluminum	_____	_____
Ge Germanium	_____	_____
As Arsenic	_____	_____

12. A charged body will be surrounded by an _____ .

Chapter 2

SOURCES OF ELECTRICITY

Every young man is familiar with the story of Benjamin Franklin and his kite. For centuries before Franklin, scientists and philosophers had observed the supernatural manifestations of lightning. It was through the experimentation and research of Dr. Franklin that the relationship between lightning and static electricity was confirmed. What is electricity and where does it come from? Years before the discovery of the electron theory by J. J. Thomson, it was suggested by Dr. Franklin that electricity consisted of many tiny particles or electric charges. He further theorized that electrical charges were created by the distribution of electrical particles in nature.

In the previous lesson, we learned that a potential difference or electromotive force (emf) was created when electrons were redistributed. A body might assume a charge; its polarity is determined by the deficiency or excess of electrons. Man since has bent his scientific interests and research on the development of machines and processes which will cause an electrical unbalance and an electrical pressure.

THE BATTERY

One of the more familiar sources of an electrical potential or voltage is the battery. In 1790, the Italian scientist, Galvani, observed a strange phenomena during the dissection of a frog which was supported on copper wires. Each time he touched the frog with his steel scalpel, its leg would twitch. Galvani reasoned that the frog's leg contained electricity. As a result of these experiments, Alesandra Volta, another Italian scientist, invented the electric cell, named in his honor, called the Voltaic Cell. The unit of electrical pressure, the volt, is also named in his honor. Volta discovered that when two dissimilar elements were placed in a chemical which acted upon them, an electrical potential was built up between them.

The student may construct several voltaic cells to demonstrate this action. Cut a one inch square of blotting paper and soak it in a strong salt solution. Place the wet paper between a penny and a nickel as shown in Fig. 2-1. If a

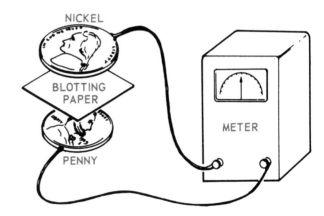

Fig. 2-1. A simple cell is produced by a nickel, a penny and a salt solution.

sensitive meter is connected to the coins, it will indicate that a small voltage is present. In Fig. 2-2 electricity is created by a grapefruit. Small cuts are made in the skin of a grapefruit. In one cut, place the penny; in the other, the nickel. Once again a meter will indicate a voltage. A

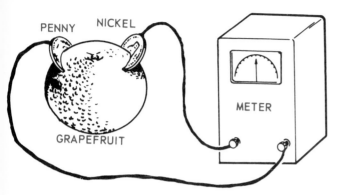

Fig. 2-2. A grapefruit will produce enough electricity to operate a small transistor radio.

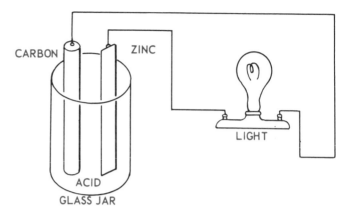

Fig. 2-3. An experimental cell made by zinc, carbon and acid.

better cell may be made by placing a carbon rod (these may be removed from an old dry cell) and a strip of zinc in a glass jar containing an acid and water as illustrated in Fig. 2-3.

LESSON IN SAFETY: When mixing acid and water, always pour acid into water. Never pour water into acid. Acid will burn your hands and eat holes in your clothing. Wash your hands at once with clear water if you have spilled some acid on them. Acid may be neutralized with baking soda. See your instructor for first aid treatment!

If the polarity of the carbon rod is tested, it will be found to be positive. The zinc will be negative. If a wire is connected between these elements or electrodes, a current will flow. A VOLTAIC CELL may be described as a means of converting chemical energy into electrical energy.

The student in electronics also needs a basic knowledge of chemistry. In the above mentioned example of a voltaic cell, sulfuric acid, (H_2SO_4) and water (H_2O) may be used as the liquid or electrolyte. When the electrodes are placed in this acid electrolyte, a chemical action takes place. The sulfuric acid breaks down into positive ions (H_2) and negative ions (SO_4). The negative ions move toward the zinc electrode, and combine with it by making zinc sulfate

($ZnSO_4$) and the positive ions move toward the carbon electrode. This action creates a potential difference between the electrodes. The zinc will have an excess of electrons and will be negative. The carbon will be positive. Such a cell will develop about 1.5 volts.

If a load, such as a light, is connected to the cell, a current will flow and the light will glow, as shown in Fig. 2-3. As the cell is used, the chemical action continues until the zinc electrode is consumed. The chemical equation for this action would be:

$$Zn + H_2SO_4 + H_2O \rightarrow ZnSO_4 + H_2O + H_2 \uparrow$$

Zinc plus sulfuric acid plus water chemically reacts to form zinc sulfate and water and free hydrogen gas. This cell cannot be recharged because the zinc has been consumed. It is called a PRIMARY CELL. The chemical action cannot be reversed.

DEFECTS IN PRIMARY CELLS

One might think that the chemical action of the voltaic cell would continue to develop a voltage as long as the active ingredients of the cell were present. In studying the equation for the discharge of the cell, you will observe the formation of free hydrogen gas. Since the carbon electrode does not enter into chemical action, the hydrogen forms bubbles of gas which collect around the carbon electrode. As the cell continues to discharge, an effective insulating blanket of bubbles will form around the carbon, which reduces the output and terminal voltage of the cell. The cell is said to be POLARIZED. The action is called POLARIZATION. To overcome this defect in the simple voltaic cell a DEPOLARIZING AGENT may be added. Compounds which are rich in oxygen, such as magnanese dioxide (MnO_2), are used for this purpose. The oxygen in the depolarizer combines with the hydrogen bubbles and forms water. This chemical action appears as:

$$2 MnO_2 + H_2 \rightarrow Mn_2O_3 + H_2O$$

The free hydrogen has been removed, so the cell will continue to produce a voltage.

One might assume that when current is not being used from the cell, the chemical action would also stop. However, during the smelting of zinc ore, not all impurities are removed. Small particles of carbon, iron and other elements remain. These impurities act as the positive electrode for many small cells within the large cell, but this chemical action adds nothing to the electrical energy produced at the cell terminals. This action is called LOCAL ACTION. It may be reduced by using pure zinc for the negative electrode, or by a process called amalgamation. A small quantity of mercury is added to the zinc during manufacturing. As mercury is a heavy liquid, any impurities in the zinc will float on the surface of the mercury, causing them to leave the zinc surface. This process increases, to a great extent, the life of a primary cell.

THE DRY CELL

Although the action of the primary cell has been described as a liquid cell or wet cell, it is not in common use. It is more convenient to make the primary cell a DRY CELL. This avoids the danger of spilling liquid acids. The common dry cell known as the LeClanche cell, is shown in Fig. 2-4. The dry cell consists of a zinc container which acts as the negative electrode. In the center is a carbon rod which is the positive electrode. Surrounding the carbon rod is a paste of ground carbon, manganese dioxide and sal ammoniac, (ammonium chloride),

mixed with water. The depolarizer is the MnO_2. The ground carbon increases the effectiveness of the cell by reducing its internal resistance. During discharge of the cell, water is formed. You may recall having difficulty removing dead cells from a flashlight. The water caused them to expand. Although this complaint has been removed by improved manufacturing techniques, it is still not advisable to leave cells in your flashlight for long periods of time. You should keep fresh cells in your flashlight, so it will be ready for emergency use.

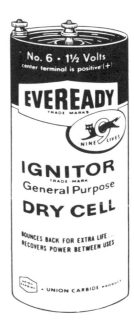

Fig. 2-4. The popular dry cell still serves many purposes in home and industry. (Union Carbide Co.)

THE MERCURY CELL

A relatively new development in the dry cell is shown in Fig. 2-5. It is called a mercury cell. It develops a voltage of 1.34 volts by the chemical action between zinc and mercuric oxide. It is expensive to manufacture, but has a decided advantage of producing approximately five times more current than the conventional dry cell. It also maintains its terminal voltage under load for longer periods of operation. The mercury cell has found wide application in powering field instruments and portable communications systems.

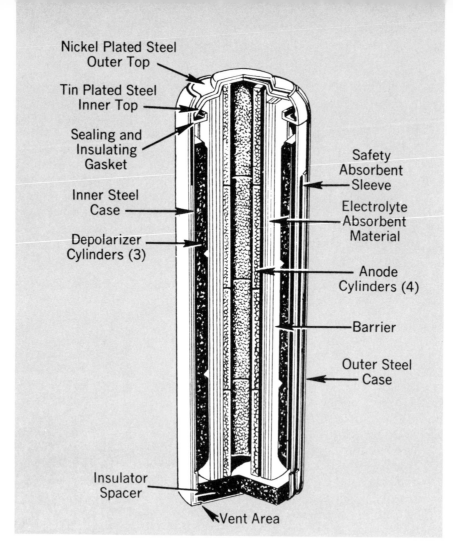

Fig. 2-5. Mercury cell battery which develops voltage by chemical action between zinc and mercuric oxide.

Labels on figure:
- Nickel Plated Steel Outer Top
- Tin Plated Steel Inner Top
- Sealing and Insulating Gasket
- Inner Steel Case
- Depolarizer Cylinders (3)
- Safety Absorbent Sleeve
- Electrolyte Absorbent Material
- Anode Cylinders (4)
- Barrier
- Outer Steel Case
- Insulator Spacer
- Vent Area

NICKEL-CADMIUM RECHARGEGABLE CELL

Another new development in the field of dry cells is the rechargeable cell. These are essentially, nickel-cadmium alkaline batteries using a paste rather than a liquid for the electrolyte. One brand uses plates of perforated nickel-plated steel on which a porous body of nickel has been sintered. The plates are impregnated with nickel hydrate for the positive and cadmium hydrate for the negative plates. Among the many advantages of these cells is their ability to be recharged. Several of these rechargeable cells are illustrated in Fig. 2-6. Other advantages include long life, high efficiency, compactness and light weight. This nickel-cadmium cell will produce a high discharge current due to its low internal resistance. A flashlight using one of these rechargeable cells is

Fig. 2-6. Rechargeable cells using nickel and cadmium hydrate are becoming quite popular. (Gulton Industries, Inc.)

shown in Fig. 2-7. Other uses include the powering of transistor radios, burglar alarm system, photo-flashers, missiles and aircraft instruments.

Fig. 2-7. This battery powered flashlight may be recharged by plugging into any 115-volt receptacle in your home. (Gulton Industries, Inc.)

THE BATTERY

Frequently a single cell is called a battery. By strict definition, a battery consists of two or more cells connected together. These cells are enclosed in one case. In the study of electricity it is important to understand the purpose and results of connecting cells in groups. First, consider the series connection. In this method the positive terminal of one cell is connected to the negative terminal of the second cell. In Fig. 2-8, four cells are connected in SERIES. The output voltage will equal,

$$E_{out} = E_{one\,cell} \times n,$$

when n equals the number of cells, so

$$E_{out} = 1.5 \text{ volts} \times 4 = 6 \text{ volts}$$

Notice that the voltage has increased four times, however, the capacity of the battery to supply a current is the same as one cell. Cells are connected in this manner to supply higher voltages for many applications. Your flashlight may use two or more cells in series. B batteries for portable radios have many cells in series to produce 22 1/2, 45 or 90 volts.

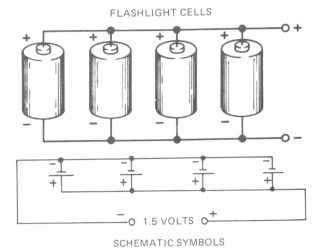

Fig. 2-9. The pictorial and schematic diagram of four cells connected in parallel.

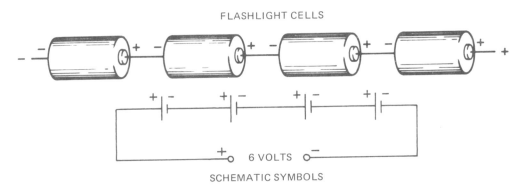

Fig. 2-8. The pictorial and schematic diagram of four cells connected in series.

19

In Fig. 2-9, the positive terminals have been connected together, and the negative terminals have been connected together. These cells are connected in PARALLEL. The total voltage across the terminals of the battery is the same as one cell only. Although the voltage has not increased, the life of the battery has been increased because the current is drawn from all cells instead of one. Generally speaking, if the load applied to the battery is a low resistance load, it is better to use the parallel connection. For a high resistance load, use the series connection.

It is also possible to connect cells in a mixed grouping. In Fig. 2-10, two groups of batteries with 6 volts terminal voltage are connected in parallel. Total voltage is still 6 volts, but the capacity has been increased by this SERIES-PARALLEL method of connecting cells.

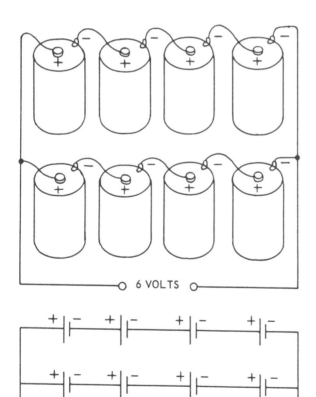

Fig. 2-10. The pictorial and schematic diagram of two groups of cells connected in series and the groups connected in parallel.

SECONDARY CELLS

The secondary cell may be recharged or restored. Chemically speaking, the reaction which occurs on discharge may be reversed by forcing a current through the battery in the opposite direction. This charging current must be supplied from another source, which may be a generator or the power company. Fig. 2-11 shows one type of battery charger used by garages and filling stations for recharing automobile batteries. An alternating current, which will be studied in a later lesson, must be RECTIFIED to a direct current for charging the battery.

Fig. 2-11. A quick charger for the automotive battery found in many service stations. (Exide Sales, Electric Storage Battery Co.)

THE LEAD ACID CELL

The more familiar example of the lead-acid cell is the automotive storage battery. A storage battery does not store electricity. It stores chemical energy which produces electrical energy. The active ingredients in a fully charged lead-acid cell consist of lead peroxide (PbO_2) which acts as the positive plate, and pure spongy lead (Pb) for the negative plate. The liquid electrolyte is sulfuric acid (H_2SO_4) and water (H_2O). One may distinguish the positive plates by their reddish-brown color. The negative plates are gray.

Although the chemical reaction is rather involved, the student may study the following equation, with the purpose of understanding, the general equation is:

Spongy Lead + Lead Peroxide + Sulfuric Acid $\rightleftharpoons$ Lead Sulfate + Water

or

$$Pb + PbO_2 + 2H_2SO_4 \rightleftharpoons 2PbSO_4 + 2H_2O$$

CHARGED   DISCHARGED

Notice that during discharge, both the spongy lead and the lead peroxide plates are being changed to lead sulfate and the electrolyte is being changed to water. When the cell is recharged, the reverse action occurs. The lead sulfate changes back to spongy lead and the lead peroxide. The electrolyte changes to sulfuric acid.

LESSON IN SAFETY: During the charging process of a storage battery, highly explosive hydrogen gas may be present. Do not light matches near charging batteries. Charge only in a well ventilated room. Batteries should be first connected to the charger, before the power is applied. Otherwise, the sparks made during connection might ignite the hydrogen gas and cause an explosion.

The electrolyte of a fully charged battery is a solution of sulfuric acid and water. The weight of pure sulfuric acid is 1.835 times heavier than water. This is called its SPECIFIC GRAVITY. The specific gravity of water is 1.000. The acid and water mixture in a fully charged battery has a specific gravity of approximately 1.300 or less. As the electrolyte changes to water when the cell discharges, the specific gravity becomes approximately 1.100 to 1.150. Therefore, the specific gravity of the electrolyte may be used to determine the state of charge of a cell. The instrument used to measure the specific gravity is a HYDROMETER, as shown in Fig. 2-12.

Fig. 2-12. At your garage or service station, the mechanic will use the hydrometer to check the state of charge of your car battery. (Exide Sales, Electric Storage Battery Co.)

Readings may be taken from the floating bulb and scale. The principle of the hydrometer is based on Archimedes principle in physics, which states that a floating body will displace an amount of liquid equal to its own weight. If the cell is in a fully charged state, the electrolyte liquid is heavier, so the float in the hydrometer will not sink so far. The distance that it does sink, is calibrated in specific gravity on the scale, and can be read directly as the state of charge of the cell.

Referring to Fig. 2-13, the plates of the modern lead-acid cell are made of a grid of

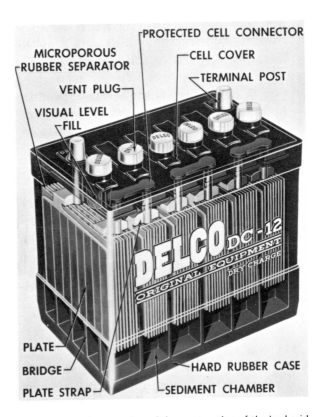

Fig. 2-13. A phantom view of the construction of the lead-acid storage battery. (Delco-Remy)

special lead-antimony alloy upon which is pressed a paste of the chemically active material. The grid holds the material in place and adds mechanical strength. When the paste on the plate grids is hard and dry, each plate is formed to its correct chemical material by placing it in an electrolyte, and passing a current through it in the proper direction. One direction makes a positive plate; the opposite direction, a negative plate. In Fig. 2-13 is shown the construction of a cell, which consists of alternate negative and positive plates grouped together. All negative plates are connected together and all positive plates are connected together by a lead plate strap at the top. The plates are insulated from each other by means of separators. These separators must be thin, porous layers of insulation and are usually made of wood, rubber or spun glass.

LESSON IN SAFETY: Sulfuric acid used in automotive storage batteries will burn your hands, eat holes in your clothes and generally destroy other materials which it contacts. When testing a cell with a hydrometer, be certain that small drops of acid at the end of the rubber tube do not drip on you or on nearby objects of value. Hold the hydrometer over the battery filler hole while taking readings. Store hydrometer in hanging position in glass or rubber jar.

LESSON IN SAFETY: Expensive storage batteries may be destroyed by excessive vibration and rough handling. Chemicals may break off from the plates, and cause internal short circuits and dead cells. Handle a battery gently and be sure it is securely clamped and bolted in your car.

In the 12 volt automotive battery, six of these cells are placed in a molded hard rubber case. Each cell has its own individual compartment. At the bottom of each compartment a space sediment chamber is provided, so that particles of chemicals broken from the plates

due to chemical action or vibration, may collect. Otherwise, these particles might short-out the plates and make a dead cell.

The individual cells are connected in series by lead alloy connectors. The entire battery is then sealed with a battery-sealing compound. Each cell is provided with a filler cap, so that distilled water may be added to the electrolyte as it evaporates. Special care of a battery includes a periodic check of the liquid level in each cell. It should be maintained at about 1/4 in. above the top edges of the plates. If the electrolyte does not completely cover the plates, the ends of the plates exposed to air will become seriously sulfated or coated with a hard inactive chemical compound. Badly sulfated batteries have a short life, so protect your investment by giving your battery good care.

LESSON IN SAFETY: Storage batteries are heavy. Use the proper carrying strap when moving a battery. Get help from your fellow worker when lifting a battery into a car. Do not strain yourself by improper and thoughtless lifting.

BATTERY CAPACITY

So that batteries may be purchased and used more intelligently, you should understand the term "capacity." The capacity of a battery is its ability to produce a current over a certain period of time. It is equal to the product of amperes supplied by the battery and the time. Capacity is measured in ampere-hours (AH). The description of an automotive battery might indicate a capacity of 100 ampere-hours. This would mean that the battery theoretically would supply,

100 amps for 1 hr.   100 x 1 = 100 amp hrs.
50 amps for 2 hrs.   50 x 2 = 100 amp hrs.
10 amps for 10 hrs. 10 x 10 = 100 amp hrs.
1 amp for 100 hrs.   1 x 100 = 100 amp hrs.

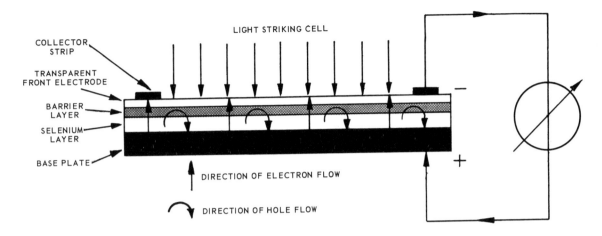

Fig. 2-14. Typical cell structure of a photovoltaic cell.
(International Rectifier Corp.)

A battery will not perform exactly by this theoretical schedule, as the rate of discharge must always be considered. A rapidly discharged battery will not give its maximum ampere-hour rating. A slowly discharged battery may exceed its rated capacity. The Society of Automotive Engineers (SAE) has set standards for the rating of automotive batteries. A manufacturer must meet these standards in order to advertise a battery as a specific ampere-hour capacity.

Several factors determine the capacity of a storage battery:

1. The number of plates in each cell. An increased number of plates provides more square inches of surface area for chemical action. Automotive batteries are commonly made with 13, 15, and 17 plates per cell. The number of plates is a determination of the life and quality of a battery.
2. The kind of separators used will have some effect on the capacity and life of a battery.
3. The general condition of the battery in respect to its state of charge, age and care will influence the capacity rating of any given battery.

ELECTRICAL ENERGY FROM LIGHT

For many years scientists have attempted to transform the light energy from the sun into useful quantities of electrical power. Although certain experimental equipment has been successful, general use of sun-power is still in its infancy. Just as light may be produced by an electric current running through a resistance or filament in a light bulb, light may also produce an electric current. The PHOTOVOLTAIC CELL is one device which will make this conversion. This may be described as a device which will develop an electrical potential between its terminals, when the surface of the cell is exposed to light. Referring to Fig. 2-14, a selenium photovoltaic cell is constructed of a layer of selenium compound, a barrier layer, and transparent front electrode. When light strikes the surface, electrons from the selenium compound cross the barrier, causing a potential difference between the base plate and the front electrode. The electrons cannot return to the selenium compound because the barrier permits conduction in one direction only. It is a unidirectional conductor.

Fig. 2-15 shows sun batteries used by experimenters. The batteries may be connected in series and parallel for higher power requirements.

Other interesting applications of photo cells include the familiar light meter used in photography. In this application, a sensitive meter responds to the current generated by the cell. The brighter the scene, the greater deflection of

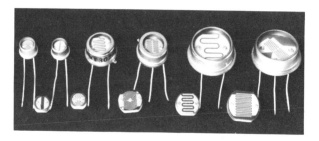

Fig. 2-15. Selenium photovoltaic cells, mounted and un-mounted, with and without leads. (Vactec Inc.)

the meter. In Fig. 2-16 are some photosensitive cells which are used to control the amount of light reaching the film. Photocells are used in industry for operations of counting, sorting and automatic inspection.

Fig. 2-16. Photocells. Various sizes — hermetic packages and plastic coated. (Vactec Inc.)

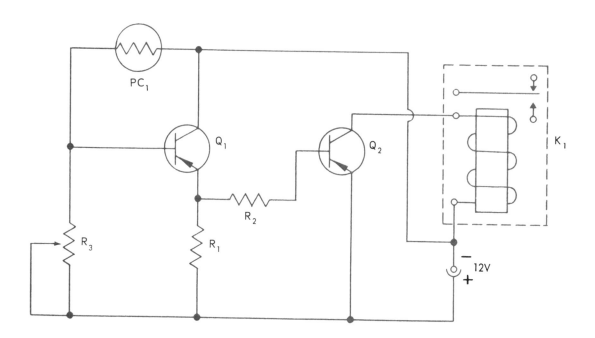

PHOTOELECTRIC CONTROL

PARTS LIST

R_1, R_2 — 100 Ohms, 1/2 W
R_3 — 1000 Ohms, Potentiometer
Q_1, Q_2 — 2N408, Transistors
K_1 — 4.6 mA, Relay
PC_1 — Photocell, (Sylvania 8143)

Fig. 2-17. Schematic and parts list for photoelectric control.

PHOTOELECTRIC CONTROL

As a student of electronics, you should build several devices which utilize the phenomenon of electrical energy from light. Fig. 2-17 shows a photoelectric control device using a photocell. When light shines on the cell, the resistance of the circuit changes. A cadmium-sulphide photocell is a light variable resistor which is most sensitive in the green to yellow portion of the light spectrum. With it you can use light to control many electronic devices. Photocells are used in counting operations, burglar alarms, door opening mechanisms and in many other devices.

ELECTRICAL ENERGY FROM HEAT

A simple device used to indicate and control the heat of electric ovens and furnaces is shown in Fig. 2-18. This is called a THERMOCOUPLE. When two dissimilar metals in contact with each other are heated, a potential difference will develop between the metals.

This simple experiment will demonstrate the electrical principle. Take a piece of iron stove pipe wire and a piece of copper wire. Twist the ends of the wire tightly together and connect the other ends to a sensitive galvanometer. Light a match and hold the flame under the

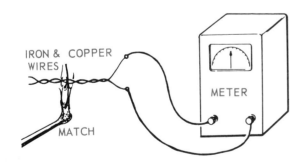

IRON & COPPER WIRES

METER

MATCH

Fig. 2-18. The basic principle of the thermocouple may be demonstrated by heating two dissimilar wires twisted together.

twisted joint. As the joint becomes hot, you will notice that a small voltage will indicate on the meter. The output voltage of the thermo-couple may be amplified and used to activate large motors, valves, controls and recording devices. Commercial types of thermocouples employ various kinds of dissimilar metals and alloys such as nickel-platinum, chromel-alumel and iron-constantan. These unfamiliar names apply to special alloys developed for thermocouples. The combination indicating device including meter and thermocouple is called a pyrometer. For an instrument which must be extremely sensitive to temperature change, a larger number of thermocouples may be joined in series. Such a group is known as a thermopile.

ELECTRICAL ENERGY FROM MECHANICAL PRESSURE

Many crystalline substances such as quartz, tourmaline and Rochelle salts have a peculiar characteristic. When a voltage is applied to the surfaces of the crystal, the crystal will become distorted. The opposite is also true. If a mechanical pressure or force is applied to the crystal surface, a voltage will be developed. The crystal microphone, Fig. 2-19, is a familiar example. Sound waves striking a diaphragm, which is mechanically linked to the crystal surfaces, cause distortion in the crystal which in turn develops a voltage across its surfaces. Thus, sound waves are converted to the electrical energy. Another familiar application is the phono cartridge which is installed in the pick-up arm of the record player. The phonograph needle is caused to move by the sound grooves in the record. The moving needle applies mechanical pressure to the crystal in the cartridge and an electrical potential is developed. The potential will vary in frequency and magnitude according to the movement of the needle. The varying potential is amplified and connected to a speaker. You hear the voice or music. Creating electricity by the mechanical distortion of a crystal is known as PIEZOELECTRIC EFFECT. Crystals may be cut for particular operating characteristics. In a later chapter, the use of crystals as frequency contols for radio transmitters will be discussed.

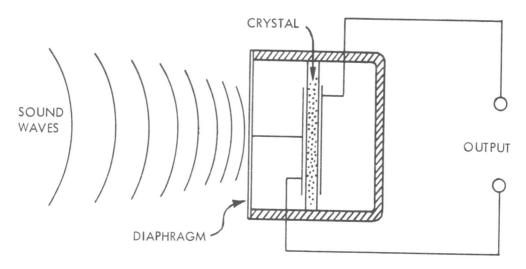

Fig. 2-19. Crystal microphone which converts sound waves to electrical energy.

ELECTRICITY FROM MAGNETISM

A primary source of electrical energy today is the dynamo or generator. Generators prove that magnetism can produce electricity.

A generator is a rotating machine that converts mechanical energy into electrical energy. This source of electricity requires detailed study, so Chapter V is devoted to the subject.

ACTIVITIES

I.   Construct a cell using a penny, a nickel and blotting paper soaked in salt solution. Measure the voltage with a multimeter.

II. Construct a thermocouple by twisting the ends of a copper and an iron wire together. Connect the wires to a voltmeter. Then heat the junction with a match. Observe the voltage developed.

III. Secure this material from the supply room.

    4 flashlight cells.
    4 cell holders.
      Voltmeter.

A.   Connect the cells in series as per diagram I.

What is the voltage between A and B? A and C? A and D? A and E?

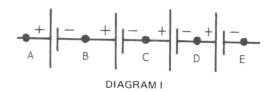

DIAGRAM I

What conclusion do you draw from this experiment?

B.   Connect the cells in parallel as per diagram II.

What is the voltage between A and B?

Remove cell No. 4. Measure voltage from A to B.

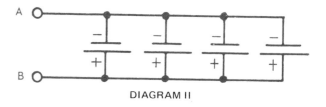

DIAGRAM II

Remove cells No. 3 and 4. Measure the voltage from A to B.

Remove cells No. 2, 3, and 4. Measure the voltage from A to B.

What conclusion do your draw from this experiment?

IV. Secure a sun battery from supply.

A. Connect voltmeter leads to battery terminals. Observe meter reading. Next, shine a flashlight on the cell and note higher meter reading. Finally, place cell in sunshine and note still higher meter meter.

TEST YOUR KNOWLEDGE

1. Name five sources of electricity.
2. What safety precautions should be observed when mixing acid and water?
3. Draw a diagram of four 1.5V cells in series. What is the terminal voltage?
4. Draw a diagram of four 1.5V cells in parallel. What is the terminal voltage?
5. Using six 1.5V cells, connect in series-parallel to produce a terminal voltage of 3 volts.
6. What are the chemical compositions of the plates of a storage battery?
7. What is the composition of the electrolyte in a lead-acid storage battery?
8. What is meant by the "specific gravity" of a liquid?
9. Name and explain two common defects of the primary voltaic cell.
10. Name the materials and draw a rough sketch of a dry cell.
11. What is the major advantage of the nickel-cadmium cell?
12. What precautions should be observed when lifting or carrying a storage battery?
13. Name several factors which determine the capacity of a storage battery.
14. Name several applications using piezo-electric effect.

Chapter 3

CIRCUITS AND POWER

Electrical circuits usually consist of a source of voltage, and one or more devices or components connected to this source by means of wires or conductors. Also, a means to turn the flow of electricity on or off is provided, by using a device called a SWITCH. A diagram showing the various connections to components, devices and switches is called a SCHEMATIC DIAGRAM. As the architect draws a plan for constructing a home, the technician or engineer will draw a schematic plan for the connections of a circuit. The student may consider the conductors as a means of distributing, or directing, the electrons around the circuit. They are the roads to accommodate the electron traffic.

The flow of electron traffic can more easily be understood by comparison to automobile traffic. Consider the size of the road. A large, smooth, four lane highway will carry more traffic than a narrow and rough country road, as illustrated in Fig. 3-1. This is also true of the electrical conductor. RESISTANCE is the opposition to electric current in a conducting wire or material. A large wire can carry more

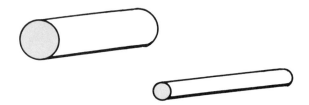

Fig. 3-2. A large conductor has less resistance than a small conductor and can carry a larger current.

electrons than a small wire. It has less resistance to the flow, because of its larger size. Fig. 3-2.

LESSON IN SAFETY: Your body is a good conductor of electricity. An electric current flowing from one hand across your chest and heart and out of your other hand can be dangerous, if not fatal. The wise technician uses only one hand when required to work on high voltage circuits. Keep one hand in your pocket! The operators of switchboards and power stations stand on insulated platforms so that there is no direct path to ground in case of accidental contact with a high voltage wire.

Conductors are conveniently arranged according to their size by the American Wire Gage System, see Fig. 3-3. The larger the gage number, the smaller its diameter and cross-sectional area. For example, a No. 14 wire is larger than a No. 20 wire.

The standard unit of measure for the cross-sectional area of a wire is the CIRCULAR MIL. A mil represents one thousandth of an inch. It is a convenient method of expressing the

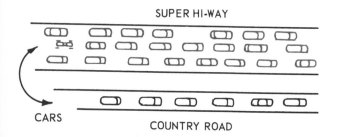

Fig. 3-1. A superhighway will carry more traffic than a country road. It has less resistance to the flow of automobiles.

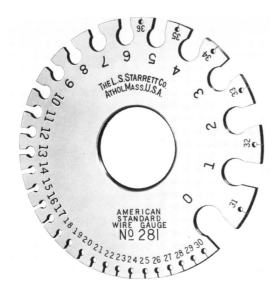

Fig. 3-3. A gauge used to determine wire size.
(L. S. Starrett Co.)

diameter of a conductor, rather than using fractions of an inch. THE CIRCULAR MIL IS THE CROSS-SECTIONAL AREA OF A WIRE WHICH IS ONE MIL IN DIAMETER. A wire having a diameter of 50 mils would have a cross-sectional area of 2500 circular mills. For purposes of comparison, simple rules of geometry may be applied. The area of a circle is equal to πR^2. Computing the area in square inches of a wire fifty thousandths (.050") of an inch in diameter or .025 inches in radius,

Cross-Sectional Area = 3.14 x (.025)²
 = 3.14 x .000625
 = .00196 sq. inches.

Notice that in the above formula $R = \frac{D}{2}$ or $\frac{.050}{2}$ or the radius is equal to one half the diameter.

Cross-sectional area could be represented by:

$$A = \pi \left(\frac{D}{2}\right)^2 \quad \text{or} \quad \pi \frac{D^2}{4} \text{ or}$$

$$\frac{3.14}{4} D^2 = .7854 D^2$$

and if D is in mils, A would equal .7854D² sq. mils. Now the circular mil was defined as the area of a wire having the diameter of one mil which would equal:

$$A = .7854 \text{ x } (1^2) = .7854 \text{ sq. mils}$$

Therefore, in order to find the circular mil area of a conductor, it is necessary to divide its square mil area by .7854 which is the square mils area in one circular mil. To simplify these mathematics, note that:

$$A = \frac{.7854 D^2 \text{ (sq. mil area)}}{.7854 \text{ (area of circular mil)}}$$
$$= D^2 \text{ circular mils}$$

THE CIRCULAR MIL CROSS-SECTIONAL AREA OF A CONDUCTOR IS EQUAL TO THE SQUARE OF ITS DIAMETER IN MILS.

To change the size of a wire from square mils to circular mils, it would be necessary to divide by .7854. To change a wire size given in circular mils to square mils, one would multiply by .7854. This information is important when comparing the current carrying capacity of round and square wires, see Fig. 3-4.

Summarizing the above discussion, one should remember that the RESISTANCE TO THE FLOW OF ELECTRICITY BY A WIRE OR CONDUCTOR WILL VARY INVERSELY WITH ITS SIZE OR CROSS-SECTIONAL AREA.

A LESSON IN SAFETY: The wires used to conduct electricity around your home have been selected with sufficient size to take care of your needs. Most circuits for lighting purposes use either No. 12 or No. 14 wire rated at 15 amperes. The wires for your range or water heater may be No. 6 to No. 10. Always use wires of the correct size and avoid trouble.

It was mentioned in Chapter I, that some materials are good conductors of electricity and others are poor conductors. So, THE KIND OF MATERIAL USED IN THE MANUFACTURING OF WIRE ALSO AFFECTS ITS ABILITY TO CONDUCT OR ITS RESISTANCE.

GAGE NO.	DIAM. MILS	CIRCULAR MIL AREA	OHMS PER 1,000 FT. OF COPPER WIRE AT 25°C	GAGE NO.	DIAM. MILS	CIRCULAR MIL AREA	OHMS PER 1,000 FT. OF COPPER WIRE AT 25°C
1	289.3	83,690	0.1264	21	28.46	810.1	13.05
2	257.6	66,370	0.1593	22	25.35	642.4	16.46
3	229.4	52,640	0.2009	23	25.57	509.5	20.76
4	204.3	41,740	0.2533	24	20.10	404.0	26.17
5	181.9	33,100	0.3195	25	17.90	320.4	33.00
6	162.0	26,250	0.4028	26	15.94	254.1	41.62
7	144.3	20,820	0.5080	27	14.20	201.5	52.48
8	128.5	16,510	0.6405	28	12.64	159.8	66.17
9	114.4	13,090	0.8077	29	11.26	126.7	83.44
10	101.9	10,380	1.018	30	10.03	100.5	105.2
11	90.74	8,234	1.284	31	8.928	79.70	132.7
12	80.81	6,530	1.619	32	7.950	63.21	167.3
13	71 96	5,178	2.042	33	7.080	50.13	211.0
14	64.08	4,107	2.575	34	6.305	39.75	266.0
15	57.07	3,257	3.247	35	5.615	31.52	335.0
16'	50.82	2,583	4.094	36	5.000	25.00	423.0
17	45.26	2,048	5.163	37	4.453	19.83	533.4
18	40.30	1,624	6.510	38	3.965	15.72	672.6
19	35.89	1,288	8.210	39	3.531	12.47	848.1
20	31.96	1,022	10.35	40	3.145	9.88	1,069.

Fig. 3-4. Copper Wire Table.

LENGTH OF WIRE AND RESISTANCE

A third consideration in the selection of a conductor is its length. If one foot of wire has a certain resistance, then ten feet of the same wire will have ten times more resistance and fifty feet will have fifty times more. Remember that this resistance uses power and creates losses in the line. Devices connected to long conductors may have insufficient voltage for proper operation due to line losses.

A LESSON IN SAFETY: When using long extension cords to operate lights and tools, be sure that they have sufficient size so that there will be little line loss or voltage drop. Small wire extension cords may heat up and burn. Motor driven tools at the end of a small, long extension cord will heat and operate inefficiently. Also, it is very wise to equip all your tools with a three wire cord. The third wire is a ground and is connected to the case and handle of the tool. This precaution has saved many lives.

RESISTANCE VARIES WITH

1. SIZE OF WIRE
2. LENGTH OF WIRE
3. KIND OF WIRE
4. TEMPERATURE

Table 3-1

RESISTANCE

IS MEASURED IN OHMS
OHMS IS REPRESENTED BY Ω
ITS LETTER SYMBOL IS R
ITS SCHEMATIC SYMBOL IS ⌁
KILOHM MEANS 1000 OHMS
MEGOHM MEANS 1,000,000 OHMS

Table 3-2

TEMPERATURE AND RESISTANCE

A fourth consideration is temperature. Most metals used in conductors, such as copper and aluminum, increase in resistance as the temperature increases. In many electronic circuits careful design is necessary to insure proper ventilation and radiation of heat from current carrying devices. When wires are enclosed in metallic coverings or conduit, the National Electric Code will specify that larger wires should be used so that heat, due to losses in the line, will not change the current carrying ability of the conductors.

The factors affecting resistance are summarized in Table 3-1. <u>MEMORIZE THEM!</u>

STRANDED WIRES

Not every conductor is a solid wire. Frequently it is necessary to use several smaller wires twisted together to form a cable. This method allows greater flexibility and ease of installation. If a wire is subject to movement or vibration, such as occurs in a lamp cord, a stranded wire produces greater strength and less danger of breaking. Stranded wires are usually made up of 7, 19, or 37 separate wires twisted together. A further improvement is made by twisting several stranded wires into one cable, see Fig. 3-5. Do not confuse this stranded cable with the multi-wire conductor used in many electronic applications. The multi-conductor has several wires, each insulated from the other. These wires are color-coded so that correct connections may be made.

A LESSON IN SAFETY: When making connections with stranded wires, the stranded ends should be tightly twisted together and coated with solder. This is called <u>tinning.</u> Fine wires separated from the connection is a mark of poor workmanship and may cause dangerous short circuits.

OHM'S LAW

One of the fundamental laws of electrical circuits was derived from experimentation done by George Simon Ohm, the German scientist and philosopher, during the 19th century. To

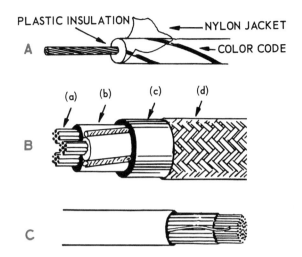

Fig. 3-5. A—Stranded hookup wire used in electronic circuits. B—Power cable, (a) Three stranded conductors, (b) Insulation and jute fillers, (c) Tough, flexible jacket, (d) Bronze protective armor. C—Wire for a telephone switchboard. Numerous conductors are insulated from each other and are color coded.

honor the achievements of Mr. Ohm, the standard unit of measurement for resistance is called the OHM. It is frequently represented by the Greek letter "omega" as Ω. If you see on a diagram 1000 Ω, it means 1000 ohms. In electronic circuits the use of the kilohm and megohm is very common. The Greek prefix "kilo" means 1000. The prefix "meg" means one million. These are summarized in Table 3-2.

Ohm's Law is stated as: The current in amperes in a circuit is equal to the applied voltage divided by the resistance, or,

$$I \text{ (in amperes)} = \frac{E \text{ (in volts)}}{R \text{ (in ohms)}}$$

Notice the letter symbols used in this equation:

I = intensity of the current in amperes
E = electromotive force in volts
R = resistance in ohms

In non-mathematical language this formula means:

As voltage is increased   — current increases
As voltage is decreased   — current decreases
As resistance is increased   — current decreases
As resistance is decreased — current increases

A mathematical explanation: A certain device which has a resistance of 100 ohms is connected to a 100V source of electricity. How much current will flow in the circuit? See Fig. 3-6.

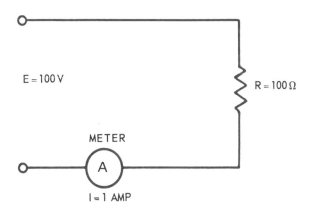

Fig. 3-6. E = 100V, R = 100 Ω, I = 1 amp.

Solution:

$$I = \frac{E}{R} \text{ or } I = \frac{100V}{100 \, \Omega} = 1 \text{ ampere}$$

If the voltage is increased to 200 volts, how much current will flow? See Fig. 3-7.

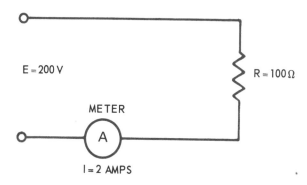

Fig. 3-7. E = 200V, R = 100 Ω, I = 2 amps.

Solution:

$$I = \frac{200V}{100 \, \Omega} = 2 \text{ amperes}$$

Returning the voltage to 100 volts, but increasing the resistance to 200 ohms, then: Fig. 3-8.

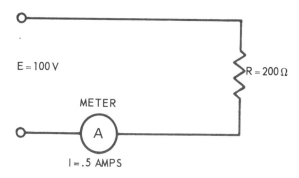

Fig. 3-8. E = 100V, R = 200 Ω, I = .5 amps.

$$I = \frac{E}{R} \text{ or } I = \frac{100V}{200 \, \Omega} = .5 \text{ amperes}$$

Finally, decrease the resistance to 50 ohms, then Fig. 3-9.

$$I = \frac{E}{R} \text{ or } I = \frac{100V}{50 \, \Omega} = 2 \text{ amperes}$$

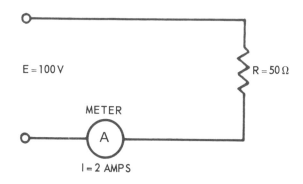

Fig. 3-9. E = 100V, R = 50 Ω, I = 2 amps.

Compare this computation to the previous discussion, and notice that these statements have been proved true. By simple transposition of symbols in the Ohm's Law equation, it is possible to solve for any one unknown quantity, if the other two are known.

Ohm's Law may be written three ways:

$$I = \frac{E}{R} \qquad E = I \times R \qquad R = \frac{E}{I}$$

Problem: A current of .5 amperes is flowing in a circuit which has 100 ohms resistance. What is the applied voltage? See Fig. 3-10.

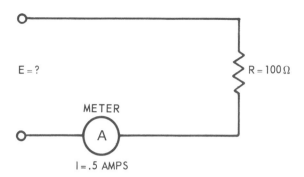

Fig. 3-10. R = 100 Ω, I = .5 amps, E = 50V.

E = IR = .5 x 100 = 50 volts

Problem: A circuit has an applied voltage of 50 volts and the current is measured at .5 amperes. What is the resistance of the circuit shown in Fig. 3-11?

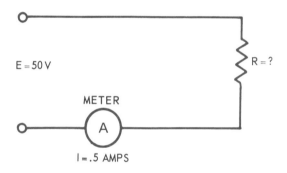

Fig. 3-11. E = 50V, I = .5 amps, R = 100 Ω.

$$R = \frac{E}{I} \text{ or } R = \frac{50}{.5} = 100 \,\Omega$$

Practice with problems is the best way to become thoroughly acquainted with Ohm's Law. A device to help you solve Ohm's Law problems is illustrated in Fig. 3-12.

Caution: When working problems involving Ohm's Law, be certain that all quantities are in their fundamental form, that is, volts in volts, current in amperes, and resistance in ohms. If one or more of these quantities appears with a prefix such as kilo, mega, or milli, a conversion must be made before solving the equation.

Fig. 3-12. A memory device to help you solve Ohm's Law problems. Cover the unknown quantity with your finger and the remaining letters give the correct equation. For example, E is unknown. Cover E, and I x R remains.

Problem: A circuit has a resistance of 10 kilohms and the current is measured as 100 milliamperes. What is the applied voltage?

E = I x R   or
E = .1 ampere x 10,000 = 1000V

For methods of making conversions to basic electrical units, see Appendix 3.

POWER

In any electrical circuit, the only component in the circuit that uses electrical power is resistance. POWER is the time rate of doing work. In the physics of machines it is discovered that when a force moves through a distance, work is done. For example, if a 10 lb. weight is lifted one foot, the work done equals,

F (force) x D (distance) or
10 x 1 = 10 ft. lb. of work.

See Fig. 3-13. No reference is made to time in this equation. One might take five seconds or 10 minutes to lift the 10 pound weight. However, if one lifted the 10 pound weight at the rate of once each second, then the power expended would be 10 ft. lb. per second. To carry the example one step further, if one lifted the 10 pound weight in one-half or .5 second, then the power expended would equal:

$$\frac{10}{.5} \text{ or 20 ft. lb. per second.}$$

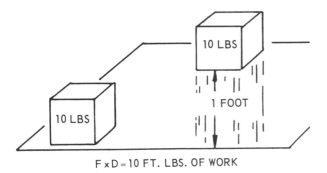

Fig. 3-13. When a 10 lb. weight is lifted one foot, 10 foot-pounds of work is done.

Fig. 3-14. A memory device to help you solve power problems. Cover the unknown quantity with your finger and the remaining letters give the correct equation.

A machine that works at a rate of 550 ft. lb./sec. is equivalent to <u>one horsepower</u>. Also 33,000 ft. lb./min. equals one horsepower.

In electricity the unit of power is the watt, named in honor of James Watt, who is credited with the invention of the steam engine. When one volt of electrical pressure moves one coulomb of electricity in one second, the work accomplished is equal to one watt of power. Recall the definition of one ampere — when one coulomb of electrons moves past a given point in a circuit in one second. So power in an electrical circuit is equal to:

$$P \text{ (watt)} = E \text{ (volts)} \times I \text{ (amperes)}$$

Use these figures when comparing electrical power to mechanical power:

746 watts = 1 horsepower

The power formula, sometimes called Watt's Law, can be arranged algebraically, so that if two quantities are known, the third unknown may be found. Use the device in Fig. 3-14 to help you.

$$P = I \times E \qquad I = \frac{P}{E} \qquad E = \frac{P}{I}$$

Example: A circuit with an unknown load has an applied voltage of 100 volts. The measured current is 2 amperes. How much power is consumed?

$$P = I \times E \text{ or } 2 \text{ amps} \times 100V = 200W$$

Example: An electric toaster rated at 550 watts is connected to a 110 volt source. How much current will this appliance use?

$$I = \frac{P}{E} \text{ or } \frac{550W}{110V} = 5 \text{ amps}$$

A LESSON IN SAFETY: An electric current produces heat when it passes through a resistance. Heaters and resistors will remain <u>hot</u> for some time after the power is removed. Handle them carefully. Burns should be treated immediately. See your instructor.

OHM'S LAW AND THE POWER LAW

It is possible to combine these laws to produce simple formulas which will permit you to solve any unknown, if two quantities are known. In Table 3-3, these equations are given and the explanation of each. Refer to the numbers of these equations.

These formulas are conveniently arranged in a wheel-shaped memory device for ready reference. Refer to Fig. 3-15. To use the device, find the equation that uses the two known quantities in your circuit.

For example: If current and resistance are known and power is unknown, use equation 11.

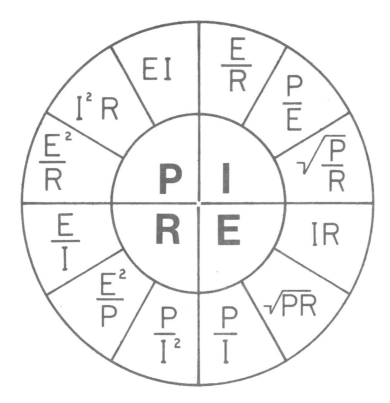

Fig. 3-15. A memory device combining Ohm's Law and Power Laws. Use it to help you solve your problems.

1.	$E = I \times R$	Ohm's Law
2.	$E = \dfrac{P}{I}$	Watt's Law
3.	$E = \sqrt{PR}$	By transposing equation 12 and taking the square root.
4.	$I = \dfrac{E}{R}$	Ohm's Law
5.	$I = \dfrac{P}{E}$	Watt's Law
6.	$I = \sqrt{\dfrac{P}{R}}$	By transposing equation 9 and taking the square root.
7.	$R = \dfrac{E}{I}$	Ohm's Law
8.	$R = \dfrac{E^2}{P}$	By transposing equation 12.
9.	$R = \dfrac{P}{I^2}$	By transposing equation 11.
10.	$P = I \times E$	Watt's Law
11.	$P = I^2 \times R$	By substituting $I \times R$ from equation 1, for E.
12.	$P = \dfrac{E^2}{R}$	By substituting $\dfrac{E}{R}$ from equation 4, for I.

Table 3-3

If power and voltage are known and current is unknown, use equation 5.

RESISTORS

Resistance units permit the design engineer to use electricity in many interesting and profitable ways. Because the resistance is the only component of a circuit that uses power, most of the useful effects and labor saving devices provided by electrical power use some form of resistance. Resistance may purposely be introduced into a circuit to reduce voltages and limit current flow in electronic devices. Fig. 3-16, shows several forms of the common molded composition type resistors. They are manufactured in a variety of shapes and sizes. The chemical composition, which causes the resistance, is accurately controlled in the industrial laboratory. They are made in many ohmic values from one ohm to several million ohms. Also the physical size of the resistor is rated in watts, which means its ability to dissipate the heat caused by the resistance. The larger sizes provide a greater radiation surface to dissipate this heat. Common sizes used in electronic work are the 1/2 watt and one watt size. One may purchase, for example, a 1000 ohm resistor in a 1/2 watt, 1 watt or 2 watt size. In each size the resistance would be the same.

Another type of small wattage resistors is the thin film resistor. They are similar to the molded composition resistor in appearance and function. However, thin film resistors are made from depositing a resistance material on a glass or ceramic tube. Leads with caps are fitted over each end of the tube to make the body of the resistor. Thin film resistors are usually color coded.

For higher current applications, resistance units are wirewound. On a ceramic core is wound a wire which has a specific resistance. The whole resistance component is insulated by a coat of vitreous enamel. Several of these resistors are illustrated in Fig. 3-17. These are manufactured in sizes from 5 watts to 200 watts, depending upon the heat dissipation required during operation.

A variation to the wirewound resistor provides an exposed surface to the resistance wire on one side to which is attached an adjustable third terminal. Such resistors, sometimes with two or more adjustable taps, are used extensively as voltage dividers in power supplies and other apparatus. These adjustable resistors are shown in Fig. 3-18.

Most electronic equipment requires variable resistance components which can be controlled by a knob on the front panel. The volume control on a radio is an example of this type of control. These are generally the rotary type. The variation in resistance is provided by a sliding contact arm on a molded composition ring as the resistive element. This type is shown in Fig. 3-19. Connections may be made through terminals to both ends of the resistance as well as to the siding contact arms. This component is called a POTENTIOMETER. For higher current and power applications, these potentiometers

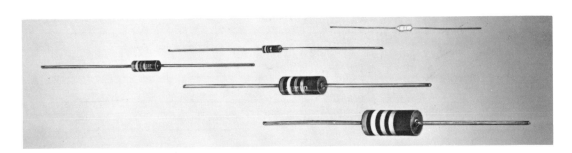

Fig. 3-16. Composition Resistors. 1/10, 1/4, 1/2 and 1 and 2W sizes.
(Ohmite Mfg. Co.)

Fig. 3-17. Resistors are made in many types, shapes and sizes to meet specific circuit requirements.

may be wirewound resistance wire instead of the molded composition. The wirewound types are shown in Fig. 3-20.

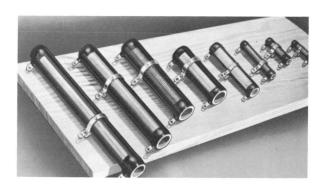

Fig. 3-18. These adjustable resistors provide a sliding tap for voltage divider applications.

There are, also, many types and sizes of resistors designed for specific applications. These include extremely accurate precision resistors, and heavy and rugged types for high power applications.

In your television set there may be a TV fuse resistor. This is a special 7 1/2 ohm resistor which acts as a slow-blow fuse or protective device in televisions which have series connected filaments.

One relatively new form of resistor, which is rapidly becoming popular in transistor circuitry, is the THERMISTOR. These components are used to offset the effect of temperature change

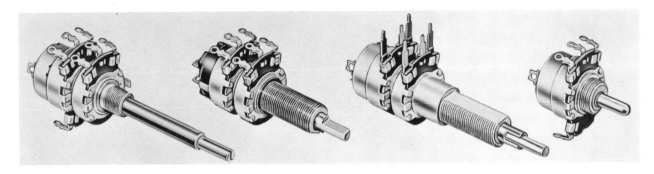

Fig. 3-19. Potentiometers used in electronic circuitry for fixed and variable resistance. (Centralab)

Fig. 3-20. A wire-wound potentiometer for higher current and power applications. (Ohmite Mfg. Co.)

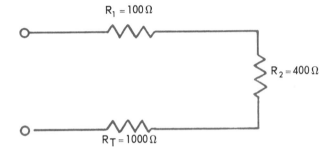

Fig. 3-21. The schematic diagram of three resistors in series.

in a circuit. They are so constructed that as the temperature increases, their resistance will decrease. Advanced studies in semiconductor characteristics will include detailed explanations of thermistor theory and application.

SERIES CIRCUITS

If a circuit is so arranged that all of the current flowing in the circuit will pass through all components, these components are connected in SERIES. The components are connected "in-line" or "end-to-end." In Fig. 3-21 three resistors are connected in series. Other components, which will be studied later, may also be connected in series. (In the following discussions, the resistance of the connecting wires in the circuits will be neglected.) If resistors are connected in series, the total resistance of the circuit is equal to the sum of the resistors. In Fig. 3-21, $R_1 = 100 \, \Omega$, $R_2 = 400 \, \Omega$, and $R_3 = 500 \, \Omega$. Then, $R_T = R_1 + R_2 + R_3$ or

$$100 + 400 + 500 = 1000 \, \Omega$$

If 100 volts were applied to the terminals of this circuit, the current flowing in the circuit would equal,

Ohm's Law: $I = \dfrac{E}{R}$ or $I = \dfrac{100V}{1000 \, \Omega} = .1$ amp.

As current passes through a resistor, a certain amount of energy is used and a certain amount of pressure or voltage is lost. The voltage loss across each resistor may be calculated by Ohm's Law, using formula, $E = I \times R$, Fig. 3-22.

For R_1, $E_{R_1} = .1 \times 100 = 10V$

For R_2, $E_{R_2} = .1 \times 400 = 40V$

For R_3, $E_{R_3} = .1 \times 500 = 50V$

The symbol such as E_{R_1} means the voltage across resistor R_1. One should particularly note that the sum of the voltage losses or <u>voltage drops</u> is equal to the source voltage. In this example,

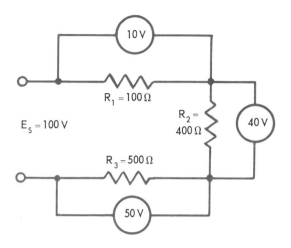

Fig. 3-22. The voltage loss or drop across each resistor may be computed by the formula, E = I x R.

$$E_{R_1} + E_{R_2} + E_{R_3} = E_{source} \text{ or}$$
$$10V + 40V + 50V = 100V$$

Because the voltage drop or loss across a resistor is computed by using Ohm's Law, E = I x R, the technician will frequently refer to this loss as the <u>IR drop</u>, which means the same as the <u>voltage drop</u>.

Since these resistors are connected in series, all the current flowing in the circuit must pass through each resistor. Therefore, current is the same in all parts of the circuit, Fig. 3-23.

The above facts concerning series circuits may be summarized in laws known as KIRCHHOFF'S LAWS.

1. The sum of the voltage drops around a series circuit will equal the source voltage.

2. The current is the same when measured at any point in the series circuit.

POWER IN A SERIES CIRCUIT

As the voltage overcomes the resistance, work is done and power is the time rate of doing work. Referring to the section on power, earlier in this chapter, you will discover that in electricity, P = I x E. The energy lost in overcoming the resistance takes the form of heat. This must be dissipated by the resistor into the surrounding air. This explains why some resistors are larger than others. They must be large enough to provide radiation surface for heat dissipation.

Referring to Fig. 3-24, the total power used in this circuit would equal:

P = I x E or .1 x 100V or 10W.

To calculate the power consumed by individual resistors, the formula $P = I^2 R$ may be used:

For R_1, $P = (.1)^2$ x 100 = 1W

For R_2, $P = (.1)^2$ x 400 = 4W

For R_3, $\underline{P = (.1)^2 \text{ x } 500 = 5W}$

For R_T, $P = (.1)^2$ x 1000 = 10W

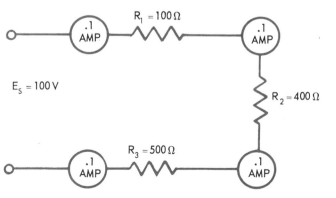

Fig. 3-23. The current measured at any point in the series circuit is the same.

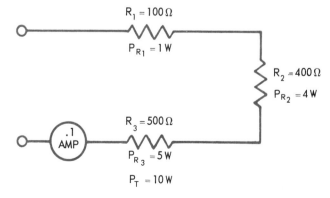

Fig. 3-24. The power used by each resistor may be computed by the formula, P = I x E.

Notice that the sum of the individual powers used by each resistor is equal to the total power used by the circuit.

PARALLEL CIRCUITS

When several components are connected to the same voltage source, the components are connected in parallel or "side by side." Multiple paths for current flow are provided by a parallel circuit, because each resistor constitutes a path of its own. In Fig. 3-25, three separate paths for electricity are provided by R_1, R_2 and R_3. One must realize, that when components are con-

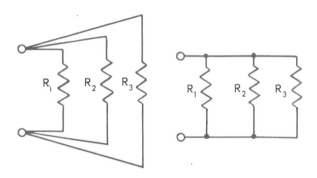

Fig. 3-25. The schematic diagram of three resistors in parallel.

nected in this manner, the total resistance of the circuit is decreased every time another component is added. Compare a parallel circuit to a system of pipes carrying water. Two pipes will carry more water than one pipe. If a third pipe is added, the three pipes will carry more water than two pipes. As more pipes are added, the total resistance to the flow of water is decreased, Fig. 3-26.

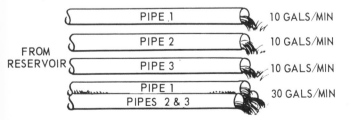

Fig. 3-26. Electricity may be compared to the flow of water. Increasing the number of pipes will decrease the resistance to water flow.

In Fig. 3-27, resistor R_1 = R_2 = R_3 so each resistor will carry the same amount of current, and the three resistors together will carry three times more current than only one. Therefore, the total resistance of the circuit must be one-third of the resistance of a single resistance. Assigning values to R_1 = 30 Ω , R_2 = 30 Ω, and R_3 = 30 Ω , then:

$$R_T = \frac{30\,\Omega}{3} \text{ or } 10\,\Omega$$

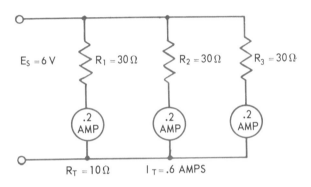

Fig. 3-27. Total current equals the sum of branch currents.

When equal resistors are connected in parallel, the total resistance of the parallel group or network is equal to any one resistor divided by the number of resistors in the network.

Total resistance R_T

$$= \frac{R}{N} \frac{\text{value of one resistor}}{\text{number of resistors in network}}$$

The applied voltage is the same for each resistor, because each is connected across the same voltage source.

Therefore, the currents are, using Ohm's Law:

For R_1, $I = \dfrac{6V}{30\,\Omega} = .2$ amps

For R_2, $I = \dfrac{6V}{30\,\Omega} = .2$ amps

For R_3, $I = \dfrac{6V}{30\,\Omega} = .2$ amps

The total current flowing through the network would be the sum of the individual branch currents or:

$$I_{R_1} + I_{R_2} + I_{R_3} = I_T \quad \text{or}$$

$$.2 + .2 + .2 = .6 \text{ amps}$$

This may be proved by applying Ohm's Law, using the total resistance,

$$I_T = \frac{E}{R_T} \quad \text{or} \quad I_T = \frac{6V}{10\,\Omega} = .6 \text{ amps}$$

Summarizing again, there are these laws concerning parallel circuits.

1. The voltage across all branches of a parallel network is the same.

2. The total current is equal to the sum of the individual branch currents.

The differences between these laws and those which apply to series circuits should be compared.

It is not unusual to find two unequal resistors connected in parallel. In this case, the total resistance may be found by using the formula,

$$R_T = \frac{R_1 R_2}{R_1 + R_2}$$

In Fig. 3-28, resistance $R_1 = 20\,\Omega$ and $R_2 = 30\,\Omega$. Apply the formula:

$$R_T = \frac{20 \times 30}{20 + 30} \quad \text{or} \quad \frac{600}{50} = 12\,\Omega$$

With an applied voltage $E_s = 6V$, then the total current in the circuit will be,

$$I_T = \frac{E}{R_T} \quad \text{or} \quad \frac{6}{12} \quad \text{or} \quad .5 \text{ amps}$$

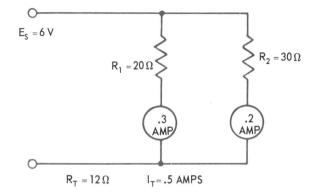

Fig. 3-28. Schematic of two unequal resistors in parallel.

This equation may be proved by finding the sum of individual branch current,

$$\text{For } R_1 \quad I = \frac{6V}{20\,\Omega} = .3 \text{ amps}$$

$$R_2 \quad I = \frac{6V}{30\,\Omega} = .2 \text{ amps}$$

$$I_1 + I_2 = .3 + .2 = .5 \text{ amps or } I_T$$

Two observations should be carefully studied.

1. The total resistance of any parallel circuit must always be less than the value of any resistor in the network. Note in Fig. 3-28. $R_T = 12\,\Omega$ which is less than R_1 which is $20\,\Omega$.

2. The branch of the circuit containing the greatest resistance conducts the least current.

When three of more resistors are connected in parallel, it is convenient to use the conductance method of finding the total resistance. As resistance in ohms is the ability to resist the flow of current, its opposite is CONDUCTANCE, which is the ability of a component to conduct. Resistance is measured in OHMS; Conductance is measured in MHOS (ohm spelled backwards). The symbol representing conductance is G and is equal to the reciprocal of resistance.

To illustrate: A resistance of $100\,\Omega$ is equal to conductance of $\frac{1}{100}$ or .01 mhos.

Ohm's Law states that $R = \frac{E}{I}$, then $\frac{1}{R} = \frac{I}{E}$ or $G = \frac{I}{E}$ and may be defined as amperes-per-volt, or

$$I = GE$$

Current in a circuit is equal to its voltage times its conductance in mhos.

For example: A circuit has an applied voltage of $E = 100V$ and a resistance of $50\,\Omega$.

By Ohm's Law: $I = \frac{100}{50} = 2$ amps

By conductance methods:

$$G = \frac{1}{R} = \frac{1}{50} = .02 \text{ mhos}$$

$$I = GE \text{ or } .02 \times 100 = 2 \text{ amps}$$

Referring to Fig. 3-29, three unequal resistors are conneced in parallel. $R_1 = 5\,\Omega$, $R_2 = 10\,\Omega$ and $R_3 = 30\,\Omega$. To find the total resistance; find the sum of the individual branch currents. Assuming a voltage of 30 volts then:

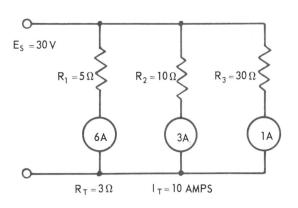

Fig. 3-29. Three unequal resistors connected in parallel.

For Branch R_1, $\quad I_1 = \frac{E}{R_1} = \frac{30}{5} = 6$ amps

For Branch R_2, $\quad I_2 = \frac{E}{R_2} = \frac{30}{10} = 3$ amps

For Branch R_3, $\quad I_3 = \frac{E}{R_3} = \frac{30}{30} = 1$ amp

$$I_T = I_1 + I_2 + I_3 = 6 + 3 + 1 = 10 \text{ amps}$$

By Ohm's Law:

$$R_T = \frac{E}{I} \text{ so } R_T = \frac{30}{10} = 3\,\Omega$$

As I_T is the sum of all of the currents, and is equal to $\frac{E}{R_T}$ (Ohm's Law) then,

$$\frac{E}{R_T} = \frac{E}{R_1} + \frac{E}{R_2} + \frac{E}{R_3}$$

Divide both sides of the equation by E, then

$$\frac{1}{R_T} = \frac{1}{R_1} + \frac{1}{R_2} + \frac{1}{R_3}$$

This equation shows that the conductance of a circuit is equal to the sum of the conductances of the branch circuits. To solve for the total resistance of a circuit in Fig. 3-29, use this formula:

$$\frac{1}{R_T} = \frac{1}{5} + \frac{1}{10} + \frac{1}{30} = \frac{10}{30}$$

Inverting both sides of the equation,

$$R_T = \frac{30}{10} = 3\,\Omega$$

This conductance formula is used in the computation of the total resistance of parallel circuits. You should practice using it by completing the problems at the end of this chapter.

SERIES AND PARALLEL COMBINATION CIRCUITS

The technician is frequently required to compute the total resistance of combination series and parallel circuits. This total resistance is referred to as the <u>equivalent resistance</u> of the circuit. It is only necessary to solve in logical order by using the series and parallel resistance formulas of the previous discussion. Referring to Fig. 3-30, R_1 = 7Ω, R_2 = 5Ω, R_3 = 10Ω, R_4 = 20Ω, R_5 = 10Ω. The applied voltage E_s = 30 volts.

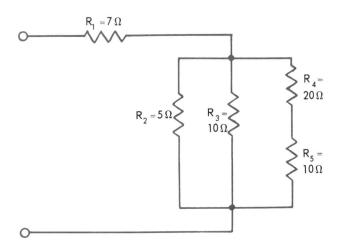

Fig. 3-30. Combination circuit of series and parallel resistors.

Step I. Combined R_4 and R_5, Fig. 3-31. They are in series, so $R_4 + R_5$ = 20 + 10 = 30Ω.

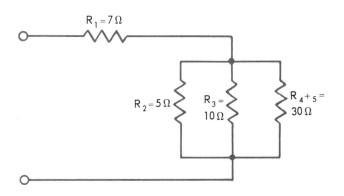

Fig. 3-31. Two series resistors in one branch of the parallel network have been combined.

Step II. Find the total resistance of the parallel circuit, Fig. 3-32.

$$\frac{1}{R_T} = \frac{1}{R_2} + \frac{1}{R_3} + \frac{1}{R_{4\&5}} \quad \text{or}$$

$$\frac{1}{5} + \frac{1}{10} + \frac{1}{30} = \frac{10}{30} \quad R_T = \frac{30}{10} = 3\,\Omega$$

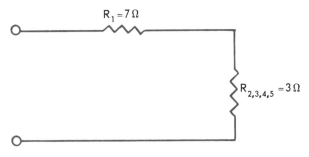

Fig. 3-32. The parallel network is combined to the single equivalent resistance.

Step III. This equivalent resistance or 3 ohms is in series with R_1 of 7 ohms, Fig. 3-33. So the R_T of the circuit equals 3 ohms + 7 ohms or 10 ohms.

The current flowing in the total circuit is;

$$I_T = \frac{E}{R_T} = \frac{30}{10} = 3 \text{ amps}$$

The voltage drop across R_1 is,

$$E_{R_1} = IR \text{ or } 3 \times 7 = 21V$$

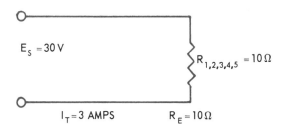

Fig. 3-33. Total resistance of the combination circuit is equivalent to a single resistance.

If 21 volts is lost by R_1, then the remaining part of the applied voltage appears across the network of parallel resistors or 9V.

The current through the three branches must equal the total current of 3 amps in the circuit, Fig. 3-34.

$$I_{R_2} = \frac{E}{R_2} = \frac{9V}{5\,\Omega} = 1.8 \text{ amps}$$

$$I_{R_3} = \frac{E}{R_3} = \frac{9V}{10\,\Omega} = .9 \text{ amps}$$

$$I_{R_{4\&5}} = \frac{E}{R_{4\&5}} = \frac{9V}{30\,\Omega} = .3 \text{ amps}$$

$$I_T = I_{R_2} + I_{R_3} + R_{R_{4\&5}}$$

$$= 1.8 + .9 + .3 = 3 \text{ amps}$$

Knowing the current and voltage of this circuit, the power consumed would be:

$$P = I \times E \text{ or } 3 \text{ amps} \times 30V = 90W$$

The sum of the powers dissipated by the individual resistors should equal 90 watts. Using the formula $P = I^2 R$ then:

$$P_{R_1} = (3)^2 \times 7 = 63.0W$$
$$P_{R_2} = (1.8)^2 \times 5 = 16.2W$$
$$P_{R_3} = (.9)^2 \times 10 = 8.1W$$
$$P_{R_4} = (.3)^2 \times 20 = 1.8W$$
$$P_{R_5} = (.3)^2 \times 10 = \underline{.9W}$$
$$P_T = 90.0W$$

To the beginning student in electronics the solution of series and parallel circuits involving the use of Ohm's and Watt's Laws requires

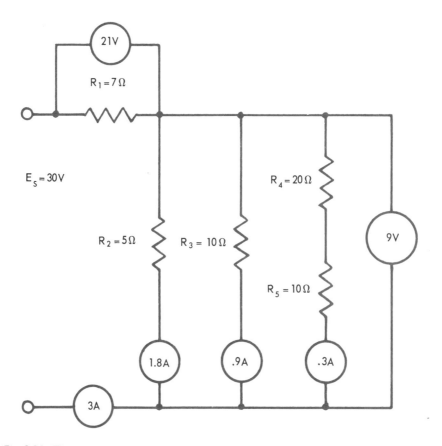

Fig. 3-34. The complete analysis of the combination circuit, showing voltages and currents.

considerable practice. Practice is the only way to assure understanding. Quick recognition of known and unknown quantities and a logical approach to the solution will assure accurate results. Always prove your answer by using other forms of Ohm's Law.

Fig. 3-35. Two electric stoves.

ACTIVITY AND EXPERIMENTATION

ELECTRIC STOVES

I. In Fig. 3-35 two simple electric stoves are illustrated which are satisfactory projects to demonstrate the heat produced by a current flowing through resistance. The little stoves are quite satisfactory for coffee and food warmers. The bases are made of 20 gage aluminum and can be shaped according to your own design. They have been fastened together with sheet metal screws. The top must be constructed of <u>insulating non-inflammable</u> material. In this project, asbestos board was used.

The line cord enters the stove through a 3/8 in. rubber grommet and is secured on the inside with a knot. The heating elements are connected to the line through single pole toggle switches.

Selecting the proper length of nickel-chrome resistance wire is a problem for you to solve. This wire is supplied in a number of gage sizes, each with a specified ohm's resistance per foot. A 330 watt stove operating on 110 volts would require,

$$I = \frac{P}{E} \text{ or } \frac{330}{110} = 3 \text{ amps current} \quad \text{and}$$

$$R = \frac{E}{I} \text{ or } \frac{110}{3} = 36.6 \,\Omega \text{ resistance}$$

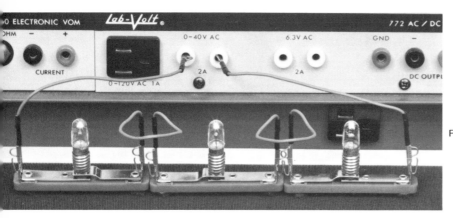

Fig. 3-36. Electrodemonstrator, wired to show series circuit.

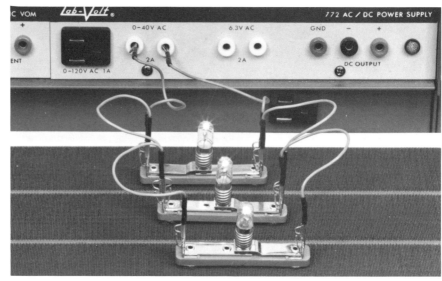

Fig. 3-37. Lights are connected in parallel.

NOTE: Information on the Electrodemonstrator will be found in Appendix 6.

If a nickel-chrome wire having a resistance of 3 ohms per ft. were used, a length of $\frac{36.6}{3}$ or 12.2 ft. is required. The wire, cut to length, is then tightly wound into a spring and fastened to the stove top as the heating element. If nickel-chrome wire is not readily available, elements may be purchased in appliance repair shops.

After the stove is operating, measure the actual power being used with a watt-meter and compare the meter reading with your own computation.

ELECTRODEMONSTRATOR

II. In Fig. 3-36, an electrodemonstrator is used to show a series circuit. Each light bulb is a resistance unit. A 6 volt power supply is used as a power source.

Procedure:

A. Connect one light only to the power source and observe its brillance.

B. Connect two lights in series and observe their billiance. Why are they dimmer?

C. Connect the three lights in series and observe that all lights are now quite dim.

D. Remove one light from the socket. Do the other lights continue to glow?

E. Measure the source voltage with a voltmeter. Measure the voltages across each light. Does the sum of these voltages equal the source voltage?

III. In Fig. 3-37 the lights from the electro-demonstrator are connected in parallel across the 6 volt power source.

Procedure:

A. Connect two lights in parallel and observe their brilliance.

B. Connect three lights in parallel and observe their brilliance. Is there a

difference? Compare to the three lights in series.

C. Unscrew one light bulb. Do the others continue to glow?

D. Using a voltmeter, measure voltage across the terminals of each light and compare to the source voltage.

E. Which circuit uses the most power? The three light series circuit or the three light parallel circuit? Explain.

PROBLEMS TO SOLVE

1.

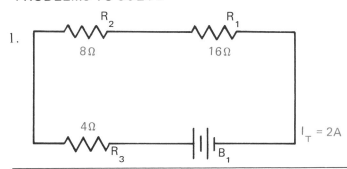

Find: R_T = _____

E_T = _____

E_{R_1} = _____

E_{R_2} = _____

E_{R_3} = _____

P_T = _____

2.

Find: R_T = _____

E_T = _____

E_{R_1} = _____

I_{R_1} = _____

I_{R_2} = _____

I_{R_3} = _____

P_T = _____

3.

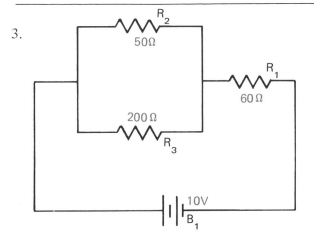

Find: R_T = _____

I_T = _____

E_{R_1} = _____

E_{R_2} = _____

E_{R_3} = _____

I_{R_1} = _____

I_{R_2} = _____

I_{R_3} = _____

P_T = _____

Using the PIRE wheel in Fig. 3-15, solve the following problems:

1. E = 100V, I = 2 amps, R = _____.

2. E = 50V, R = 1000 ohms, I = _____.

3. I = .5 amps, R = 50 ohms, E = _____.

4. E = 10V, I = .001 amps, R = _____.

5. I = .05 amps, R = 1000 ohms, E = _____.

6. P = 10W, I = 2 amps, E = _____.

7. E = 100V, I = .5 amps, P = _____.

8. P = 500W, E = 250V, I = _____.

9. I = .01 amps, R = 100 ohms, E = _____.

10. P = 100W, I = 2 amps, R = _____.

11. E = 10V, P = 10W, R = _____.

12. E = 500V, I = 2 amps, R = _____.

13. E = 100V, R = 1000 ohms, P = _____.

14. I = .5 amps, R = 50 ohms, P = _____.

15. I = 4 amps, R = 10 ohms, P = _____.

16. I = 10 mA, E = 50V, P = _____.

17. I = 20 mA, E = 100V, R = _____.

18. P = 10W, I = 1 amp, R = _____.

19. E = 1000V, R = 1000 ohms, I = _____.

20. I = 100 mA, R = 100 ohms, E = _____.

21. I = 100 mA, R = 100 ohms, P = _____.

22. P = 500W, E = 100V, I = _____.

23. E = 100V, R = 100 ohms, P = _____.

24. E = 50V, R = 10 kilohms, I = _____.

25. P = 10W, R = 10 ohms, E = _____.

26. P = 50W, R = 2 ohms, I = _____.

TEST YOUR KNOWLEDGE

1. State Ohm's Law three ways.
2. State Watt's Law three ways.
3. In a series circuit,

$$R_T =$$

4. What do you know about voltages and current in a series circuit?
5. In a parallel circuit of several equal resistors,

$$R_T =$$

6. In a parallel circuit of two unequal resistors,

$$R_T =$$

7. In a circuit of several unequal resistors in parallel,

$$R_T =$$

8. What do you know about the voltages and currents in a parallel circuit?
9. In a series circuit,

$$P_T =$$

10. In a parallel circuit,

$$P_T =$$

11. In series resistors,

$$R_T =$$

Chapter 4

MAGNETISM

For centuries the mystery of magnetism has baffled the scientist. Shepherds, tending their flocks in ancient days, were mystified by small pieces of stone, which attracted the iron tip on the shepherd's staff. The ancient Chinese navigators discovered that a small piece of this peculiar stone attached to a string would always turn in a northerly direction. These small stones were iron ore. They were called magnetite by the Greeks, because they were found near Magnesia in Asia Minor. Since mariners used these stones in the navigation of their ships, the stones became known as "leading stones" or LODESTONES. These were the first forms of natural magnets.

Today, a magnet may be defined as a material or substance which has the power to attract iron, steel and other magnetic materials. By laboratory experiments, it was discovered that the greatest attractive force appeared at the ends of a magnet. These concentrations of magnetic force are called <u>magnetic poles</u>; each magnet has a NORTH POLE and a SOUTH POLE. Further experimentation revealed that existing between the North and South Pole were many invisible lines of magnetic force. Each line was an independent line and not crossing or touching its adjacent line. The photo, Fig. 4-1, shows the magnetic lines of force. This field picture was created by placing a bar magnet beneath a sheet of paper, upon which was sprinkled some fine iron filings. Note the pattern of lines existing between the poles. Note the concentration of lines at each end of the magnet or its poles. Each line of force from the north pole goes to the south pole through

space and returns to the north pole through the magnet. These closed loops of the magnetic field may be described as <u>magnetic circuits</u>. You may compare the magnetic circuit to the electrical circuit, where the magnetizing force may be compared to voltage, and the magnetic lines may be compared to current.

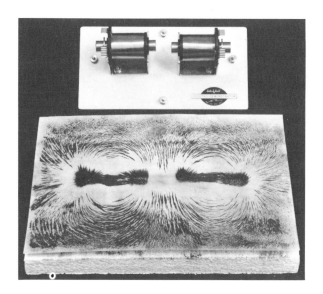

Fig. 4-1. The electrodemonstrator is used to show the magnetic field of two permanent magnets. See Activities at the end of this chapter, page 59.

Further scientific investigation has proved that the EARTH acts as one enormous magnet, with its poles close to the North and South geographical poles. Referring to Fig. 4-2, you will observe that magnetic north and the North geographic Pole do not coincide and a compass would not necessarily point toward true north. This angle between true and magnetic north is called the <u>angle of declination or variation</u>.

There is, however, an imaginary line around the earth where the angle of declination is zero. When standing on this line your compass would point to true north as well as magnetic north. At all other locations on the surface of the earth, a correction must be applied to the compass reading to find true north.

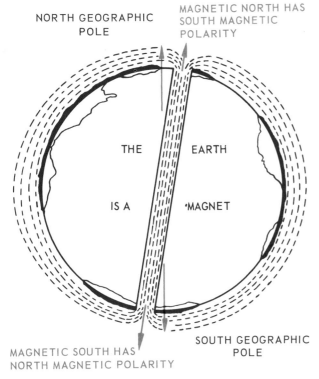

NORTH GEOGRAPHIC POLE

MAGNETIC NORTH HAS SOUTH MAGNETIC POLARITY

THE   EARTH

IS A   'MAGNET

MAGNETIC SOUTH HAS NORTH MAGNETIC POLARITY

SOUTH GEOGRAPHIC POLE

Fig. 4-2. The earth is an enormous magnet and is surrounded by a magnetic field.

LAWS OF MAGNETISM

The power of a magnet to attract iron has already been discussed. In Fig. 4-3, two bar magnets have been hung in wire saddles and are free to turn. The student will notice that when the N pole of one magnet is close to the S pole of the other that an attractive force will bring the two magnets together. On the other hand, if the magnets are turned so that two N poles or two S poles are close to each other, there will be a repulsive force between the two magnets. This proves two of the Laws of Magnetism:

UNLIKE POLES ATTRACT EACH OTHER

LIKE POLES REPEL EACH OTHER

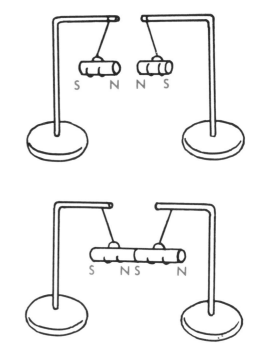

S N N S

S N S N

Fig. 4-3. These permanent magnets demonstrate the Laws of Magnetism.

By placing the magnets under a sheet of paper and using the iron filings to detect the invisible fields, you will observe in Fig. 4-4, that in the attractive position (N and S together) the invisible field is very strong between the two poles. However, when the magnets are placed with like poles together, the repulsive force is indicated by no invisible lines between the poles.

CAUTION: Certain precautions should be observed in the handling and storage of permanent magnets. Since the magnetism is a result of the molecular alignment, any rough handling such as dropping and pounding would upset the molecular alignment and destroy the magnet.

What causes a substance to become magnetized? An acceptable theory states that the molecules in an iron bar act as tiny magnets, and are arranged in-line so that their North and South poles are together. When the iron is demagnetized, the molecules are in random positions. This is illustrated in Fig. 4-5. This conclusion is substantiated by breaking a piece

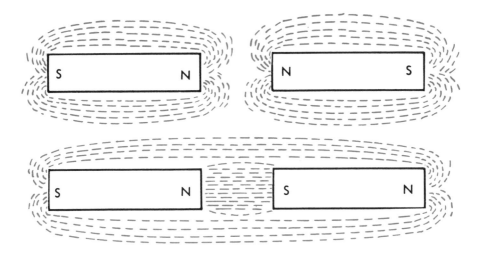

Fig. 4-4. These sketches show the magnetic fields of attracting and repelling magnets.

of magnetized iron into several pieces. Then each piece is a separate magnet. Fig. 4-6 shows a broken magnet. The conclusion is further justified by the way a magnet is made. For example, take an unmagnetized bar, and rub it several times in the same direction with a permanent magnet. A test will show that the bar is now magnetized. The rubbing with the magnet lined-up the molecules and caused the iron to become magnetized. Commercially, magnets are made by placing the material to be magnetized in a very strong magnetic field.

Permanent magnets are manufactured in a variety of shapes and sizes for particular applications in homes and industry, Fig. 4-7. Very small magnets may be used to temporarily attach letters and figures to a bulletin board or visual teaching aid, Fig. 4-8. The resistance symbols in this illustration have small magnets glued to their backs. They are held in place on a sheet iron bulletin board by magnetism. Industry uses these "boards" for traffic control, production flow, temporary notices and other announcements.

CAUTION: Magnets should be stored in pairs with north and south poles together to insure long life. A single horseshoe magnet may be preserved by placing a small piece of soft iron called a "keeper" across its poles.

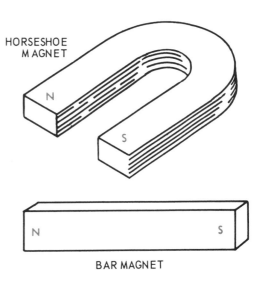

HORSESHOE MAGNET

BAR MAGNET

Fig. 4-7. Magnets are manufactured in a variety of styles, shapes and sizes.

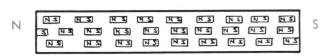

Fig. 4-5. In a permanent magnet the molecules are in line.

Fig. 4-6. A long magnet may be broken into several smaller magnets.

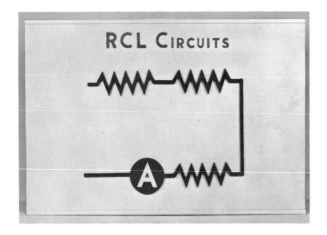

Fig. 4-8. Small magnets are used to attach symbols to the metal board. This idea has hundreds of applications.

CAUTION: Heat will destroy a magnet. Heat energy causes an increase in molecular activity and expansion which permits the molecules to return to their random positions of the unmagnetized piece of steel.

MAGNETIC FLUX

The many invisible lines of magnetic force surrounding a magnet are called the MAGNETIC FLUX. If it is a strong magnet, the lines will be more dense. So the strength of a magnetic field may be determined by its flux density, or the number of lines per square inch or per square centimetre. Flux density is expressed by the equation:

$$B = \frac{\Phi}{A}$$

where B equals flux density, Φ (phi) represents the number of lines and A is the cross-sectional area in either square inches, or square centimetres. If A is measured in square centimetres, then B is the number of lines per square centimetre or gauss.

THIRD LAW OF MAGNETISM

A simple experiment will demonstrate the third law of magnetism. Place one bar magnet on a table, then slowly approach it with a second bar magnet. If the polarities are opposite, they will attract each other when close enough together. This experiment proves that the attractive force increases as the distance between the magnets decreases. Actually this magnetic force, either attractive or repelling varies inversely as the square of the distance between the poles. For example, if the distance between two magnets with like poles is increased to twice the distance, the repulsive force reduces to one-quarter of its former value. This is valuable information. It explains the reason for accurately setting the gap in magnetic relays and switches. The term gap is the distance between the core of the electromagnet and its armature.

ELECTRIC CURRENT AND MAGNETISM

Scientists, during the eighteenth and nineteenth centuries, directed much of their research to the relationship between electricity and magnetism. The Danish physicist, Hans Christian Oersted, discovered that a magnetic field existed around a conductor carrying an electric current. The proof of this statement is seen in Fig. 4-9, where a current carrying

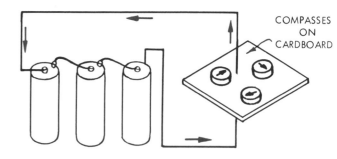

Fig. 4-9. Compasses line up to show circular pattern of magnetic field around current carrying conductor.

conductor is passed through a sheet of cardboard. Small compasses placed close to the conductor, point in the direction of the magnetic lines of force. Reversing the current will also reverse the direction of the compasses by 180 deg. This shows that the direction of the magnetic field depends upon the direction of current flow. The left hand rule for a conductor may be used to reveal the direction of the magnetic field. Grasp conductor with your left hand, extending your thumb in direction of current flow. Your fingers around conductor will indicate circular direction of field.

Fig. 4-10. These conventions are used to show the relationship between current flow and the magnetic field.

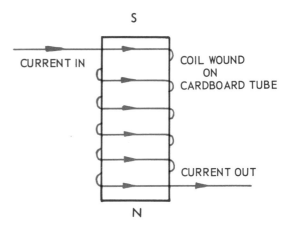

Fig. 4-11. A wire wound into a coil is a solenoid and has a polarity determined by the direction of current flow.

In Fig. 4-10, the dot in the center of the conductor at left is the point of an arrow. It shows that current is flowing toward you. Circular arrows give the direction of the magnetic field. This principle is especially important when electrical wires carry alternating currents. Then, placement of wires, or "lead dress," has certain influence on the behavior of the circuit.

THE SOLENOID

When a current carrying conductor is wound into a coil or SOLENOID, the individual magnetic field encircling the conductors tend to merge or join together. A solenoid will appear as a magnetic field with a N pole at one end, and a S pole at the opposite end. This solenoid is shown in Fig. 4-11. The LEFT HAND RULE FOR A COIL may be used to determine the polarity of the coil. Grasp the coil with your left hand in such a manner that your fingers encircle the coil in the direction of current flow. Your extended thumb will point to the N pole of the coil.

The strength of the magnetic field of the solenoid depends upon the number of turns of wire in the coil and the current in amperes flowing through the coil. The product of the amperes and turns is called the AMPERE-TURNS of a coil and is the unit of measure-

ment of field strength. If, for example, a coil of 500 ampere-turns will produce a required field strength, any combination of turns and amperes totaling 500 will be satisfactory, such as,

50 turns x 10 amps = 500 amp-turns

100 turns x 5 amps = 500 amp-turns

500 turns x 1 amp = 500 amp-turns

MAGNETIC CIRCUITS

Although a detailed study of magnetic circuits is beyond the scope of this text, you should be familiar with the terms used to describe quantities and characteristics of magnetic circuits. It has already been mentioned that a similarity exists between magnetic circuits and electrical circuits. Comparing circuits, it is found that the number of lines of magnetic flux, Φ (in electricity this is current, I) is in direct proportion to the force producing them. This force (F) is in units of magnetomotive force called GILBERTS. (In electricity this is voltage, E.) Also there is resistance in a magnetic circuit which is called RELUCTANCE and its unit of measurement is the REL ($\mathcal{R}$). The equation then becomes,

$$\Phi = \frac{F}{\mathcal{R}} \text{(Rowland's Law)}$$

where,

Φ = total number of lines of magnetic force

F = force which produces the field in gilberts

$\mathcal{R}$ = resistance to the passage of lines of force

The relationship between F and ampere-turns is,

F = 1.257 IN

where,

F = gilbert, I is the current and N = the number of turns.

Other terms encountered in the study of magnetism are as follows,

H = Intensity of a magnetizing force through a unit of length and is usually expressed in gilberts per centimetre. Mathematically,

$$H = \frac{F}{\ell} \text{ or } \frac{1.257 \text{ IN}}{\ell}$$

The term ℓ is expressed in centimetres.

The term reluctance has been mentioned as the unit of measurement of the resistance to the passage of magnetic lines of force. This may be further defined as the resistance offered by one cubic centimetre of air. Experimentation proves that as the flux path increases in length, the reluctance increases, and as the cross-sectional area of the flux path decreases, then the reluctance increases. The reluctance is also dependent upon the material used as the path. The ability of a material or substance to conduct magnetic lines of force is called PERMEABILITY. The permeability of air is considered as 1. The term is expressed by the Greek letter μ, pronounced mu. The factors determining the resistance in a magnetic circuit may

be compared to the factors affecting the resistance of electron flow in a conductor. Refer to Chapter 3. For the magnetic circuit:

$$\mathcal{R} = \frac{\ell}{\mu A} \quad \text{where,}$$

$\mathcal{R}$ equals rels, ℓ equals length of path, μ equals the permeability of the material in the path, and A equals the cross-sectional area of the path.

There is a definite relationship between the density of a magnetic field, the force producing the field, and the permeability of the material in which the field is produced. Mathematically,

$$\text{Permeability } \mu = \frac{\begin{array}{c}\text{B Flux lines per Cm}^2 \\ \text{or gauss}\end{array}}{\begin{array}{c}\text{H gilberts per Cm} \\ \text{of length}\end{array}}$$

The B, H and μ of common magnetic materials may be found in an electricity handbook.

ELECTROMAGNETS

The construction of a solenoid coil has already been discussed. The student should realize that in the case of the solenoid, air only is the conductor in the magnetic circuit. Other substances will conduct magnetic lines of force better than air. These materials would be described as having greater PERMEABILITY. To prove this, a soft iron core may be inserted in the solenoid coil, Fig. 4-12. The strength of the magnetic field will be greatly increased. Two reasons may be given. First, the magnetic lines have been confined or concentrated into the smaller cross-sectional area of the core and

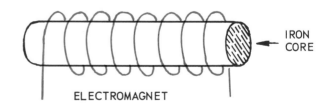

ELECTROMAGNET

IRON CORE

Fig. 4-12. The coil with an iron core is described as an electromagnet.

secondly, the iron provides a far better path (greater permeability) for the magnetic lines. Such a device is known as an ELECTRO-MAGNET. The rules for the determination of the polarity of an electromagnet are the same as those for the solenoid. Use the left hand rule. You may wish to construct a simple electro-magnet as illustrated in Fig. 4-13.

The electromagnet is made by selecting a prewound solenoid coil and inserting the soft iron core in the center. Wire according to the diagram. Experiment with this coil by using a compass. Connect the ends of the coil to a 6-volt battery and notice the polarity of the coil. Reverse the connections to the battery and observe the compass action. Remembering the direction of electron flow in a circuit, does this experiment prove the Left Hand Rule?

Continue this experiment by using the electromagnet to pick up several small nails. As the nails are held by the magnetic force, open the switch. The nails will fall. Leaving the switch still open, try to pick up some fine iron filings. Notice that some filings will be attracted to the core. In other words, the core has retained a small amount of its magnetism. This is called RESIDUAL MAGNETISM. As very little magnetism remained, the core would be considered as having LOW RETENTIVITY. Now replace the core of the magnet with a steel core and repeat the experiment. Once the steel is magnetized by the coil, it will be found that it retains its magnetism. A permanent magnet has been made because steel has a HIGH RETENTIVITY.

SOLENOID SUCKING COIL

Referring to Fig. 4-14, the experimentation is continued to demonstrate the sucking action of the solenoid. First, energize the solenoid coil by closing the switch, then bring the iron core close to one end of the coil. Notice the pulling magnetic force. Release the core from your

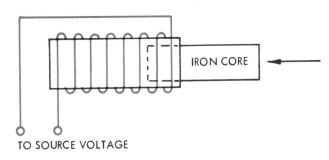

Fig. 4-14. When the solenoid is energized, the core is sucked into the center.

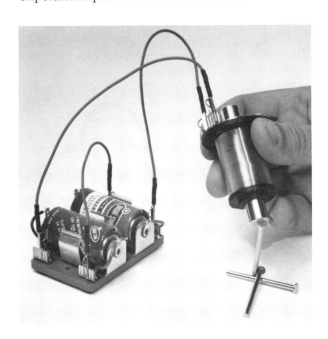

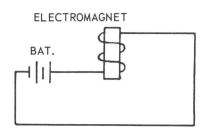

Fig. 4-13. The electrodemonstrator is used to show the principles of an electromagnet.

fingers and it will be sucked into the center of the coil and come to rest at the center of the coil. Is this not converting magnetism into mechanical motion? The solenoid sucking coil has numerous applications in industry for the electrical control of mechanical action.

A simple and practical door chime may be built by arranging the coil and parts as in the photo, Fig. 4-15. The door chime uses the

Fig. 4-15. In this application, the electrodemonstrator is used as a door chime. The same principles and construction may be used for other worthwhile and useful projects.

sucking coil principle. When the coil is energized, the core is sucked into the coil and strikes the chime.

THE RELAY

The relay is a device used to control a large flow of current by means of a low voltage low current circuit. It is a magnetic switch. Assemble the relay, as shown by the diagram, Fig. 4-16. A flashlight cell is used to energize the electromagnet and this circuit is controlled by the switch. When the coil is magnetized, its attractive force pulls the lever arm, called an ARMATURE, toward the coil. The contact points on the armature will open or close depending upon the arrangement, and control the larger high voltage circuit. In this circuit a light bulb is used which is connected to a 115 volt power source. The advantages of this device are apparent.

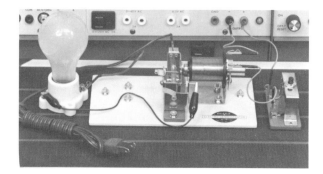

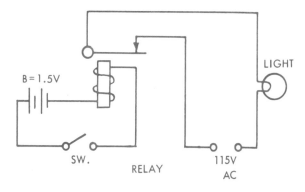

Fig. 4-16. This circuit uses the electrodemonstrator as a relay.

1. From the safety point of view, the operator touches only a harmless low-voltage circuit, yet controls perhaps several hundred volts by means of the relay.
2. Heavy current machines may be controlled from a remote location without the necessity of running heavy wires to the controlling switch.
3. Switching action by means of relays may be very rapid and positive.

There are hundreds of applications of relays at home and in industry to control motors and machines. In the automobile the relay is used in the current voltage regulator, in controlling the starting motor, the horn and the headlights. In electronic equipment and radio transmitters a relay is used in numerous high voltage circuits and provides a sensitive, positive means of control. Relays of this type are shown in Fig. 4-17. By consulting a radio parts catalog, you

will notice a large variety of relays designed for specific applications, see Fig. 4-18. In the selection of a relay for a special purpose, consideration should be given to the number of switching contacts required and the current carrying ability of these contacts. Well designed relays have points made of silver, silver alloys, tungsten and other alloys. A relay may be selected for either opening or closing a circuit and its "at rest" position would be "normally closed" or "normally open." The coil is probably the most important specification. Relays are designed with coils which operate on a direct current, or an alternating current. Some are very sensitive and require a milliampere or less to energize. The coil should be selected that will produce sufficient magnetic field at its rated voltage to insure positive contact of switch points at all times.

THE REED RELAY

The reed relay is another similar application of magnetism. Referring to Fig. 4-19, you will notice two magnetically sensitive switch contacts are enclosed in a glass tube. If a permanent magnet is brought close to the glass tube, it will cause the switch contacts to close. The reed relay can also be operated by an electromagnet. The operating coil is placed around the reed relay.

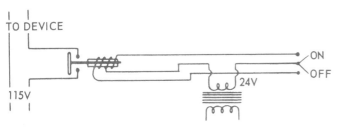

Fig. 4-17. Basic circuit using low-voltage relay, as used to control lighting circuit in the home.

Fig. 4-18. Type of relay used in electronic control circuits.

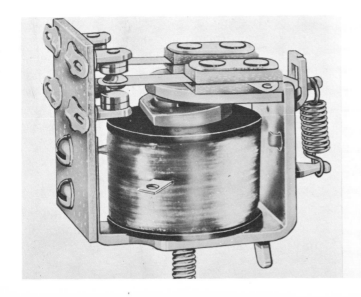

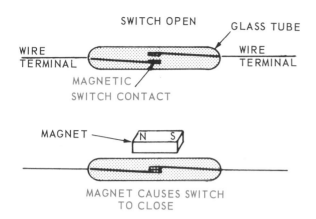

Fig. 4-19. A sketch of a reed relay. The magnet causes the reed contacts to close.

CIRCUIT BREAKER

In reference to electromagnets, the strength of a magnetic field is dependent upon the ampere-turns. If a coil is made with a fixed number of turns, then its field strength may be varied by controlling the current in the coil. Upon this principle is devised the Circuit

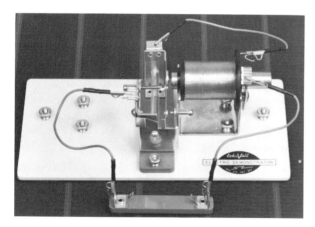

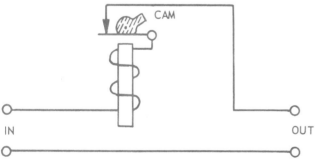

Fig. 4-20. Above. Electrodemonstrator assembled as a circuit breaker. Below. Circuit breaker circuit.

Breaker. Assemble the coil and armature parts as in Fig. 4-20. Compare this device to the relay and notice that in the circuit breaker, the coil and armature points are in series and the whole device is connected in series in the line carrying the load. If the line current exceeds a predetermined value, the coil will build up sufficient magnetic force to overcome the spring tension of the armature. The armature releases the points and will open and "break" the circuit. The contact points must be "reset" or closed manually. This is a protective device and prevents a circuit from being overloaded beyond the safe carrying capacity of the wires. Modern homes and industry use the circuit breaker in place of the fuse.

LESSON IN SAFETY: The circuit breaker is installed in the home for protection. It prevents the overloading of circuits and the danger of fire. The breaker should never be disabled. If the breaker repeatedly opens the circuit, discover the reason at once and take immediate steps to remedy it.

DOOR BELL AND BUZZER

Probably the most familiar of all devices using electromagnets is the buzzer. Arrange the coils and armature as in Fig. 4-21. Notice that the coil and the contact points are again connected in series. When the coil is energized, the armature is attracted toward the coil and the contact points open and "break" the circuit. The attraction of the coil falls to zero and the armature spring pulls the contact points closed. This once again energizes the coil and opens the contacts. This action continues as long as a voltage is applied. The device "buzzes" due to vibration of the armature. An extension may be placed on the vibrating armature on which a striker is attached. The striker hits a bell. This is the principle of the doorbell.

Specific examples of electromagnetic applications will be discussed in the later chapters of this text. You must thoroughly understand the

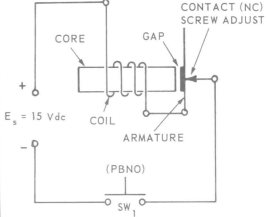

Fig. 4-21. Left. Electrodemonstrator assembled as a buzzer. Right. Buzzer circuit.

$E_s = 15$ Vdc

action of magnetic fields. You will find that principles of magnetism serve as the basis for many electrical and electronic phemonena.

MAGNETIC SHIELDS

A magnetic field has no boundaries. To the beginning student, it is difficult to understand that the force of the magnet will pass through any kind of material whether it be a concrete wall, glass, wood or any material one might mention. It is, nevertheless, true. However, an instrument or a circuit may be shielded from magnetic lines of force. This is possible by the application of the permeability of some materials. If a piece of iron is placed in a magnetic field, the lines of force will follow through the iron rather than through the air, because iron has a much greater permeability. The iron acts as a low resistance path for the magnetic lines. In Fig. 4-22 this action may be observed. By this method, magnetic lines may be conducted around an instrument. This is called "shielding." Many shields are used in electronic equipment to prevent magnetic fields from affecting the operation of circuits.

ACTIVITIES

I. The electrodemonstrator is set-up as in Fig. 4-1. The coils are used in this experiment only as a holding device for the two permanent

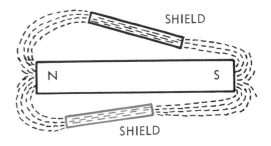

Fig. 4-22. Magnetic lines of force may be conducted around device by a low permeability material used as a shield.

magnets. Suspend the cardboard over the magnets on long bolts fastened to the base. Fix the magnets in the coils by wedging them with a pointed match stick or toothpick.

A. Place magnets so that poles attract. Sprinkle iron filings on cardboard and observe pattern.

B. Place magnets so that poles repel. Observe pattern.

C. Place magnets so that ends touch and observe pattern. What conclusions can you draw from these observations?

II. Make two wire saddles as shown in Fig. 4-3, and suspend permanent magnets in free space.

A. Place so that north poles are next to each other. Observe.

B. Place so that south poles are next to each other. Observe.

C. Place so that unlike poles are next to each other. Observe.

Make your conclusions and state the Laws of Magnetism in your own words.

III. Construct and operate the electro-magnet, the buzzer, the door chime and the circuit breaker. Use your electrodemonstrator. Ask your instructor to check each activity BEFORE power is applied.

MAGNETIC PUTTING GREEN

Here is an inexpensive and practical project which demonstrates several principles of electrical theory. It will also give you an opportunity to develop some other skills including the bending and shaping of metals, using taps and dies, working with wood, winding coils and following a blueprint. This project, Fig. 4-23 is a wintertime putting green. When the snow is deep and the weather cold, the putting green may be placed on the living room carpet. Here you may develop those golfing skills which will

help your game next summer. If you make your "putt," the ball will be <u>magnetically returned to you.</u>

The plans given should be used only as a general guide to construction. The putting green shown in the photo and drawings represents a satisfactory working model, but you are encouraged to change the overall appearance of the device. You may wish to use other materials. You may wish to devise other mechanical arrangements for the activation of the solenoid.

These suggestions may help.

1. The coil is mounted on adjustable brackets. The proper setting will require some experimentation, so the solenoid will kick the ball back to you at the proper moment.

2. The length of the plunger and the tension of the spring will determine the force of the return of the ball. You may want the ball to return several feet.

3. The spring brass contact in the switch must be so adjusted that the weight of a golf ball

Fig. 4-23. Magnetic putting green. See next page for additional drawings.

will close the switch and trigger the solenoid. Actual movement of the switch should only be about 1/32 in.

4. The switch plunger should protrude less than 1/16 in. so the ball will push down on it. If it protrudes too far, the ball will be stopped.

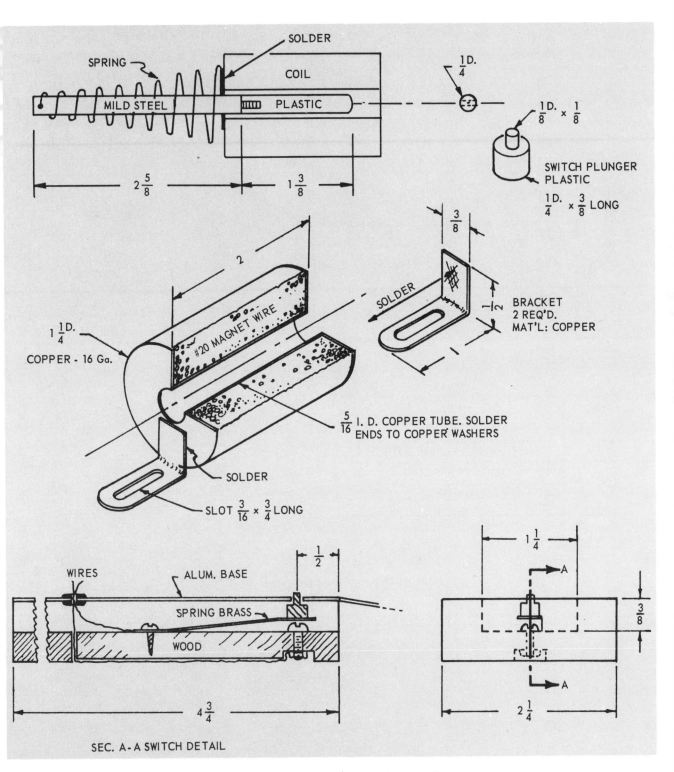

Fig. 4-23. Magnetic putting green (continued).

5. The coil is wound with No. 20 enameled copper or cotton-covered wire. Other sizes may be used. The 1 1/4 in. copper spool is filled with windings. Since the coil is connected across 115 volt ac line, it should have enough reactance to limit current flow.

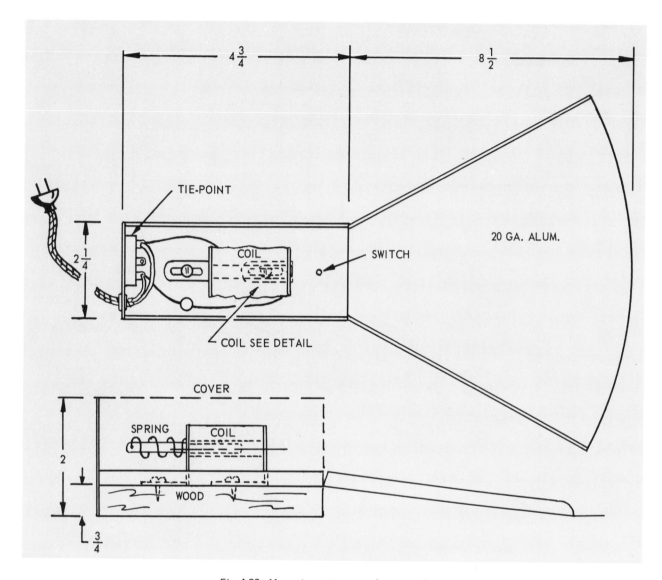

Fig. 4-23. Magnetic putting green (continued).

TEST YOUR KNOWLEDGE

1. State the first Law of Magnetism.
2. Why were early natural magnets called Lodestones?
3. Does the compass point toward true north? Explain.
4. Compare a magnetic circuit to an electrical circuit. Use Rowland's Law.
5. State the Third Law of Magnetism.
6. State in your own words the meaning of the following letter symbols.

   B   =

   Φ   =

   F   =

   H   =

   μ   =

   IN   =

7. State the Left Hand Rule for determining the polarity of an electromagnet.
8. What is meant by permeability?
9. What is the relationship between IN and Gilberts?
10. Explain the term residual magnetism.
11. Why are relays used in electrical circuits?
12. Explain the operation of a magnetic circuit breaker.

Chapter 5

GENERATORS

ELECTRICAL ENERGY FROM MECHANICAL ENERGY

In the studies of magnetism and the construction of the solenoid coil in Chapter 4, the discoveries of Hans Christian Oerstead were cited as proof that an electric current will produce a magnetic field. In 1831 another great scientist, Michael Faraday, pondered this question: If electricity would produce magnetism; can magnetism produce electricity? Based on the research and discoveries of Mr. Faraday, the electric dynamo was developed. He is known as the "father of the dynamo."

A simple experiment may be performed to illustrate the principles of Faraday's discoveries. Referring to Fig. 5-1, the coil is attached to a galvanometer. This sensitive meter indicates the flow and direction of an electric current. Insert a permanent bar magnet into the hollow coil.

Notice that as the magnet moves into the coil, the indicating needle of the meter will show a current in one direction. As the magnet is withdrawn from the coil, the needle will indicate a current in the opposite direction. Also observe that when the magnet is not moving, no current is produced. Repeat this same demonstration by holding the magnet in a fixed position, and move the coil up and down over the magnet. The conclusions drawn from this experiment are the fundamental principles of the operation of a generator (dynamo). In order to produce an electric current, there must be a magnetic field, a conductor and RELATIVE MOTION BETWEEN THE FIELD AND THE CONDUCTOR. A GENERATOR is a device which converts mechanical energy into electrical energy.

The more convenient and practical method of producing relative motion between a magnetic field and a conductor is to suspend a

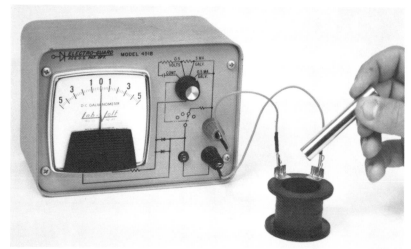

Fig. 5-1. Using large coil of electro-demonstrator, and a permanent magnet, for Faraday's experiment.

rotating coil within the field. This coil is called the generator ARMATURE. Fig. 5-2 illustrates simple generator action when a single turn coil is rotated in a field. Note the four positions during rotation. In A, both wires of the coil are moving parallel to the field so no voltage is induced. In position B, both sides of the coil are cutting the field at right angles and maximum voltage is induced. Position C is similar to

position A and the voltage has dropped to zero. In position D, once again the coil is cutting the field at right angles and maximum voltage is induced but in the opposite direction or polarity. Fig. 5-3 is a graph representing the instantaneous voltages induced in the coil during one revolution. Some explanation of this induced voltage is necessary. Fig. 5-4 shows single conductors passing through a magnetic field, but in opposite directions. Current flow direction is indicated by the arrows. In each case the induced current in the conductor forms a magnetic field around the conductor which opposes or is repelled by the fixed field. This is known as LENZ'S LAW. This opposition to the moving conductor must exist and some form of mechanical force must be applied to overcome this opposition. Water and steam power are used to turn generators in a large power plant.

The strength of the induced voltage in a rotating coil depends upon:

1. The number of magnetic lines of force cut by the coil.

2. The speed at which the conductor moves through the field.

Therefore, when a single conductor cuts across 100,000,000 (10^8) magnetic lines in one second, one volt of electrical pressure will be produced. The voltage can be increased by winding the armature with many turns of wire and also by increasing its speed of rotation. Reviewing Chapter 4 on Magnetism, this relationship may be expressed by the mathematical equation:

$$E = \frac{\Phi N}{10^8}$$

where, E equals the induced voltage

Φ equals the lines of magnetic flux

N equals revolutions per second

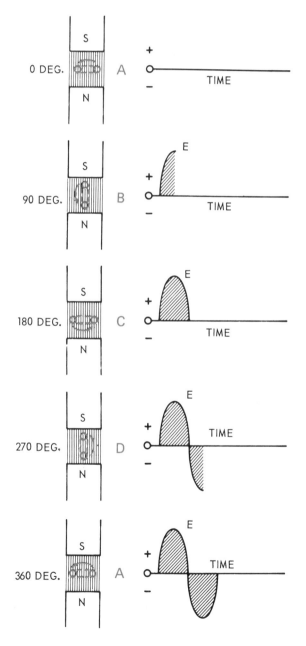

Fig. 5-2. The step-by-step development of the induced voltage during one revolution of armature in magnetic field.

For example, if the fixed magnetic field consisted of 10^6 lines of magnetic flux and a single

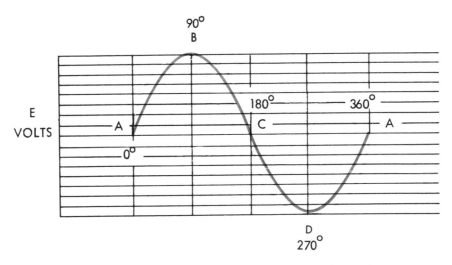

Fig. 5-3. A graph of voltage induced during one revolution of coil. Notice change in polarity.

conductor cut across the field 50 times per second, the induced voltage would equal:

$$E = \frac{10^6 \times 50}{10^8} = 50 \times 10^{-2} = .5 \text{ volts}$$

(The use of scientific notation or powers of ten is explained in the appendix.)

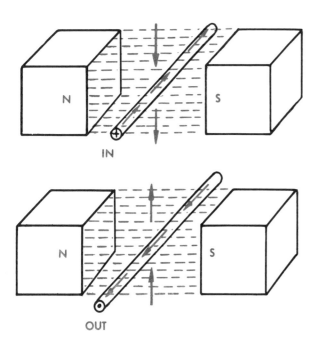

Fig. 5-4. The direction of current flow as a conductor conducts across a magnetic field.

CONSTRUCTION OF A GENERATOR

Improvements must be made on the simple generator described. In order to produce a strong magnetic field, the permanent magnets may be replaced by electromagnets. These are represented by field coils placed over pole pieces or shoes fastened to the steel frame or case of the generator. The revolving armature is suspended in the case with proper bearings. The single coil is replaced by coils of many turns of wire on the armature. External connections to the rotating armature are made through com-mutator sections and brushes. COMMUTATOR is a new word in this discussion.

Referring to the previous explanation, it will be seen that the induced current in the rotating coil flows first to maximum in one direction, then reverses and flows to maximum in the opposite direction. This defines an alternating current, and will be studied later in this chapter. However, it is desired to have a direct current at the output terminals of the dc generator. The alternating current in the rotating armature is converted to a pulsating direct current in the external circuit by means of the commutator. Examination of the diagrams in Fig. 5-5, will explain the action of the commutator and brushes.

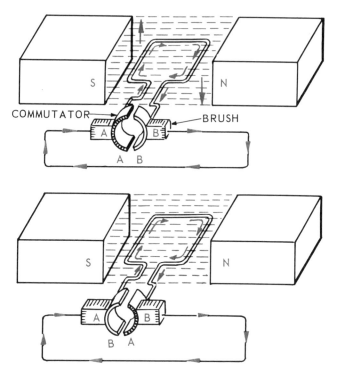

Fig. 5-5. The commutator changes the alternating current in the armature to a pulsating direct current in the external circuit.

Note that the current in the external circuit is always flowing in one direction. The output of this generator would appear as in the graph in Fig. 5-6. It would rise and fall from zero to maximum to zero, but always in the same direction. Following the action in Fig. 5-5, brush A is in contact with commutator section A, and brush B is in contact with section B. The first induced wave of current flows through the armature out of brush B around the external circuit and into brush A, completing the circuit. When the armature revolves one half turn, the induced current will reverse its direction, but the commutator sections have also turned with the armature. The induced current flowing out of section A is now in contact with brush B and

flows through the external circuit to brush A into section B, completing the circuit. The commutator has acted as a switch and reversed the connections to the rotating coil when the direction of the induced current was reversed. The current in the external circuit is pulsating direct current.

The output of this generator is not a smooth direct current. It is a pulsating or surging current. Further improvements may be made on the generator by increasing the number of rotating coils on the armature, and providing commutator sections for each set of coils. To help you understand the addition of coils to the armature, a simplified generator with two coils at right angles to each other is shown in Fig. 5-7. Each coil has its own induced current and as the current starts to fall off in one, it is replaced by the induced current in the next coil

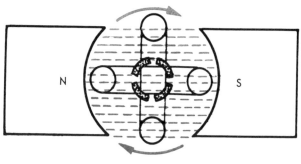

Fig. 5-7. Two coils at right angles are rotated in magnetic field.

as it cuts across the magnetic field. The graph of the output would appear as in Fig. 5-8. Notice that it is still a pulsating current, but the pulsations are twice as frequent and not as large. The output of the two coil generator is much smoother. By increasing the number of coils, the output will approach a pure direct current with only a ripple variation.

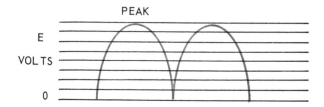

Fig. 5-6. A graph representing the generator output in volts.

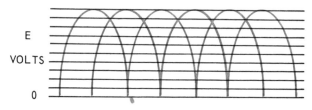

Fig. 5-8. A graph showing output of the generator in Fig. 5-7.

GENERATOR LOSSES

All of the mechanical power used to turn the generator is not converted into useful electrical power. There are some losses. Reviewing the studies of electrical conductors will emphasize that all wires have some resistance, depending on their size, material and length. Resistance uses power. In the generator coils there are many feet of copper wire. The resistance of this wire must be overcome by the induced voltage. Voltage used in this manner contributes nothing to the external circuit and only produces heat in the generator windings. Power loss in any resistance is equal to:

$$P = I^2 R$$

This is a particularly aggravating situation, because the loss increases as the square of the current. To be specific, if the current in the generator doubles, the power loss is four times or 2^2 more. The limiting factor in the generator output is usually the size of the wire and current carrying capacity of the armature windings. Losses resulting from resistance in the windings are classified as COPPER LOSSES or the $I^2 R$ loss.

The armature windings in the generator are wound on an iron core, which is slotted to hold the coils. If a moving conductor cuts across a magnetic field, a voltage is induced and the armature core may be also considered as a conductor. If the core were a solid block of iron, voltages would be induced in the core causing an alternating current to flow in the core. These alternating currents in the core are known as EDDY CURRENTS. They produce heat which is a loss of energy. To reduce this loss to a negligible value, the core is built up of thin sections or laminations. Each lamination is insulated from the others by lacquer or frequently by only oxide or rust. Laminating the core reduces voltage in the core, and increases the resistance to the eddy currents. The student should also reason that eddy currents will increase in the core, as the speed of rotation increases or the field density increases. When iron cores are used in rotating machines and transformers, they are laminated to reduce eddy current losses.

A third loss occurring in a generator is more difficult to understand. It is called HYSTERESIS LOSS. It is sometimes called molecular friction. As the armature rotates in the fixed magnetic field, many of the magnetic particles in the armature core remain in a line-up position with the fixed field. This causes a rotation of these particles in respect to those not in alignment with the field. This rotation causes internal friction between the magnetic particles and creates a heat loss. Generator manufacturers now use a silicon steel which has a low hysteresis loss. Heat treating of the core by annealing further reduces this loss.

TYPES OF GENERATORS

Basic generator types include: independently excited field, shunt, series and compound.

INDEPENDENTLY EXCITED FIELD GENERATOR

Generator output is determined by the strength of the magnetic field and the speed of rotation. The strength of the field is measured in ampere-turns, so an increase in current in the field windings will increase the field strength. Since output voltage is in direct proportion to field strength times the speed of rotation, most output regulating devices depend upon varying the current in the field. The field windings may be connected to an independent source of dc voltage as illustrated in Fig. 5-9. With the speed constant, the output may be varied by controlling the exciting voltage of the dc source. This is accomplished by inserting resistance in series with the source and field windings. This type of generator is called an independently excited field generator.

SHUNT GENERATOR

The inconvenience of a separate dc source for field excitation led to the development of the shunt generator, in which a part of the generated current is used to excite the fields. The field windings consist of many turns of relatively small wire and actually use only a small part of the generated current. The total current generated must, of course, be the sum

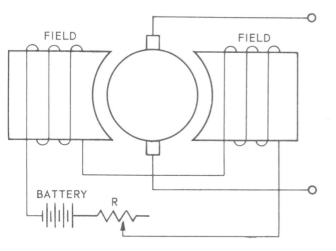

Fig. 5-9. The diagram of an independently excited generator.

of the field excitation current and the current delivered to the load. Thus, for practical purposes, the output current can be considered as varying according to the applied load. The field flux does not vary to a great extent, so the terminal voltage remains constant under varying load conditions. This type of generator is considered as a constant voltage machine. If the generator excites its own field, how does the generator start? The initial generator voltage is produced when the armature windings cut

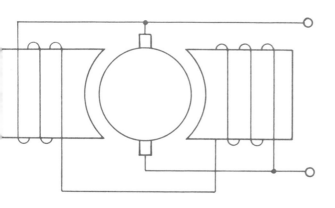

Fig. 5-10. The diagram of a shunt generator.

across a small magnetic field caused by the residual magnetism in the pole shoes or field coil cores. The conventional diagram of a shunt generator is shown in Fig. 5-10.

All machines are designed to do a certain amount of work. If overloaded their lives are shortened. A generator is no exception. With an overload the shunt generator terminal voltage will drop rapidly and the excessive current will cause the armature windings to heat and finally break down.

SERIES GENERATOR

The field windings of a generator may be placed in series with the armature and the load. Such a generator is diagramatically sketched in Fig. 5-11. It is seldom used.

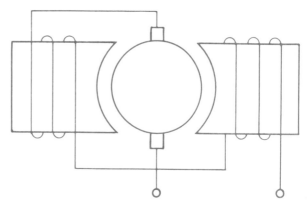

Fig. 5-11. A diagram of a series wound generator.

COMPOUND GENERATORS

The compound generator employs both series and shunt windings in the field. The series windings are usually a few turns of relatively large wire and are mounted on the same poles with the shunt windings. Both windings contribute to the field strength of the generator. If both act in the same direction or polarity, an increase in load would cause an increase of current in the series coils, thus increasing the magnetic field and increasing the terminal voltage of the output. The fields would be <u>additive</u> and the resultant field would be the sum of both coils. However, the current through the series winding can produce magnetic saturation of the core and result in a decrease of voltage as the load increases.

The manner by which the terminal voltage behaves is dependent upon the <u>degree of compounding</u>. A compound generator which maintains the same voltage either at no-load or full-load is said to be FLAT-COMPOUNDED. An OVER-COMPOUNDED generator, then, will increase the output voltage at full-load, and an UNDER-COMPOUNDED generator will have a decreased voltage at full-load current. A variable resistance may be inserted in parallel with the series winding to adjust the degree of compounding. Fig. 5-12 shows the <u>schematic</u> diagrams of the compound, the series and the shunt generator for comparison.

VOLTAGE AND CURRENT REGULATION

The regulation of a power source, whether it is a generator or a power supply, may be defined as the percentage of voltage drop between no-load and full-load. Mathematically, it may be expressed as:

$$\frac{E_{no\text{-}load} - E_{full\text{-}load}}{E_{full\text{-}load}} \times 100 = \% \text{ Regulation}$$

To explain this formula, assume that the voltage of a generator with no-load applied is 100 volts. Under full-load the voltage drops to 97 volts, then,

$$\frac{100V - 97V}{97V} = \frac{3}{97} \times 100 = 3.1\% \text{ approx.}$$

It is desirable in most applications to maintain the output generator voltage at a fixed value under varying load conditions. At this point, you must understand that the output voltage of the generator depends upon the field strength; the field strength depends upon the field current.

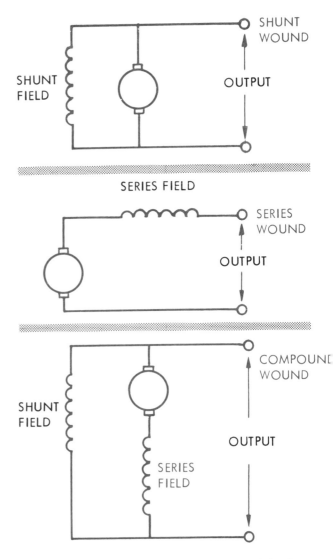

Fig. 5-12. A comparison of the wiring diagrams of the shunt, series and compound generator.

Current, according to Ohm's Law, varies inversely with resistance. Therefore, a device which would vary the resistance in the field circuit would also vary the voltage output of the generator. This regulator is drawn in Fig. 5-13, and may be used in the automobile. The generator output terminal G is connected to the battery and to the winding of a magnetic relay. The voltage produced by the generator causes a current to flow in the relay coil. If the voltage exceeds a predetermined value, the increased current will cause sufficient magnetism to open the relay contacts. Notice that the generator field is grounded through these contacts. When

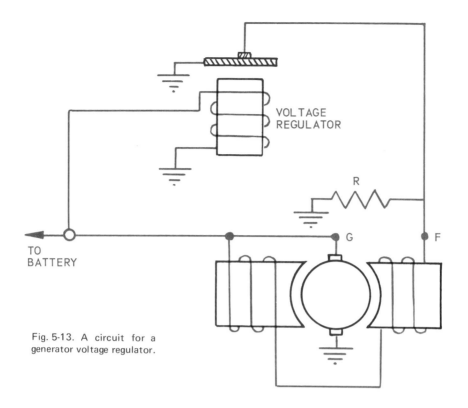

Fig. 5-13. A circuit for a generator voltage regulator.

they open, then the <u>field current</u> must pass through resistance R to ground. This reduces the current which reduces the field strength and reduces the terminal voltage. When the voltage is reduced, the relay contact closes, permitting maximum field current, and the terminal voltage rises. In operation these contact points vibrate, alternately cutting resistance in and out of the field circuit and maintaining a constant voltage output of the generator.

Mechanical-magnetic relays have served this purpose for many years, but electronic devices are now appearing on many models of cars. An electronic regulator using transistors for the switching functions is shown in Fig. 5-14. A study of transistor circuitry will be made in a later chapter of this text.

Fig. 5-14. A transistorized current-voltage regulator now used on many automobiles. (Delco-Remy)

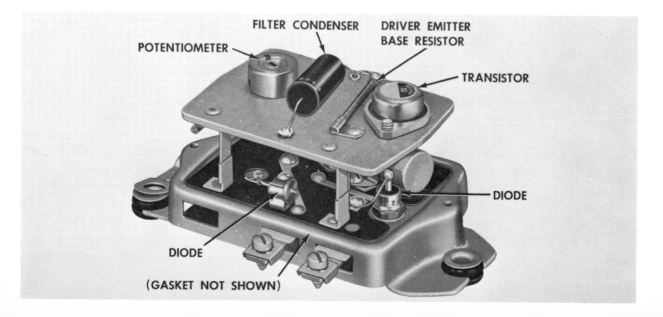

ALTERNATING CURRENT

In comparing direct current to alternating current, dc flows in only one direction, while ac periodically reverses its direction of flow in the circuit. In dc, the source voltage does not change its polarity. In ac, the source voltage changes its polarity between positive and negative. The graph in Fig. 5-15 illustrates the magnitude and polarity of an ac voltage. Commencing at zero, the voltage rises to maximum in the positive direction, then falls back to zero,

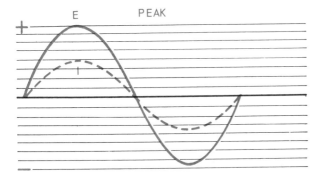

Fig. 5-15. Graph of current and voltage of alternating current.

then rises to maximum with the opposite polarity and returns to zero. The current wave is also plotted on the graph showing the flow of current and the direction of flow. Above the zero line, current is flowing in one direction; below the zero line, the current is flowing in the opposite direction. The graph represents instantaneous current and voltage at any point in the cycle. What is a CYCLE? A cycle may be defined as a sequence or chain of events occurring in a period of time. An ac cycle may be described as from zero to maximum negative to zero. The alternating current in your home changes direction sixty times per second, or has a frequency of sixty cycles per second (60 cps). FREQUENCY measured in cycles per second or hertz (Hz), is the number of complete sequences or chains of events occurring per second. If sixty cycles occur in one second, then the time period for one cycle is 1/60 of a second or .0166 seconds. This is the PERIOD of

the cycle. Referring again to Fig. 5-15, the maximum rise of the waveform represents the AMPLITUDE of the wave and the PEAK voltage or current.

In the study of the dc generator, we learned that the current induced in a rotating loop of wire in a magnetic field flowed first in one direction and then in the opposite direction. It was an alternating current. Points to remember:

1. The frequency of this cycle of events increases with an increase in speed of rotation.

2. The amplitude of the induced voltage depends upon the strength of the magnetic field.

In the solution of problems involving alternating currents, VECTORS are used to represent the magnitude and direction of a force. A vector is a straight line drawn to a convenient scale representing units of force. An arrowhead on the line shows the direction. The development of an ac wave is shown in Fig. 5-16, as a single coil armature, represented by the rotating vector, revolving one revolution through a magnetic field. Assuming that the peak induced voltage is 10 volts, using a scale of 1 inch equals 5 volts, the vector is two inches or 10 volts long. Vectors of this nature are assumed to rotate in a counterclockwise direction. The time base at the right in Fig. 5-16, is a line using any convenient scale, representing the period of one cycle or revolution of the vector. The time line is divided into intervals representing the time for specified degress of rotation during the cycle. For example: At 90 deg. rotation, one quarter of the time period is used; at 270 deg. rotation, three quarters of the time period is used. The wave is developed by plotting amplitude of voltage at any instant of revolution against the time interval. The developed wave is called a sine wave because the instantaneous induced voltages are proportioned to the sine of the angle θ (theta) that the vector makes with the horizontal. (Refer to Appendix for explanation of trigonometric functions and

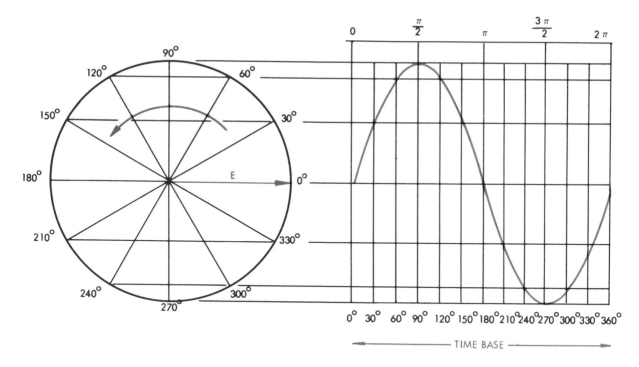

Fig. 5-16. The development of a SINE WAVE.

tables.) Then, the instantaneous voltage may be found at any point of the cycle by making use of the following equation:

$$e = E_{max}\text{SIN } \theta$$

To apply this equation: Assume that an ac generator is producing a peak voltage of 100 volts. What is the instantaneous voltage at 45 deg. of rotation?

$$e = 100 \text{ x sin } 45^{o}$$

$$e = 100 \text{ x } .707 = 70.7\text{V}$$

A study of an ac wave by comparison with direct current raises the important question of the actual value of the ac wave. The voltage and current are constantly varying and are at peak value only twice during a cycle.

Frequently it is necessary to find the AVERAGE VALUE of the wave. This is the mathematical average of all the instantaneous values during one half-cycle of the alternating current. The formula you would use for computing the average value from the peak value (max.) of the ac wave is:

$$E_{avg} = .637 \text{ } E_{max}$$

If E_{avg} is known the conversion to find E_{max} may be made by,

$$E_{max} = 1.57 \text{ } E_{avg}$$

A more meaningful value of an alternating current is found by comparing it to the heating effect equivalent to a direct current. This has been named the EFFECTIVE VALUE and may be found by the formula,

$$E_{eff} = .707 \text{ } E_{max}$$

If E_{eff} is known, the conversion to find E_{max} may be made by,

$$E_{max} = 1.414 \text{ } E_{eff}$$

The effective value is frequently called the rms (root-means-square) value because it represents the square root of the average of all currents squared between zero and maximum of the wave. The currents are squared so the power produced may be compared to direct current. Watt's Law states: $P = I^2 R$. By using the .707 factor the value of a direct current may be found which will equal the alternating current. For example; a 5 ampere ac will produce the same heating effect in a resistance as,

$$I_{eff} = .707 \times 5 = 3.53 \text{ amps dc}$$

Note that average and effective values may be applied to either voltage or current waves.

PHASE DISPLACEMENT

Several wave forms may be drawn on the same time base to show the phase relationship between them. In Fig. 5-17, E and I represent the voltage and current in a given circuit. The

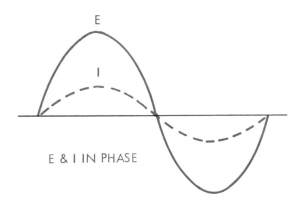

Fig. 5-17. Current and voltage are in phase.

current and voltage rise and fall simultaneously and cross the zero line at the same point. The current and voltage are IN PHASE. The "in phase" condition only exists in the purely resistive circuit as will be discovered in later instruction. Many times the current will lead or lag the voltage as in Fig. 5-18, creating a PHASE DISPLACEMENT between the two waves. This displacement is measured in degrees and is equal to the angle θ between the two polar vectors.

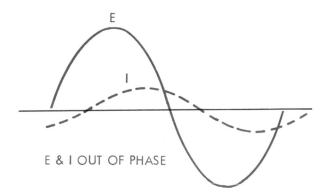

Fig. 5-18. Current and voltage are out of phase.

The waves are <u>out of phase</u>. Several out of phase currents and voltages may be displayed in this manner.

ALTERNATING CURRENT GENERATOR

The ac generator is similar in many respects to the dc generator with one important exception. The commutator is omitted. The ends of the armature coils are brought out to SLIP RINGS. Brushes sliding on the slip rings provide constant connection to the coils at all times and the current in the externally connected circuit is an alternating current. In large commercial type generators the magnetic field is rotated and the armature windings are placed in slots in the stationary frame or stator of the generator. This method allows the generation of large currents in the armature without the necessity of conducting these currents through moving or sliding rings and brushes.

The rotating field is excited through slip rings and brushes by means of an externally connected dc generator called the EXCITER. The dc voltage is necessary for the magnetic field. Commercial power generators are turned by water power and steam.

THE ALTERNATOR

Some of the new models of cars have replaced the dc generator with an ac generator called the ALTERNATOR. Fig. 5-19, shows the internal construction of this machine. The

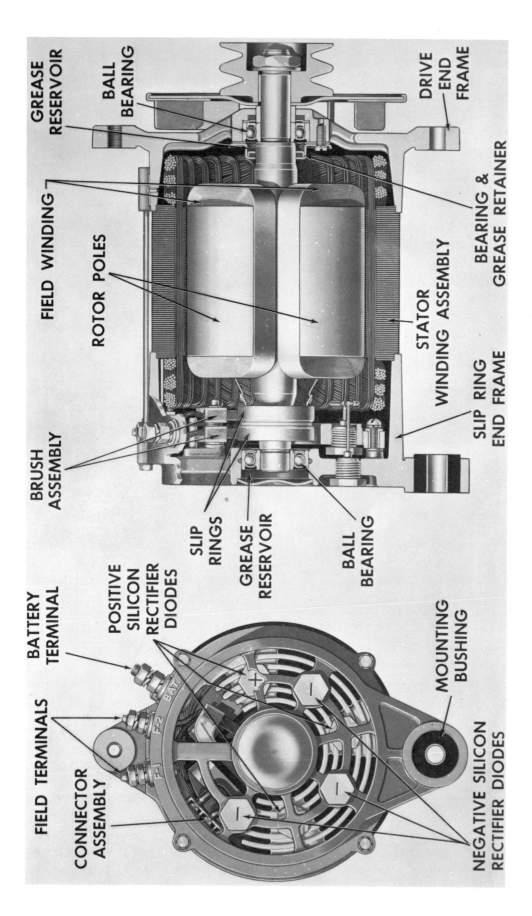

GREASE RESERVOIR

BALL BEARING

DRIVE END FRAME

FIELD WINDING

ROTOR POLES

STATOR WINDING ASSEMBLY

BEARING & GREASE RETAINER

SLIP RING END FRAME

BRUSH ASSEMBLY

SLIP RINGS

GREASE RESERVOIR

BALL BEARING

BATTERY TERMINAL

POSITIVE SILICON RECTIFIER DIODES

FIELD TERMINALS

CONNECTOR ASSEMBLY

MOUNTING BUSHING

NEGATIVE SILICON RECTIFIER DIODES

Fig. 5-19. The ac generator used on many automobiles. It is called an "alternator." (Delco-Remy)

output is <u>rectified</u> to direct current for charging the battery and other electrical devices in the car. Manufacturers claim certain advantages of the alternator over the dc generator, among which are higher output at lower speeds and trouble-free service.

The principle of the generator may be demonstrated in the useful and practical rpm meter illustrated in Fig. 5-20. It is simple to construct and provides an opportunity for the individual design of the case. A small dc motor, such as found in toy automobiles, is used as the generator. The shaft of the motor is wrapped

with friction tape or rubber. This friction wheel is held against a rotating shaft or wheel. The generated emf is measured on the 0-10 mA meter. This meter may be calibrated to read directly in rpms. The switch permits measurements in either clockwise or counterclockwise rotation. The "adjust screw" is used as a calibrator or range setting control.

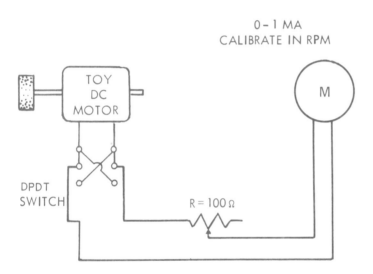

Fig. 5-20. A simple and practical rpm meter or tachometer.

TEST YOUR KNOWLEDGE

1. Write a brief description of Faraday's experimentation and discovery.
2. The strength of induced voltage depends upon _____ and _____.
3. What is the purpose of the commutator?
4. Name and describe three losses associated with the generator.
5. Why are the cores of armature and transformers constructed of laminated sheets of iron?

6. Name three types of generators and draw a schematic diagram of each.
7. A generator has a no-load voltage of 25 volts. When load is applied, terminal voltage drops to 24 volts. What is the percent of regulation?
8. How may output of generator be controlled when operating at constant speed?
9. What is the purpose of the cut-out in an automobile generator circuit? Explain

operation and draw schematic diagram.

10. Define: cycle, frequency, period, amplitude of an ac wave.

11. A generated voltage has a peak value of 100 volts. Using the vector scale of 1 inch = 50 volts, construct a sine wave. You will need a drawing board and instruments.

12. In problem 11, what is the instantaneous voltage at 69 degrees?

13. What is the effective value of the generated voltage in problem 11?

14. What is the peak value of the alternating current used in your home. (115 volts)

15. Draw waveforms showing the following current and voltage relationships:
    a. In phase.
    b. Current lagging by 45 deg.
    c. Current leading by 30 deg.

Application of electronics is shown in a modern machine shop. Programmed tape is being used to control cutting operation of a milling machine.
(Cincinnati Milacron)

Chapter 6

INDUCTANCE AND RL CIRCUITS

The study of electricity and electronics revolves around the characteristics of a circuit such as inductance, capacitance, resistance and combinations of these components in series and parallel. An analysis of resistance in a circuit was made in Chapter 3. This chapter will help you to answer these questions:

1. What is an inductor and inductance?

2. What is the effect of inductance in a circuit?

3. What methods are used to measure and compute values of current and voltage in an inductive circuit?

INDUCTANCE

INDUCTANCE may be defined as the property in an electric circuit which resists a change in current. This resistance to a change of current is the result of the energy stored within the magnetic field of a coil. A coil of wire will have inductance. Set up again the experiment illustrated in Chapter 5, Fig. 5-1, using the hollow coil A connected to a galvanometer. Move the permanent magnet in and out of the coil, and notice the deflection of the needle in the meter. When the magnet moves in, current will flow in one direction, and the current will reverse direction when the magnet is withdrawn. No current will flow unless the magnet is moved. This is the same principle as demonstrated in Chapter 5, on the theory of generators. By moving the magnet a voltage was induced in the coil which caused current to indicate on the meter. There was a definite relationship between the movement of the magnet and the direction of current flow. This is an application of Lenz's Law, which states that the field created by the induced current is of such a polarity that it opposes the field of the permanent magnet.

A coil connected to a source of direct current will build up a magnetic field when the circuit is closed. The expanding magnetic field cutting across the coil windings will induce a counter emf or voltage, which will oppose the source voltage and oppose the rise in current. When the current reaches its maximum value and there is no further change, there is no longer an induced counter emf. The current is then only limited by the ohmic resistance of the wire. However, if the source voltage is disconnected, the current tends to fall to zero, but the collapsing magnet field again induces a counter emf which retards the reduction of current.

The inductance of the coil resists any change in current value. The symbol for inductance is the capital letter L and inductance is measured in a unit called a henry (h). A henry represents the inductance of a coil if one volt of induced emf is produced when the current is changing at the rate of one ampere per second. This may be expressed mathematically as,

$$E = L \frac{\Delta I}{\Delta t}$$

where E equals the induced voltage, L is the inductance in henrys, ΔI is the change of current in amperes and Δt is the change of time in seconds, The symbol Δ means a "change in." Because the strength of an induced voltage is dependent upon the strength of the field, it follows that a stronger magnetic field would produce higher induced voltage. The magnetic field may be strengthened by inserting a core in the coil. The iron core has higher permeability than air and concentrates the lines of force. This was explained in the study of magnetism.

Large inductors are usually wound on laminated iron cores and their inductance is measured in henrys. Smaller inductors, used at higher frequencies, may have powdered iron or air cores. They have inductance measured in millihenrys, (1/1000 of a henry) and microhenrys, (1/1,000,000 of a henry). An inductor used in a power supply called a choke is shown in Fig. 6-1. Radio frequency chokes using air cores are shown in Fig. 6-2.

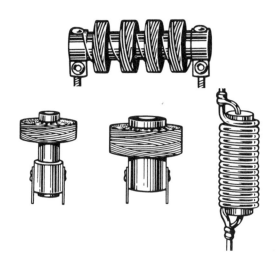

Fig. 6-2. RF or radio frequency chokes.　(J. W. Miller Co.)

tion is dependent upon the number of turns of wire in the coil, the relationship between the length of the coil to its diameter and the permeability of the core. This effect may be observed by setting up the experiment in Fig. 6-3. First, connect the light directly to the 6V power source and observe that as the switch is closed the light will burn at full brilliance instantaneously. Connect coil F in series with the light and close the switch. A slight delay in approaching full brilliance should be noted due to the inductance of the coil. Turn the light off and insert the iron core in the coil. Once again close the switch and note that the light comes to brilliance rather slowly and not to full brilliance. Why does the core increase the delay in brilliance? Repeat the above experiment, using coil D in which self-inductance is increased by the number of turns of fine wire in the coil. Move the core in and out of the coil and note the change in brilliance due to the change in self-inductance. This principle may be used in theaters to dim the houselights.

TRANSIENT RESPONSES

The response of the current and voltage in a circuit after an immediate change in applied voltage is known as a TRANSIENT RESPONSE. Referring to the diagram in Fig. 6-4, a coil is connected to a dc voltage source. When

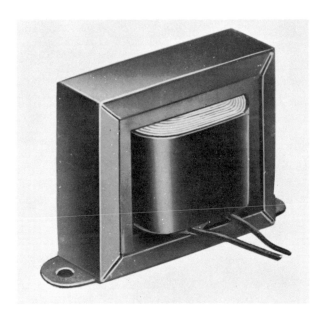

Fig. 6-1. Choke, used in filter section of power supply. (United Transformer Corp.)

A changing current through an inductor produces an expanding or collapsing magnetic field which is cutting across the wires of the coil. A counter emf is induced which opposes the change of current. This is called self-induction and the magnitude of the self-induc-

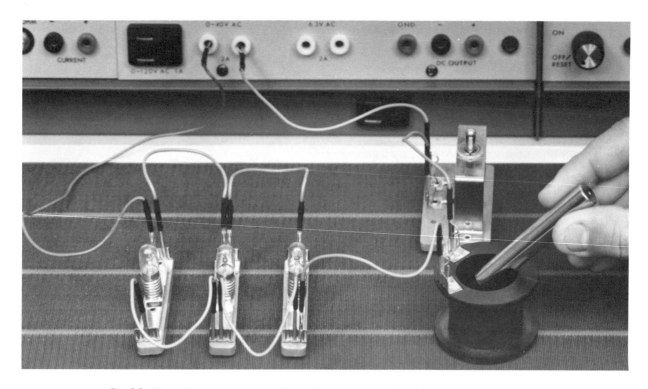

Fig. 6-3. The self-induction of a coil may be demonstrated with the electrodemonstrator. Experiment with different sizes and kinds of cores.

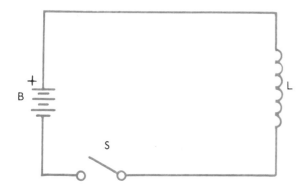

Fig. 6-4. The schematic diagram of the L circuit described in the text.

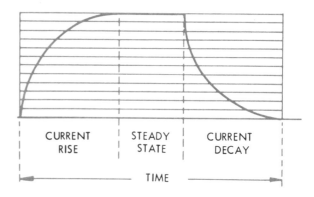

Fig. 6-5. The graph shows the transient response of the L circuit as the switch is closed and opened.

the switch is closed, the current will build up gradually, due to the self-induction of the coil and the internal resistance of the battery. When the switch is opened, the current will decay in like manner. The rise and decay of the current in this circuit is shown by the graph in Fig. 6-5. It is important to understand that the opposition to the rise or decay of current occurs only when there is a change in applied voltage. When there is no change, the current remains at its steady-state value, dependent only upon the resistance of the coil.

When a resistance is connected in series with an inductor, as illustrated in the RL circuit in Fig. 6-6, the behavior of the voltage and current should be carefully studied for increases and decreases. For example, when the switch is closed, the counter emf (E_L) of the coil equals the source voltage. Since no current has started to flow, the IR drop across R equals zero. As the current gradually builds up, the voltage E_R increases and counter emf E_L decreases until a steady state condition exists. All the voltage drop is E_R and there is no drop across L.

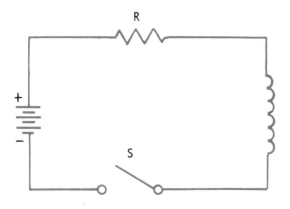

Fig. 6-6. A diagram of the RL circuit.

ered charged or discharged after the duration of five time constant periods. The following table gives the voltage at the end of each time constant, assuming E source is 100 volts.

TIME CONSTANT	CHARGING	DISCHARGING
1	63.2V	36.8V
2	86.5V	13.5V
3	95V	5V
4	98V	2V
5	99V	1V

Now if the RL circuit is shorted by another switch, S_2 as in Fig. 6-7, the stored energy in the field of L immediately develops a voltage and the circuit is discharged by current flowing through R. The graphs showing the charge and discharge of the circuit appear in Fig. 6-8. The base line or X axis of these graphs represents time and the transient response of an RL circuit does require a definite time, depending upon the values of R and L. This is called the TIME CONSTANT of the circuit and may be found by the formula,

$$t = \frac{L \text{ (in henrys)}}{R \text{ (in ohms)}}$$

when t equals the time in seconds for the current to increase to 63.2 percent of its maximum value or to decrease to 36.7 percent. For all practical purposes, the circuit is consid-

These points are plotted in Fig. 6-8 and a clear picture of the rise and decay of the voltage and current may be seen.

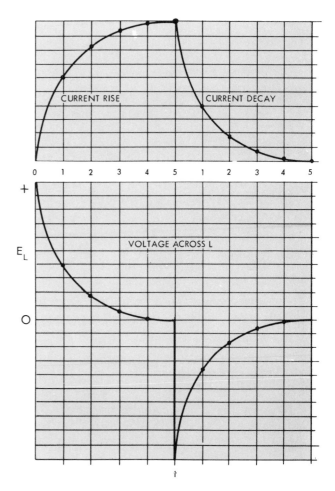

Fig. 6-8. The transient response curves for current and inductive voltage of the RL circuit.

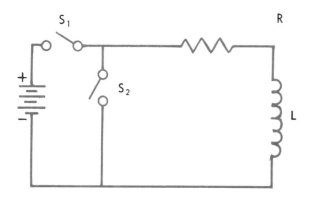

Fig. 6-7. The coil L is shorted through switch S_2 and the magnetic field collapses.

MUTUAL INDUCTANCE

In the study of generator regulation, brief mention was made of the magnetic fields created by the windings of the cut-outs as being additive or cancelling. When two coils are within magnetic reach of each other, so the flux lines of one coil will link with or cut-across the other coil, they would be considered as having MUTUAL INDUCTANCE. If they are close to each other, many magnetic lines in the flux would link with the second coil. Conversely, if they were some distance apart, there might be very little linkage. The mutual inductance of two coils may be greatly increased if a common iron core is used for both coils. The degree to which the lines of force of one coil link with the windings of the second coil is called COUPLING and if all lines of one cut across all the turns of the other it is UNITY COUPLING. Various percentages of coupling may exist due to the mechanical position of the coils. The amount of mutual inductance may be found by the formula,

$$M = k\sqrt{L_1 L_2}$$

where, M = mutual inductance in henrys

   k = the coefficient or percentage
   of coupling

   L_1 and L_2 = the inductance of
   respective coils.

The mutual inductance must be considered when two or more inductors are connected in series or parallel in a circuit.

THE TRANSFORMER

One of the familiar applications of mutual inductance is the TRANSFORMER. A transformer may be defined as a device used to transfer energy from one circuit to another by electromagnetic induction. Basically, a transformer consists of two or more coils of wire around a common laminated iron core, so the

coupling between the coils approaches UNITY. Transformers have no moving parts and require very little care. They are simple and rugged and efficient devices.

The construction of the simple transformer is illustrated in Fig. 6-9 with its schematic symbol. The first winding or input winding is called the PRIMARY and this winding receives the energy from the source. The second winding or output winding is called the SECONDARY and the output load is attached to this winding. The energy in the secondary is the result of the mutual induction between the secondary and the primary windings. The varying magnetic field of the primary cuts across the windings of the secondary and induces a voltage in the secondary. Therefore, the transformer is a device which must operate on an alternating current or a pulsating direct current. The primary field must be a moving magnetic field in order for induction to take place.

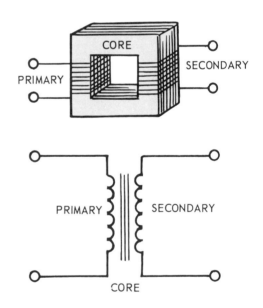

Fig. 6-9. A sketch showing transformer construction and the schematic symbol used in circuit diagrams.

In Fig. 6-10, two types of construction are shown, namely, the core type and the shell type. In the core type the coils surround the core and in the shell type the core surrounds the coil. Both types are used in general applica-

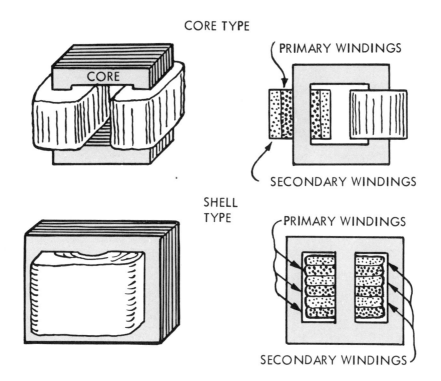

CORE TYPE

PRIMARY WINDINGS

CORE

SECONDARY WINDINGS

SHELL TYPE

PRIMARY WINDINGS

SECONDARY WINDINGS

Fig. 6-10. The construction of core and shell type transformers.

tions. In either type of construction, consideration must be given to assure maximum coupling between the primary and secondary so that very little flux loss or leakage results. In the core type, parts of both windings are wound on each leg of the core. In the shell type the windings are in alternate layers.

TURNS RATIO

One of the major uses of the transformer is the result of its ability to step-up or step-down a voltage. An example is the power transformer in Fig. 6-11, which is found in the power supply of radios and transmitters. In this transformer, several secondary windings are used to provide high and low voltages necessary for proper operation of the equipment. How may this be accomplished? Reference to the diagram in Fig. 6-12, will show that the output voltage in relation to the input voltage is the same as the ratio between the number of turns in secondary and primary. To express this mathematically,

$$\frac{E_{primary}}{E_{secondary}} = \frac{N_{primary}}{N_{secondary}}$$

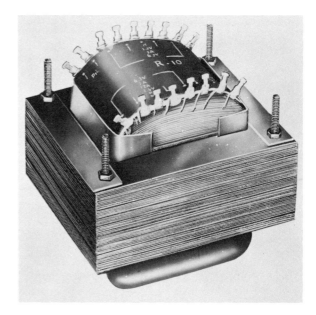

Fig. 6-11. A power transformer with multiple secondary taps used in the power supply for electronic circuits. (United Transformer Corp.)

where E is voltage and N equals the number of turns. This relationship is called the TURNS-RATIO of the transformer. An example will clarify this relationship.

83

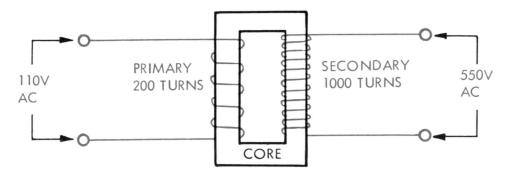

Fig. 6-12. The turns-ratio determines the output voltage.

A given transformer has 200 turns in the primary and 1000 turns in the secondary. If the applied voltage is 110 volts ac, what is the secondary voltage?

$$\frac{110 \text{ volts}}{E_s} = \frac{200}{1000}$$

Transposing this equation,

$$E_s = \frac{1000 \times 110}{200} = 550V$$

This is an example of a STEP-UP transformer.

If this transformer were constructed with a 10 turn secondary, then,

$$\frac{110V}{E_s} = \frac{200}{10}$$

$$E_s = \frac{110 \times 10}{200} = 5.5V$$

This exemplifies a STEP-DOWN transformer.

In Fig. 6-13 is shown another example of a power transformer with several taps on the secondary winding to provide a variety of voltages. A TAP is a fixed electrical connection made to a winding. If a tap is made on the secondary winding as in Fig. 6-14, two voltages can be obtained from the output with reference to the center tap. One of these two voltages is 180 deg. out of phase with the other. The tapped secondary winding of the power trans-

Fig. 6-13. A typical power supply transformer used in radios, TVs, and transmitters. (United Transformer Corp.)

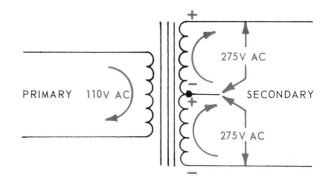

Fig. 6-14. Tap on secondary winding.

former will be further explained in full wave rectifier circuit operation in the chapter on Power Supplies.

The primary of a power transformer may be also tapped. This permits varying input voltages to be used. Sometimes this is necessary if electrical or electronic equipment is moved from one location to another where the ac voltages are not the same.

The development of the transformer explains the reason that alternating current is universally used. The student should realize that the power used in the secondary circuit must be supplied by the primary. Assuming that the transformer is 100 percent efficient, then the power $I_s \times E_s$ in the secondary must equal $I_p \times E_p$ in the primary.

Example: A step-up transformer will produce 300 volts in the secondary when 100 volts ac is applied to the primary. If a 100 ohm load is applied to the secondary, as in Fig. 6-15, a current of 3 amperes would flow.

$$I = \frac{300 \text{ volts}}{100 \text{ ohms}} = 3 \text{ amps}$$

The power used in the secondary would be:

$$P = I_s \times E_s = 3 \text{ amps} \times 300V = 900 \text{ watts}$$

Since the primary must supply this power, then

$$I_p = \frac{P_p}{E_p} \text{ or } \frac{900 \text{ watts}}{100 \text{ volts}} \text{ or } 9 \text{ amps}$$

and

$$I_s \times E_s = I_p \times E_p = 900 \text{ watts}$$

The important lesson of this transformer action is,

As voltage <u>increases</u> in the secondary, the current in the secondary <u>decreases</u>.

Why is the power company concerned with these facts? Remember, in the study of conductors, that all wires have some resistance. And also remember that the power loss in a circuit or conductor is, $P = I^2 R$. A long length of wire having a resistance of 2 ohms and

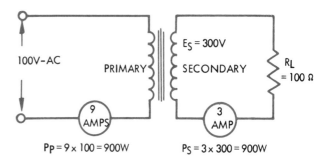

$$P_P = 9 \times 100 = 900W \qquad P_S = 3 \times 300 = 900W$$

Fig. 6-15. The relationship is shown between voltage, amperage and power in the primary and secondary of a transformer.

carrying a 10 ampere current will have a power loss of,

$$P = I^2 R = 10^2 \times 2 = 200 \text{ watts}$$

If the current is doubled to 20 amperes, then the loss would be,

$$P = I^2 R = 20^2 \times 2 = 800 \text{ watts}$$

or four times as much. Power loss varies as the square of the current. Therefore, it is more economical to raise the voltage with the associated decrease in current and reduce power loss in the transmission lines. Transformers are used for this purpose, as illustrated in Fig. 6-16.

By turning on an electric appliance in your home, which uses 10 amperes of current, the power consumed would be,

$$P = I \times E = 10 \times 120 = 1200 \text{ watts}$$

In the 12,000 volts city lines a current of .1 amp would flow,

$$1200 \text{ watts} = .1 \text{ amp} \times 12,000 \text{ volts}$$

In the cross country power lines a current of .02 amp would flow,

$$1200 \text{ watts} = .02 \text{ amp} \times 60,000 \text{ volts}$$

By raising the voltage, the power company is able to supply great cities and industries with electricity and use relatively small wires for

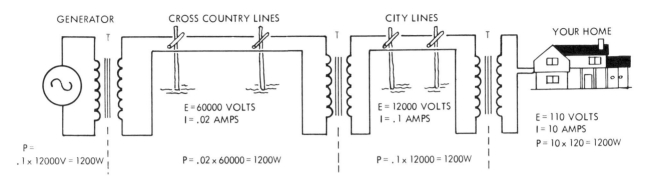

GENERATOR

CROSS COUNTRY LINES

CITY LINES

YOUR HOME

E = 60000 VOLTS
I = .02 AMPS

E = 12000 VOLTS
I = .1 AMPS

E = 110 VOLTS
I = 10 AMPS
P = 10 x 120 = 1200W

P =
.1 x 12000V = 1200W

P = .02 x 60000 = 1200W

P = .1 x 12000 = 1200W

Fig. 6-16. The use of transformers in the transmission of electric power to reduce line losses.

transmission lines. Although smaller wires do have greater resistance per foot, the loss resulting from this resistance is only in direct proportion to the current used. The power company selects a wire that has the least resistance, yet is large enough to carry the anticipated current. It must also be structurally strong to withstand high winds, ice and snow.

ISOLATION

A power or filament transformer provides magnetic coupling from the primary winding to the secondary winding even though there is no physical connection between them. Because of this effect, the secondary winding is "isolated" from the primary winding. This is an important safety factor. A person can be shocked only by touching both of the secondary wires in a transformer. With unisolated 110 volts alternating current, one lead is hot and one lead is ground. A person standing on ground and touching a hot lead with one hand will be shocked.

Autotransformers have only one winding. The primary and the secondary windings are connected and there is no isolation. In Fig. 6-17, note that autotransformers can be either step-up or step-down.

TRANSFORMER LOSSES

Three types of losses are associated with transformer construction. All losses result in heat.

1. COPPER LOSSES are the result of the resistance of the wire used in the transformer windings. These are sometimes called the I^2R losses and vary as the square of the current according to Ohm's Law and the Power Law.

2. EDDY-CURRENT losses are caused by small whirlpools of current induced in the core material. These losses are minimized by the laminated core construction. Each lamination is insulated from its adjacent layer by varnish, thus cutting the path for these currents to flow.

3. HYSTERESIS LOSS is sometimes called molecular friction, and is the result of the magnetic particles changing polarity in step with the induced voltage. Special alloys and heat treating processes are used in the manufacture of core materials, which minimize hysteresis loss.

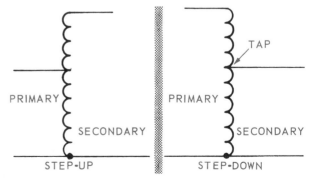

Fig. 6-17. Autotransformers are less expensive than regular transformers and the tap may be made variable so that the secondary voltage can be changed.

REASONS FOR USING A TRANSFORMER

The basic reasons for using a power or filament transformer are:

1. To step up or step down voltage or to step up or step down current.

2. To provide two voltages 180 deg. out of phase with each other.

3. To provide isolation from the primary to the secondary.

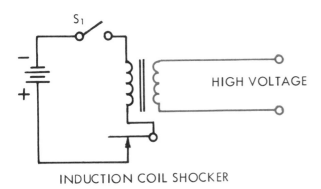

INDUCTION COIL SHOCKER

THE INDUCTION COIL

A transformer requires an alternating current or a pulsating direct current. A simple device to demonstrate transformer action is the "Electric Shocker" or induction coil constructed in Fig. 6-18. At first a buzzer is assembled. The armature and point action in the buzzer circuit cause the magnetic field of the coil to expand and collapse. It is a continually moving field. A secondary coil of many turns of fine wire is placed over the primary buzzer coil. A high voltage is induced in the secondary.

LESSON IN SAFETY: People vary in their ability to stand an electric shock. Although this project does not produce a dangerous shock to a normal healthy person, good judgment should be exercised when operating it.

A capacitor has been added to this circuit to improve its action. Capacitance will be studied in later lessons.

Fig. 6-18. The electrodemonstrator is used to show transformer action and an induction coil.

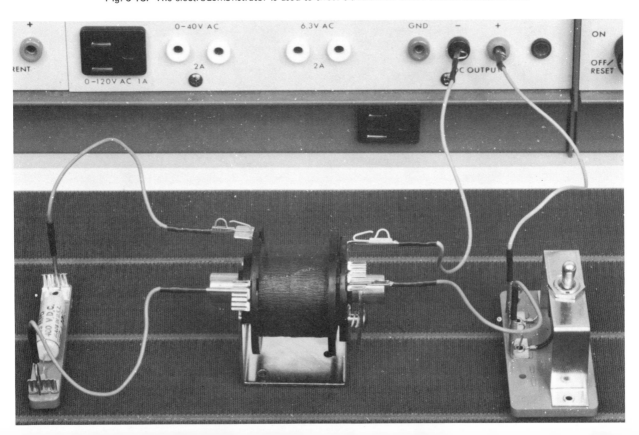

THE IGNITION SYSTEM

To make a practical application of the induction coil, observe the ignition system of the automobile in Fig. 6-19. The primary and secondary coils are clearly identified in the coil. The breaker points in the distributor are a mechanical switch operated by the camshaft in the engine and provide a means of "making and breaking" the primary circuit of the coil. A voltage of 15,000 volts and more is induced in the secondary of the coil, which is automatically distributed to the spark plugs of the engine. The high voltage causes a spark to jump across the points of the spark plug and ignite the gasoline and air mixture in the cylinder. The

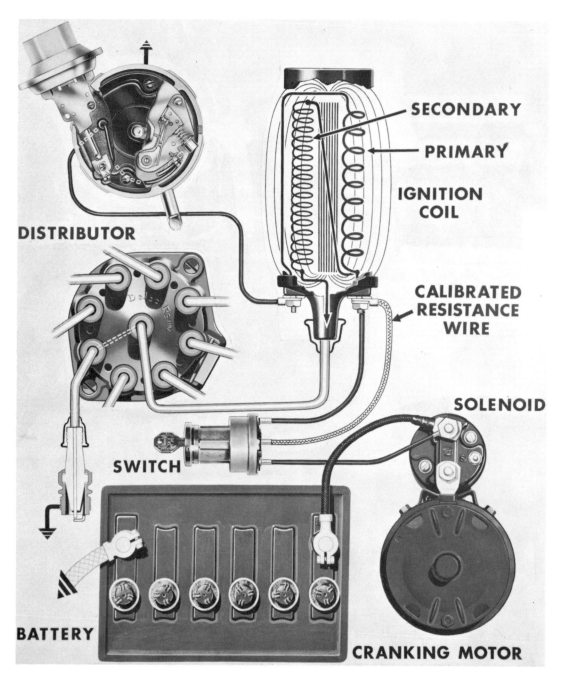

Fig. 6-19. The automotive ignition system. (Delco-Remy)

spark should occur at just the right instant. This is called "timing" the engine.

PHASE RELATIONSHIP IN TRANSFORMER

The secondary output of a transformer may be in phase or 180 deg. out of phase with the primary voltage, depending upon the direction of the windings and the method of connection. Restating Lenz's Law, the polarity of the induced voltage will be opposite to the voltage producing it. A diagram of a transformer and the waveforms of the primary and secondary voltage appear in Fig. 6-20. The direction of the

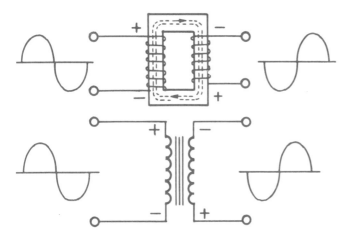

Fig. 6-20. The phase relationship between the primary and secondary of the transformer. The schematic symbol is shown.

magnetic flux is indicated by arrows within the core. A transformer of this nature actually inverts the alternating voltage wave applied to the primary.

SERIES AND PARALLEL INDUCTANCE

When coils are connected in series so that no mutual inductance exists between them, the total inductance may be found by,

$$L_T = L_1 + L_2 + L_3$$

The individual inductances add like series resistors. However, when there is mutual inductance the formula becomes,

$$L_T = L_1 + L_2 \pm 2M$$

The plus or minus sign before the M indicates that plus (+) should be used when the coils are aiding each other and the minus (−) when the magnetic fields oppose each other.

When inductors are connected in parallel, without mutual inductance,

$$L_T = \cfrac{1}{\dfrac{1}{L_1} + \dfrac{1}{L_2} + \dfrac{1}{L_3}}$$

There are many applications of mutual inductance in electricity and electronics. A skilled technician must have a complete understanding of the characteristics of circuits containing these components.

INDUCTANCE IN AC CIRCUITS

At this point only the transient response of an inductance connected to a dc voltage source has been studied. In an ac circuit the applied voltage is constantly varying and reversing polarity. Consequently the self-inductance is always present in the circuit, developing counter emfs which oppose the source voltage. This phenomenon may be observed by setting up the experiment again in Fig. 6-3. The light is first connected to a 6V dc source and its brilliance noted. Now connect the light to a 6V ac transformer and note the brilliance. Is it brighter or dimmer? The inductance indirectly has some effect upon the current flow. This type of resistance to an alternating current is called REACTANCE. Its letter symbol is X and because it is caused by an inductor, its symbol is X_L called INDUCTIVE REACTANCE. Its unit of measurement is the OHM.

INDUCED CURRENT AND VOLTAGE

A brief review will help in the understanding of Fig. 6-21. Remember that the induced voltage in a coil is the counter emf and opposes the source so therefore is 180 deg. out of phase with the source voltage. Also the greatest counter emf is induced when the current change is at maximum. Studying Fig. 6-21, you will observe that the greatest <u>rate of change</u> of the current is at the point it is crossing the zero line at 90 deg. and 270 deg. At these points also, the applied voltage must be at maximum and the induced voltage is at maximum. At the points 180 deg. and 360 deg. the <u>current change is</u>

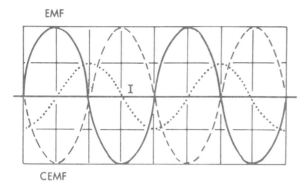

Fig. 6-21. Solid line is electromotive force. Dashed line is counter electromotive force. Dotted line is current. Greatest CEMF occurs when current is changing at its most rapid rate.

<u>minimum.</u> The current is at its maximum value, just ready to start its decrease. It is important to note that the current is 90 deg. <u>out of phase</u> with the applied voltage in a circuit containing pure inductance and the current is <u>lagging behind the voltage.</u> The magnitude of the reactive force opposing the flow of an ac is measured in Ohms and may be mathematically expressed by

$$X_L = 2\pi fL$$

where,

X_L = Inductive reactance in ohms

f = frequency in hertz

L = Inductance in henrys

π = 3.1416

Important: As frequency or inductance is increased, the inductive reactance increases. They are in direct proportion.

The behavior of an inductance in a circuit by applying ac voltages of different frequencies may be observed in the graph in Fig. 6-22. An 8 henry inductor is first connected to a dc (f = 0) and its reactance is zero. Using the formula, $X_L = 2\pi fL$, the reactance is computed for frequencies of 50, 100, 500 and 1000 Hz. The graph shows the linear increase of reactance as the frequency is increased.

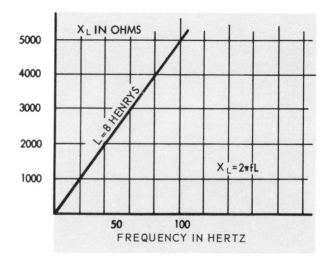

Fig. 6-22. As frequency increases; inductive reactance increases.

In the graph, Fig. 6-23, the frequency is held constant at 100 Hz and the reactance is plotted as the inductance is increased from .32 henrys to .8 henrys. Notice the linear increase in reactance as the inductance is increased.

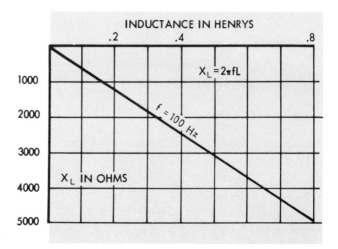

Fig. 6-23. As inductances in a circuit is increased, inductive reactance also increases.

In later studies of electronics these principles will be applied to filter and coupling circuits. The student must thoroughly understand the characteristics of reactance and how it changes as a result of inductance and frequency of applied voltage.

REACTIVE POWER

In the study of pure resistive circuits the power consumed by a circuit was found to be the product of the voltage and the current, P = I x E. This is not true, however, in an ac circuit containing inductance only. A current will flow, limited by the reactance of the circuit, but the power used to build the magnetic field will be returned to the circuit when the field collapses. This introduces two new terms. TRUE POWER represents the actual power used by a circuit. APPARENT POWER is equal to the product of the effective voltage and the effective current. Reference to Fig. 6-24 will clarify these definitions. An applied ac voltage to the inductive circuit causes a 10 ampere current, so the <u>apparent power</u> will equal,

100V x 10 amps = 1000 volt-amperes

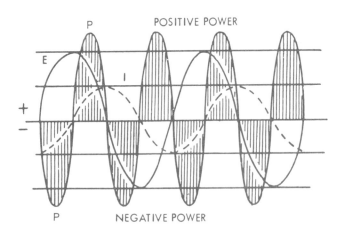

Fig. 6-24. In a theoretical circuit containing inductance only the true power is zero. The current lags the voltage by 90 deg.

The term "watts" is not used for apparent power. During one cycle equal amounts of positive power and negative power are used

which total zero. The relationship between the true power and the apparent power in an ac circuit is called the POWER FACTOR. In the example,

$$\text{Power Factor (PF)} = \frac{0}{1000 \text{ volt-amps}} = 0$$

The power factor may also be found trigonometrically by using the cosine of the phase displacement angle θ between the current and the voltage. In this example, the circuit is purely inductive and the current lags the voltage by 90 deg. The power factor is,

Cos θ or Cos 90 deg. = 0

RESISTANCE AND INDUCTANCE IN AN AC CIRCUIT

When resistance and inductance are in series in a circuit, power is used. If a circuit contains only resistance as in Fig. 6-25, the current and voltage are in phase and the power consumed is equal to I x E. Even though the polarity of the voltage changes and the current reverses, positive power is consumed. A resistance consumes

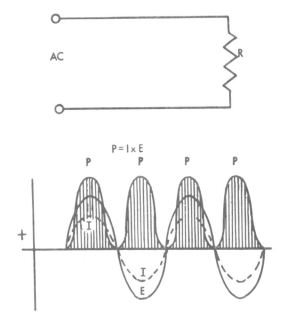

Fig. 6-25. In a pure resistive circuit, true power and apparent power are the same. The current and voltage are in phase.

power, regardless of the direction of the current. The power factor in a circuit of this type is cos 0 deg. = 1 and the apparent power is equal to the true power.

The circuit characteristics change when an inductor is added in series with the resistor. This does not necessarily have to be a resistive component, because the wire from which the coil is wound will have a certain amount of resistance. For the purpose of an example, consider Fig. 6-26. The series resistance equals

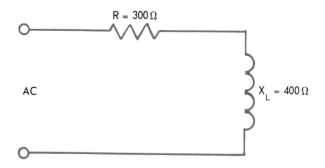

Fig. 6-26. The R L circuit described in the text.

300 ohms and the inductive reactance equals 400 ohms. This reactive component will cause the current to lag by an angle of 90 deg. or less. The forces opposing the current can be considered as the resistance and the reactance which are 90 deg. out of phase. To find the resulting opposition, the two forces must be added vectorially as in Fig. 6-27. The vectors

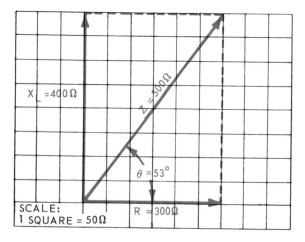

Fig. 6-27. The vector addition of X_L and R which are 90 deg. out of phase.

may be added graphically by placing the tail of the X_L vector to the arrow head of the R vector. A vector then drawn from origin 0 to the head of the vector X_L represents the magnitude and direction of the combined forces. The total opposition to an alternating current in a circuit containing resistance and reactance is called IMPEDANCE. It is measured in OHMS.

Problems of this nature are usually solved mathematically by applying the PYTHAGOREAN THEOREM for the solution of right triangles. This geometric theorem states that the hypotenuse of a right triangle is equal to the square root of the sum of the squares of the two sides. In the problem,

$$X_L = 400 \ \Omega \text{ and R} = 300 \,\Omega, \text{ then}$$

$$Z = \sqrt{300^2 + 400^2}$$

$$= \sqrt{90,000 + 160,000} = 500 \ \Omega$$

$$\text{or, } Z = \sqrt{R^2 + X_L{}^2}$$

The impedance of the circuit is 500 ohms. The angle between vector Z and vector R is the angle θ and represents the phase displacement between the current and voltage resulting from the reactive component. Now Cos θ is equal to the power factor and also equals $\dfrac{R}{Z}$ (see appendix IV) so:

$$\text{Power Factor} = \text{Cos } \theta \text{ and Cos } \theta =$$

$$\frac{R}{Z} \text{ or } \frac{300}{500} \text{ or } .6$$

The angle whose cosine is .6 is 53.10 deg. (approx.)

The current lags the voltage by an angle of 53.1 deg. The true power in this circuit will equal the apparent power times the power factor or cos. θ :

True Power = Apparent Power x Cos θ

and

$$\text{Cos } \theta = \frac{\text{True Power}}{\text{Apparent Power}}$$

In Fig. 6-28 the wave forms for current, voltage and power are drawn. Assuming an applied ac voltage of 100V, the current in the circuit will equal:

$$I = \frac{E}{Z} \text{ or } I = \frac{100}{500} = .2 \text{ amps}$$

and the apparent power can be computed by:

Apparent Power = E x I or
.2 x 100 = 20 volts-amperes

and the true power equals:

True Power = I x E x Cos θ

$$= .2 \text{ x } 100 \text{ x } \text{Cos } 53.1 \text{ deg.}$$

$$= 20 \text{ x } .6 = 12 \text{ watts}$$

This computation may be proved by

$$\text{Power Factor} = \frac{\text{True Power}}{\text{Apparent Power}} =$$

$$\frac{12}{20} = \frac{3}{5} = .6$$

This is not only theory, for in practice the power factor must be considered whenever the power company connects its power lines to a manufacturing plant. Industries must keep the power factor of their circuits and machinery within specified limits or pay the power company a premium. Power used by a reactive circuit is sometimes called "wattless power." It is returned to the circuit. In Fig. 6-28 the power in shaded area above the zero line is used, whereas the power below the line is "wattless power" and is returned to the circuit.

OHM'S LAW FOR AC CIRCUITS

In the computation of circuit values in ac circuits the familiar Ohm's Law is used with the following exception. Z is substituted for R because Z represents the total resistive force opposing the current, then:

$$I = \frac{E}{Z}, \quad E = IZ, \quad Z = \frac{E}{I}$$

The application of Ohm's Law in an ac circuit will be shown in the following problem.

A series circuit containing an 8 henry choke and a 4000 ohm resistor, is connected across a 200 volt, 60 Hz ac source. The diagram of this problem appears in Fig. 6-29.

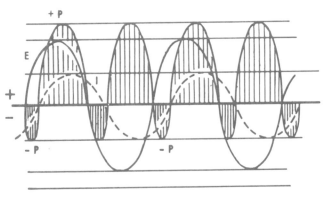

Fig. 6-28. The relationship between voltage, current and power in the example described in the text. θ = 53.1 deg.

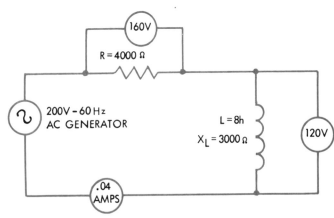

Fig. 6-29. The circuit of the problem analyzed in the text.

Step I. Find the reactance of L,

$X_L = 2\pi fL$ or

$2 \times 3.14 \times 60 \times 8 =$

   3014 Ω (3000 Ω approx.)

Step II. Find the impedance of the circuit,

$Z = \sqrt{R^2 + X_L^2}$ or $\sqrt{4000^2 + 3000^2}$ or

$\sqrt{16,000,000 + 9,000,000} =$

$\sqrt{25,000,000} = 5000\ \Omega$

Step III. Find the current in circuit,

$I = \dfrac{E}{Z}$ or $\dfrac{200}{5000} = .04$ amps

Step IV. Find the voltage drop across R and X_L

$E_R = I \times R = .04 \times 4000 = 160V$

$E_{X_L} = I \times X_L = .04 \times 3000 = 120V$

The sum of the voltage drops does not equal the applied voltage. The reason for this apparent inaccuracy is because the two voltages are 90 deg. out of phase which requires vector addition, so

$E_s = \sqrt{160^2 + 120^2} = \sqrt{25,600 + 14,400}$

   $= \sqrt{40,000} = 200$ volts (approx.)

Step V. Find the phase angle θ between I and E

$\cos\theta = \dfrac{R}{Z} = \dfrac{4000\ \Omega}{5000\ \Omega} = \begin{array}{l}.8 \text{ which is the} \\ \text{cosine of angle 36 deg.}\end{array}$

Step VI. What is the true power and apparent power?

Apparent Power $= I \times E = .04 \times 200 =$
                8 volt-amperes

True Power $= I \times E \times \cos\theta =$
           $.04 \times 200 \times .8 = 6.4$ watts

Note: Some figures have been approximated in the above problem to avoid uneven figures and to clarify the problem.

FOR DISCUSSION:

1. Why does the power company use high voltages in their cross-country transmission lines?
2. What is the difference between Apparent Power, and True Power?
3. Discuss the differences between a step-up, and a step-down transformer.
4. Power companies request industry to maintain their power factor close to one, or pay additional premiums. Why?

TEST YOUR KNOWLEDGE

1. What is inductance in a circuit?
2. Inductance is measured in _____ .
3. Convert to henrys:

   200 μH =
   50 mH =

4. Inductance in a circuit causes the current to _____ the voltage.
5. Draw a graph showing the rise and decay of current in a RL circuit.
6. The time constant of a RL circuit may be found by the formula, t = _____ .

7. Draw the symbol for a power transformer.

8. The primary of a transformer has 200 turns of wire and the secondary has 800 turns. If 115 volt ac is applied to primary, what will be the secondary voltage?

9. A load attached to the secondary of the transformer in problem 8 draws a 100 mA current. What is the primary current?

10. Name and describe the losses associated with transformers.

11. What is the purpose of the breaker points in the automotive ignition system.

12. The resistance to an alternating current resulting from circuit inductance is called _____ . It is measured in _____ . Its symbol and formula _____ .

13. The only component which uses power in a circuit is _____ .

14. In a series RL circuit, L = 2H, R = 500 ohms, E_S = 100V 60 hertz ac.
    Find:

$$X_L =$$
$$Z =$$
$$I =$$
$$E_R =$$
$$E_L =$$
$$\theta =$$
$$PF =$$

True Power =

Apparent Power =

15. On graph paper, draw vectors and sine curves representing current and voltage of Problem 14. With a red pencil, draw the power curve.

Chapter 7

CAPACITANCE IN
ELECTRICAL CIRCUITS

Early in the history of the science of electricity, the characteristics of the capacitor were investigated. One of the earlier forms was named the Leyden Jar, which consisted of a glass jar covered on the inside and outside by tinfoil. A charge was introduced into the jar by means of a rod and chain. The characteristic of capacitance is a phenomenon of electricity which should be thoroughly understood. Capacitors have thousands of applications in electricity and electronics. In this chapter will be found answers to the following questions:

1. What is a capacitor?

2. How does a capacitor behave in a circuit?

3. How does capacitance affect an ac circuit?

4. What are the results of combining capacitance and resistance in a circuit?

5. What are some of the common applications of capacitors in electronic equipment?

THE CAPACITOR

A capacitor is a simple electronic component, and may be described as two plates of conductive material separated by insulation, which is called the dielectric. The understanding of the capacitor action requires a knowledge of the electron theory. In Fig. 7-1, the capacitor is represented by two metal plates separated by air. These plates are connected to a dc source of voltage. The circuit appears to be an open circuit because the plates are not in physical

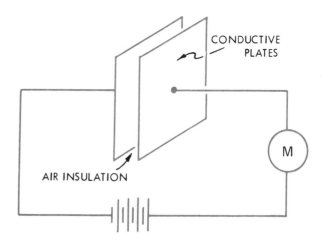

Fig. 7-1. The capacitor in its basic form.

contact with each other. The meter in the circuit, however, will show a momentary flow of current when the switch is closed. A study of Fig. 7-2, will show that as the switch is closed, electrons from the negative terminal of the source will flow to one plate of the capacitor. These electrons will repel the electrons from the second plate (like charges repel) and these

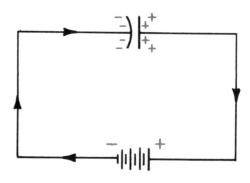

Fig. 7-2. The capacitor charges to the source voltage.

electrons will be attracted to the positive terminal of the source. The capacitor is now charged to the same potential as the source and is opposing the source voltage. The capacitor may be removed from the circuit and it will remain charged. The energy is stored within the electric field of the capacitor.

A LESSON IN SAFETY: Many large capacitors in radios, TVs and other electronic equipment retain their charge after the power has been turned off. These capacitors should be discharged by shorting their terminals to the chassis with an insulated screwdriver. If this is not done, these voltages may destroy test equipment and might give a severe shock to the technician working on the equipment.

CAPACITANCE in a circuit may be defined as that property which opposes any change in voltage. Compare this to inductance, which is the property that opposes a change in current. This capacitive effect will be explained in greater detail later in this chapter. It is important to remember that in the circuit of Fig. 7-2, no electrons flowed through the capacitor. A capacitor will block a direct current. However, one plate became negatively charged, the other positively charged, and a strong electric field exists between them. Insulating or dielectric materials vary in their ability to support an electric field. In electronics this ability is termed the dielectric constant of the material. The constants of various materials are given in comparison to the dielectric constant of dry air, which is assigned the value of one (1). Dielectric constants of some familiar materials are listed in Table 7-1.

The capacitance of a capacitor is determined by the number of electrons that may be stored in it for each volt of applied voltage. The unit of measurement of capacitance is the FARAD and represents a charge of one coulomb which raises the potential one volt. This equation states that,

Material	Dielectric Constant
Air	1
Wax paper	3.5
Mica	6
Glass (window)	8
Pure water	81

Table 7-1

$$C \text{ (in farads)} = \frac{Q \text{ (in coulombs)}}{E \text{ (in volts)}}$$

Capacitors used in electronic work have capacities measured in microfarads $\left(\frac{1}{1,000,000} \text{ of a farad}\right)$ and picofarads $\left(\frac{1}{1,000,000} \text{ of } \frac{1}{1,000,000} \text{ farad.}\right)$

The Greek Letter " μ " is used to represent micro and pico would appear as "p." A typical capacitor might be rated as 250 pF meaning 250 picofarads.

Capacitance is determined by:

1. The material used as a dielectric.

2. The area of the plates.

3. The distance between the plates.

These factors are related in the mathematical formula as,

$$C \text{ (in pF)} = .225 \frac{KA}{d}(n-1)$$

where,

K = dielectric constant

A = area of one side of one plate in square inches

d = the distance between plates in inches

n = number of plates.

SCHEMATIC SYMBOL FOR VARIABLE CAPACITOR

Fig. 7-3. Variable capacitors are made in many types and sizes.   (Hammarlund Mfg. Co.)

This formula is presented at this time so the student will realize the following facts:

1. Capacity <u>increases</u> when the area of the plates is increased or the dielectric constant is increased.

2. Capacity <u>decreases</u> when the distance between the plates is increased.

This information will be useful in later studies of capacitance in electronic circuits.

TYPES OF CAPACITORS

There are hundreds of sizes and types of capacitors manufactured today. Some of the more common ones will be illustrated and discussed.

1. VARIABLE CAPACITORS, Fig. 7-3. This group of capacitors consists of metal plates which intermesh as the shaft is turned. The stationary plates are called the <u>stator</u> and the rotating plates are the <u>rotor</u>. Many of these are used in radio transmitters and receivers. It is a component like one of these that you turn when tuning a radio. Referring to the past discussion, this capacitor is at maximum capacity when the plates are fully meshed. Notice the schematic symbol for a variable capacitor.

2. FIXED PAPER CAPACITOR, Fig. 7-4. This common variety of capacitor is made up of layers of tinfoil separated by waxed paper as the dielectric. The wires extending from the ends connect to the foil plates. The assembly is tightly rolled into a cylinder and sealed

SCHEMATIC SYMBOL FOR FIXED CAPACITOR

Fig. 7-4. A fixed value paper capacitor encapsulated in plastic. (Sprague Products Co.)

with special compounds. To provide rigidity, some manufacturers encapsulate these capacitors in plastic. These molded capacitors will withstand severe heat, moisture and shock.

3. RECTANGULAR OIL FILLED CAPACITORS, Fig. 7-5. These capacitors are hermetically sealed in metal cans. They are oil filled and have extremely high insulation resistance. This type is found in the power supply of radio transmitters and other electronic equipment.

Fig. 7-5. The oil filled capacitor.

4. ELECTROLYTICS, CAN TYPE, Fig. 7-6. This kind of capacitor employs a different method of plate construction. Some manufacturers use aluminum plates and a wet or dry electrolyte of borax or carbonate. A dc

voltage is applied during manufacture and by electrolytic action a thin layer of aluminum oxide is deposited on the positive plate, effectively insulating it from the electrolyte. The negative plate is a connection to the electrolyte; the electrolyte and positive plates form the capacitor. These capacitors are very convenient when a large amount of capacity is needed in a small space. The polarity of these capacitors must be observed. A reverse connection will destroy them. The cans may contain from one to four different capacitors. The terminals are marked by ▲ , ● , ■ and ▬ symbols. See Fig. 7-6. The metal can is usually the common negative terminal for all the capacitors. A special metal and a fiber mounting plate is supplied for easy installation on a chassis.

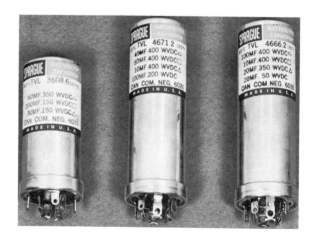

Fig. 7-6. Can type electrolytic capacitor.

5. ELECTROLYTICS, TUBULAR, Fig. 7-7. The construction of these capacitors is similar to the can type. The main advantage is the smaller size. They have a metal case which is enclosed in an insulating tube. These also are made in dual, triple and quadruple units in one cylinder.

6. CERAMIC DISC, Fig. 7-8. A very popular small capacitor which is used extensively in radio and TV work. It is made of a special ceramic dielectric on which is fixed the silver

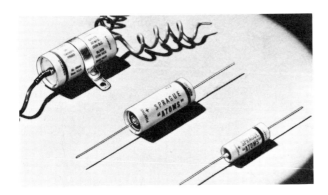

Fig. 7-7. The tubular electrolytic capacitor.
(Sprague Products Co.)

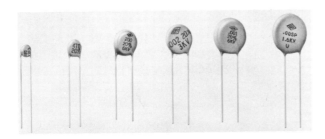

Fig. 7-8. A group of ceramic disc capacitors.
(Centralab)

plates of the capacitor. The whole component is treated with special insulation to withstand heat and moisture.

7. MICA CAPACITOR, Fig. 7-9. These small capacitors are made by stacking tin foil plates together with thin sheets of mica as the dielectric. The assembly is then molded into a plastic case.

Fig. 7-9. Mica capacitors. (See Appendix 2 on How to read the capacitor color code.)

8. TRIMMER CAPACITOR, Fig. 7-10. This is a form of a variable capacitor. The adjustable screw compresses the plates and increases the capacitance. Mica is used as a dielectric.

Fig. 7-10. Several types of trimmer capacitors.
(Centralab)

These, as the name suggests, are used where a fine adjustment of capacitance is necessary. They are used in conjunction with larger capacitors and are connected in parallel with them.

LESSON IN SAFETY: When making adjustments on trimmer and padder capacitors, the screw should be turned with a special fiber or plastic screwdriver called an alignment tool. The capacitive effect of a steel screwdriver, if used, would result in inaccurate adjustment.

TRANSIENT RESPONSE OF THE CAPACITOR

The response of the current and voltage in a circuit after an immediate change in applied voltage, is known as a TRANSIENT RESPONSE. Referring to the diagram, Fig. 7-11, a capacitor and a resistor are connected in series

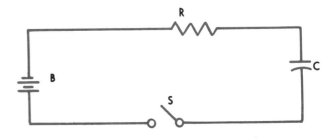

Fig. 7-11. This series RC circuit can be used to demonstrate transient response of a capacitor.

across a voltage source. When the switch is closed, maximum current will flow, gradually decreasing until the capacitor has reached its full charge, which is the same as the applied voltage. However, the voltage initially is zero across the capacitor, and when the switch is closed this voltage will gradually build up to the value of the source voltage. This is graphically illustrated in Fig. 7-12. When the switch is opened, the capacitor remains charged. Theoretically it would remain charged but there is always some leakage through the dielectric and during a period of time the capacitor will discharge itself.

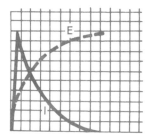

Fig. 7-12. A graph showing current and voltage in the series RC circuit.

In Fig. 7-13, the series combination of charged capacitor and resistor are short-circuited by providing a discharge path. Because there is no opposing voltage, the discharge current will instantaneously rise to maximum and gradually fall off to zero. The combined graph showing the charge and discharge of C is shown in Fig. 7-14. Consideration must also be given to the voltages appearing across R and C in this circuit. As the voltage across R is a result of current flow, E = IR, then the maximum

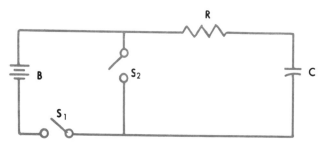

Fig. 7-13. The RC circuit is short-circuited by closing switch S_2.

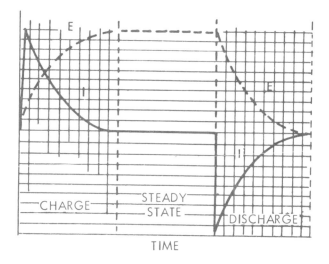

Fig. 7-14. This combination graph shows the rise and decay of current and voltage in the series RC circuit.

voltage appears across R when maximum current is flowing. This is immediately after the switch is closed in Fig. 7-11, and after the discharge switch is closed in Fig. 7-13. In either case the voltage across R drops off or decays as the capacitor approaches charge or discharge. The graph of the voltage across R is drawn in Fig. 7-15.

RC TIME CONSTANT

During the charge and discharge of the series RC network previously described, a period of time did elapse. This time is indicated along the

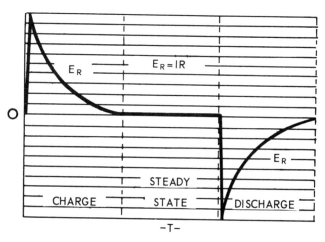

Fig. 7-15. The graph shows the voltage drop across R as the capacitor is charged and discharged.

base of x-axis of the graphs in Fig. 7-14 and 7-15. The time duration for the capacitor to charge or discharge 63.2 percent is known as the TIME CONSTANT of the circuit. It may be computed by the formula:

t (in seconds) = R (in ohms) x C (in farads)

For complete charge or discharge, five time constant periods are required. Assuming a source voltage equal to 100 volts, Table 7-2 shows the time constant, percentage and voltage.

TIME CON- STANT	PERCENT OF VOLTAGE	E CHARGING	E DIS- CHARGING
1	63.2%	63.2V	36.8V
2	86.5%	86.5V	13.5V
3	95.0%	95.0V	5V
4	98.0%	98.0V	2V
5	99+%	99+V	1V

Table 7-2

A problem will clarify this charging action. A .1 microfarad capacitor is connected in series with a megohm (1,000,000 ohm) resistor across a 100 volts source. How much time will elapse during the charging of C? (Using powers of ten is explained in the appendix.)

$$t = .1 \times 10^{-6} \times 10^6 = .1 \text{ sec.}$$

One tenth of a second is the time constant and C would charge to 63.2 volts during this time. At the end of five time constants or 5 x .1 = .5 sec., the capacitor is considered fully charged.

An interesting experiment may be performed by setting up the circuit in Fig. 7-16. A neon lamp will remain out until a certain voltage is reached. At this voltage called the "ignition or firing voltage" the lamp glows and offers negligible resistance in a circuit. In this circuit capacitor C charges through resistance R. When the voltage across C develops to the ignition

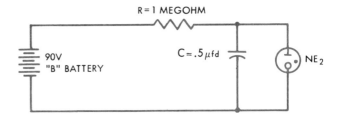

Fig. 7-16. The flashing circuit called a relaxation oscillator.

voltage, the lamp glows and C is rapidly discharged. This cycle repeats over and over, causing the neon light to flash. The frequency of the flashes may be changed by varying either the value of C or R. Time constants have many uses in electronic circuits and frequent reference will be made to them in the following chapters. Timing circuits are used in industry to control the sequence and duration of machine operations. The photo-enlarger uses a time-delay circuit to control the exposure time. These circuits provide many interesting effects for the experimenter in electronics.

WISE OWL PROJECT

The wise old owl pictured in Fig. 7-17 is called "Hooty Hoot." Because of a patented blinking feature, RC circuits cause the eyes to flash alternately. See the schematic in Fig. 7-17. This project demonstrates the principles of RC time constants which you have just covered.

Parts List for HOOTY-HOOT

    2 NE2 Neon Lamps
    2 2.2 megohm resistors
    1 .1 μ fd Paper Capacitors, 200 volts
    1 90 volt B Battery, Portable size.

The wise owl is easy to build and it makes an amusing toy and conversation piece. Assemble the parts according to the circuit diagram in Fig. 7-17. Frame the owl's body with chicken wire. Install the complete RC circuit and cover the body of the bird with paper mache.

PARALLEL AND SERIES CIRCUITS

Many times the technician is required to work with capacitors connected in series and parallel, and the effect of such connections

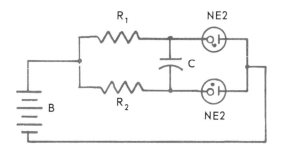

Fig. 7-17. Hotty-Hoot employs RC circuits so that his eyes will alternately flash. The blinking feature is covered by U.S. Patent No. 2,647,222, owned by Gold Seal Co., Bismarck, ND. Used by permission of patent holder.

should be explored. When capacitors are connected in parallel as in Fig. 7-18, the total capacitance is equal to the sum of the individual capacitances.

$$C_T = C_1 + C_2 + C_3 \ldots$$

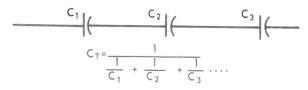

$$C_T = C_1 + C_2 + C_3 \ldots$$

Fig. 7-18. Capacitors in parallel.

When two capacitors are connected in series, then,

$$C_T = \frac{C_1 \times C_2}{C_1 + C_2}$$

and if two or more are connected in series as in Fig. 7-19:

$$C_T = \frac{1}{\frac{1}{C_1} + \frac{1}{C_2} + \frac{1}{C_3} \ldots}$$

$$C_T = \frac{C_1 C_2}{C_1 + C_2}$$

$$C_T = \frac{1}{\frac{1}{C_1} + \frac{1}{C_2} + \frac{1}{C_3} \ldots}$$

Fig. 7-19. Capacitors in series.

Up to this point the WORKING VOLTAGE of a capacitor has only been briefly mentioned. The dielectrics used for capacitors can only withstand certain voltages. If this voltage is exceeded, the dielectric will breakdown and arcing will result. This may possibly cause a short circuit and ruin other parts of the circuit in which it is connected. Increased voltage ratings require special materials and thicker dielectrics. When a capacitor is replaced, special attention should be paid to its capacitance value and also it its <u>dc working voltage</u>.

When a capacitor is used in an ac circuit, the working voltage should safely exceed the peak ac voltage. A 120 volt effective ac voltage will have a peak voltage of 120V x 1.414 = 169.7 volts.

CAPACITANCE IN AC CIRCUITS

The previous lesson has explained that a capacitor will block a <u>direct current</u>. Current only flows in the circuit during the charging or discharging time. If an ac voltage is applied to a capacitor, the plates charge to one polarity during the first half cycle. During the next half cycle they charge to the opposite polarity. An ac meter in the circuit would indicate a current flow at all times. In Fig. 7-20 this is proved experimentally by connecting the light and capacitor in series to a 6 volt dc source. Does the light glow? Now connect the same circuit to a 6 volt ac source and notice that the light

burns dimly, showing that alternating current is flowing as a result of the alternate charging of the capacitor.

For purposes of emphasis, the current and voltage in a capacitive circuit will be reviewed. Referring to Fig. 7-21, as the ac voltage starts to rise, the current flow is maximum because C is in a discharged state. As C becomes charged to the peak ac voltage, the charging current drops to zero, Point A. As the voltage begins to drop, the discharging current begins to rise in a negative direction and reaches maximum at the point of zero voltage, Point B. This phase difference continues on throughout each cycle. In a purely capacitive circuit the current leads the voltage by an angle of 90 deg. The magnitude of the current in the circuit depends upon the size of the capacitor. Larger capacitors (more capacitance) require a larger current to charge them. Also affecting the current flow is the frequency of the ac voltage. The current flow depends upon the rate of charge and discharge of the capacitor. As the frequency of the ac is increased, then current will increase. These relationships are stated mathematically by,

$$X_c = \frac{1}{2\pi fC}$$

where,

X_c = Capacitive reactance in ohms
f = Frequency in hertz
C = Capacitance in farads

Similar to inductive reactance, the resistance to the flow of an alternating current as a result of capacitance is called CAPACITIVE REAC-TANCE. It is measured in ohms, like dc resistance and X_L. The formula indicates that:

1. As the frequency increases, X_c decreases.

2. As capacitance is increased, X_c decreases.

Problem: What is the reactance of a 10 μ fd capacitor operating in a circuit at a frequency of 120 hertz (Hz)?

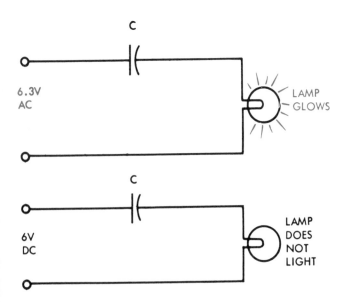

Fig. 7-20. The light will glow when connected to an ac source. The capacitor will block direct current.

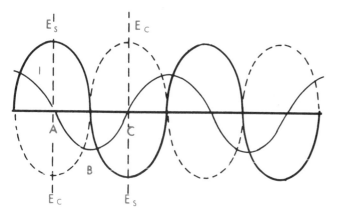

Fig. 7-21. A graph showing applied voltage, current and the voltage across C, when the circuit is connected to ac.

$$X_c = \frac{1}{2\pi \times 120 \times 10 \times 10^{-6}}$$

$$= \frac{1}{6.28 \times 120 \times 10}$$

$$= \frac{1 \times 10^5}{753.6} = .0013 \times 10^5 \text{ or } 130 \ \Omega$$

The reactance of a .1 μfd capacitor as frequency is varied may be observed in the graph of Fig. 7-22. As frequency is changed to 50, 100, 1000 and 10,000 Hz, the reactances are computed respectively by the formula,

$$X_c = \frac{1}{2\pi fC}$$

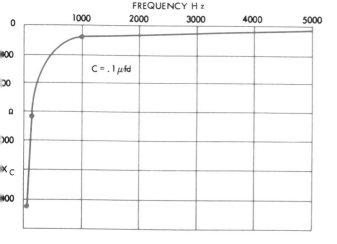

Fig. 7-22. The graph shows that as frequency increases, the capacitive reactance decreases.

In Fig. 7-23 the frequency is held constant at 1000 Hz and the reactance is plotted for capacitors of .5 mfd, .1 mfd, .05 mfd, .01 mfd and .001 mfd. These are some common sizes of capacitors used in electronic work. They are used in filtering, coupling and bypassing networks. More detailed explanations will be included in later studies.

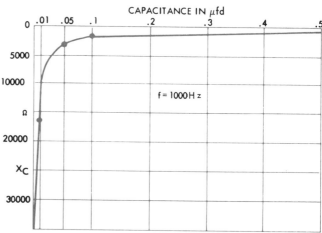

Fig. 7-23. This graph shows that as capacitance increases, the reactance decreases.

REACTIVE POWER

When a capacitor is discharged, the energy stored in the dielectric is returned to the circuit. This may be compared to inductance which returns the energy stored in the magnetic field to the circuit. In either case electrical energy is used temporarily by the reactive circuit. This power in a capacitive circuit is also called WATTLESS POWER. In Fig. 7-24 the voltage and current waveforms are drawn for a circuit containing pure capacitance. The power wave form is a result of plotting the products of the instantaneous voltage and current at selected points. The power waveform shows that equal amounts of positive power and negative power are used by the circuit, resulting in zero power consumption. The TRUE-POWER therefore is zero. (Refer to Chapter 6 to study of Reactive Power in an Inductive Circuit and compare these circuits.) The APPARENT POWER is equal to the product of the effective voltage

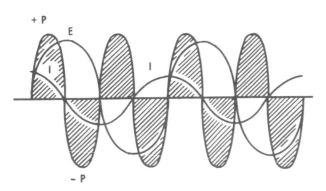

Fig. 7-24. The wave forms showing current, voltage and power in a pure capacitive circuit.

and the effective current. Referring to the circuit in Fig. 7-25 an applied ac voltage to the capacitive circuit causes a 10 ampere current. The apparent power will equal,

$$100V \times 10 \text{ amps} = 1000 \text{ volt-amperes}$$

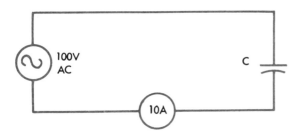

Fig. 7-25. The schematic of the theoretical capacitive circuit.

The relationship between the true power and the apparent power in an ac circuit is called the POWER FACTOR. In this example,

$$\text{Power Factor (PF)} = \frac{0 \text{ (True Power)}}{1000 \text{ volt-amps (Apparent Power)}} = 0$$

The power factor may also be found trigonometrically. It is equal to the cosine of the phase displacement angle θ between the current and the voltage. In the pure capacitive circuit, the current leads the voltage by an angle of 90 deg. and θ therefore is 90 deg.

$$\text{Cos } \theta \text{ or Cos 90 deg.} = 0$$

The true power may be found by,

$$\text{True Power} = \text{I} \times \text{E} \times \text{Cos } \theta$$

$$= 10 \times 100 \times \text{Cos 90 deg.}$$

$$(\text{Cos 90 deg.} = 0)$$

$$= 0$$

RESISTANCE AND CAPACITANCE IN AN AC CIRCUIT

When resistance is present in a circuit, power is consumed. If the circuit contains only resistance, then the voltage and current are in phase. There is no phase angle θ and the power factor is one,

$$\text{Cos 0 deg.} = 1$$

and the apparent power equals the true power.

The circuit characteristics change when capacitance is added in series with the resistor. The capacitive reactance is also a force which resists the flow of an alternating current. Because the capacitive reactance causes a 90 deg. phase displacement, the total resistance to an ac current must be the vector sum of the X_c and R. These vectors are drawn in Fig. 7-26. Assuming the circuit has 300 ohms resistance and 400 ohms capacitive reactance, then the resultant force is 500 ohms which is called the impedance Z of the circuit.

$$Z = \sqrt{300^2 + 400^2}$$
$$= \sqrt{90,000 + 160,000} = 500 \, \Omega$$

$$Z = \sqrt{R^2 + X_c^2}$$

The angle between vector Z and vector R represents the phase displacement between the current and the voltage as a result of the reactive component. This is angle θ (theta). Now cosine θ is equal to the power factor and also equals $\frac{R}{Z}$ so,

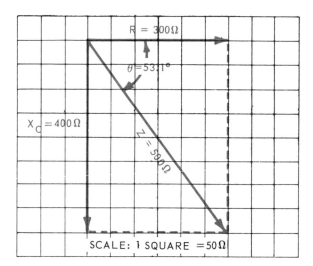

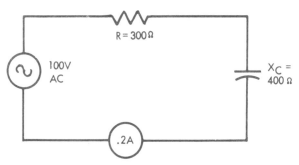

Fig. 7-26. Above. The vector relationship between R and X_C and the resultant vector Z for impedance. Below. The circuit for the problem in the text.

$$\text{Power Factor} = \text{Cos } \theta \text{ and Cos } \theta =$$
$$\frac{R}{Z} \text{ or } \frac{300}{500} = .6$$

The angle which has a cosine of .6 is 53.1 deg. (approx.). The current then leads the voltage by an angle of 53.1 deg. The true power in this circuit will equal to:

$$\text{Apparent Power x Cos } \theta = \text{True Power or}$$

$$\text{Apparent Power x Power Factor} = \text{True Power}$$

Also the power factor is the relationship between the true power and the apparent power, and this equality may be expressed as,

$$\text{Cos } \theta = \text{Power Factor} = \frac{\text{True Power}}{\text{Apparent Power}}$$

Using Ohm's Law for ac circuits, the current flowing in circuit, Fig. 7-26 is equal to, (100 volts ac is applied)

$$I = \frac{E}{Z} \text{ or } I = \frac{100}{500} = .2 \text{ amps.}$$

The apparent power is therefore,

$$I \text{ x } E \text{ or } .2 \text{ x } 100 = 20 \text{ volt-amps}$$

and,

$$\text{True Power} = I \text{ x } E \text{ x Cos } \theta$$
$$= .2 \text{ x } 100 \text{ x Cos } 53.1 \text{ deg.}$$
$$= 20 \text{ x } .6 = 12 \text{ watts}$$

Proof of this equation may be realized by,

$$\text{Power Factor} = \frac{\text{True Power}}{\text{Apparent Power}} = \frac{12}{20} = .6$$

There are three electrical properties of any circuit, R, L, and C. These will be found in many combinations including R, RL, and RC networks which have been studied in the last two chapters. In Chapter 8, RCL circuits will be considered. Always review previous lessons, if the theory discussed is not clearly understood. The nature of the study of electricity and electronics requires the student to have a firm grasp of all previous knowledge before progression into more difficult applications.

AUTOMATIC CONTROL FOR GARAGE LIGHTS

This project makes a very fine present and may prevent some accidents. It uses two principles you have already studied:

1. The photocell.

2. RC time constants.

It works like this. When the headlights from the car strike the photocell, it automatically turns on the garage lights. The lights remain on for several minutes depending on the setting of R_3.

This project is illustrated in Fig. 7-27 and 7-28. The tube on the metal box acts as a shade so the circuit will be triggered only by direct approaching light. It may be soldered or screwed onto the box. The sun battery B_2 is mounted behind the tube. The switch activates the circuit. The ac plug permits connection by extension cord to the toggle switch which operates the garage lights. It should be connected across the switch. This is in parallel with the switch.

The parts list for this project follows:

B_1 — 15-volt battery

B_2 — Sun battery, International Rectifier Corp., Type B2M or equivalent

C_1 — 1000 mfd electrolytic, 15-volt

CR_1 — Sylvania 1N34 diode or RCA SK3087

I_1 — 115-volt light bulb, wattage not greater than 100

K_1 — 1 K ohm sensitive relay, Sigma 11F-1000G or equivalent

Q_1 — Sylvania 2N1265 transistor, or RCA SK3003

Q_2 — Sylvania 2N229 transistor, or RCA SK3010

R_1 — 10 K potentiometer

R_2 — 22 K, 1/2W

R_3 — 1 meg potentiometer

R_4 — 1K, 1/2W

S_1 — Single-pole, single-throw switch; slide, toggle, or push button.

Here is how it works: When the sun cell is activated by light, transistor Q_1 conducts and charges C_1, and also Q_2 conducts through the relay coil, which closes the switch points. Garage lights go on. When the light source is removed (head lights off) the charge on C_1 keeps Q_2 conducting. The charge on C_1 gradually bleeds off through R_3 and R_2. At a predetermined point, set by R_3, the relay points open. (Garage lights turn off.) Sufficient time has passed so that you can turn on the garage lights in the usual manner.

R_1 is a sensitivity control, and should be adjusted so that the headlights will trigger the circuit. No attempt is made to show the physical wiring of this circuit. The photo only suggests on way of many to build the project. It should be compact. Care must be taken always when working with transistors.

Fig. 7-27. Automatic control for garage light.

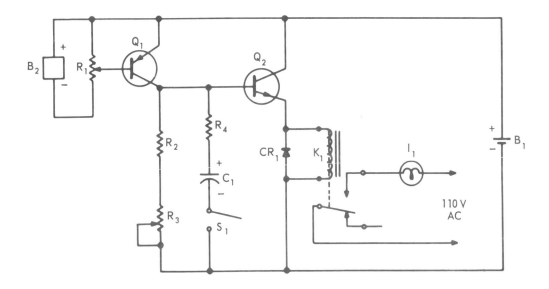

Fig. 7-28. Wiring diagram for garage light control, shown in 7-27.
(Sylvania Electric Products, Inc.)

FOR DISCUSSION

1. Why should adjustments on trimmer capacitors be made with a plastic or fiber screwdriver?
2. Why does a capacitor block dc, yet will pass an ac signal?
3. What is meant by the term TRANSIENT RESPONSE?
4. Why does a theoretical circuit containing only capacitance consume no power?
5. Should a capacitor with a working voltage of 150 volts be used in circuit with 115 volts ac supplied from your house circuit? Why?

TEST YOUR KNOWLEDGE

1. Draw a circuit of a battery, switch and capacitor in series. Explain the charging of C. Show direction of electron flow.
2. What SAFETY precaution should be followed when working on capacitors in a radio or TV?
3. Capacitance in a circuit may be defined as that property which opposes a change in _____ .

4. What is meant by the term DIELECTRIC CONSTANT?
5. Capacitance is measured in _____ .
6. What three physical characteristics determine the capacity of a capacitor?
7. Ask your instructor to show you a sample of each of the different types of capacitors illustrated in the text.
8. The time constant of an RC circuit is equal to _____ x _____ and, _____ time constant periods are required for fully charging the capacitor.
9. Capacitors in parallel add together like _____ in series.
10. State formula for C_T when two unequal capacitors are in series.
11. The opposing force to an alternating current as a result of capacitance is called _____ _____ . Its symbol is _____ . It is measured in _____ and may be found by the formula _____ .
12. As capacitance is increased, X_c _____ . As frequency in increased, X_c _____ .
13. In a theoretical circuit containing only capacitance, the current _____ the voltage by an angle of _____ or less. The power consumed is _____ .

14. A series circuit of R = 300 ohms and C = .1 μ fd is connected to a 50 volt 400 Hz source.

    Find:

    X_c

    Z

    E_{X_c}

    E_R

    I

    θ

15. A capacitor has a value of .1 μ fd. What is its reactance at,

    50 Hz – X_c =

    100 Hz – X_c =

    1000 Hz – X_c =

    10 Hz – X_c =

16. The frequency of the current in a circuit is 1000 Hz. What is reactance of a

    .5 μ fd =

    .1 μ fd =

    .05 μ fd =

    .01 μ fd =

    .001 μ fd =

Chapter 8

TUNED CIRCUITS: RCL NETWORKS

RCL NETWORKS

The effects of resistance, capacitance and inductance have been studied in previous chapters. These principles should be reviewed before progressing into combination circuits.

1. In an ac circuit containing resistance only, the applied voltage and current are in phase. There is no reactive power and the power consumed by the circuit is equal to the product of volts times amperes.

2. In an ac circuit containing inductance only, the current lags the voltage by an angle of 90 deg. They are not in phase. The power consumed by the circuit is zero.

3. In an ac circuit containing resistance and inductance, the current will lag the voltage by a phase angle of less than 90 deg. The total resistive force is the vector sum of the ohmic resistance and the inductive reactance. This is the impedance of the circuit.

4. In an ac circuit containing capacitance only, the current leads the voltage by an angle of 90 deg. The power consumed is zero.

5. In an ac circuit containing resistance and capacitance, the current will lead the voltage by an angle of less than 90 deg. The impedance is equal to the vector sum of the resistance and the capacitive reactance.

6. In an ac circuit containing resistance, inductance and capacitance in series, the resulting impedance is equal to the vector addition of R in ohms, X_L in ohms and X_C in ohms.

RESONANCE

A special condition exists in a RCL circuit, when it is energized at a frequency at which the inductive reactance is equal to the capacitive reactance, $X_L = X_C$. As inductive reactance, X_L, increases as frequency increases and capacitive reactance, X_C decreases as frequency increases, there is one frequency at which they are both equal. This is the RESONANT FREQUENCY of the circuit, (f_o). In the vector diagram of Fig. 8-1, $X_L = 100$, $X_C = 100$, and R = 50. X_L and X_C are opposing each other

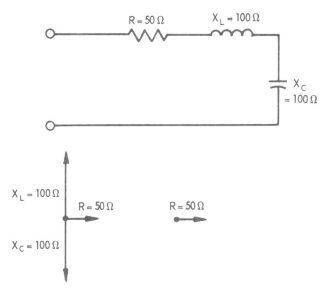

Fig. 8-1. The condition of resonance exists when $X_L = X_C$ and the circuit appears as a resistive circuit.

because they are 180 deg. out of phase, and the algebraic sum of these vectors is <u>zero</u>, and only resistance R of 50 ohms remains. The current and voltage are in phase. This particular frequency of the circuit may be computed by the resonant frequency formula:

$$f_o = \frac{1}{2\pi\sqrt{LC}}$$

where,

f_o = resonant frequency in hertz

L = inductance in henrys

C = capacitance in farads

The derivation of this formula may be understood by the following steps,

At resonance $X_L = X_c$

or

$$2\pi fL = \frac{1}{2\pi fC}$$

Transpose: $2\pi L$ to right side of the equation by dividing both sides by $2\pi L$

$$f = \frac{1}{(2\pi)^2 fLC}$$

Move f to left side by multiplying both sides by f,

$$f^2 = \frac{1}{(2\pi)^2 LC}$$

Take square root of both sides of the equation,

$$f = \frac{1}{2\pi\sqrt{LC}}$$

To simplify the equation further, $\frac{1}{2\pi}$ or $\frac{1}{6.28}$ = .159, so

$$f_o = \frac{.159}{\sqrt{LC}} \text{ (approximately)}$$

Example: What is the resonant frequency of a circuit which has 200 μH inductance and 200 pF capacitance?

Solution:

$$f_o = \frac{.159}{\sqrt{200 \times 10^{-6} \times 200 \times 10^{-12}}} = \frac{.159}{\sqrt{40,000 \times 10^{-18}}}$$

$$= \frac{.159}{\sqrt{4 \times 10^4 \times 10^{-18}}} = \frac{.159}{\sqrt{4 \times 10^{-14}}} = \frac{.159}{2 \times 10^{-7}}$$

$$= .08 \times 10^7 \times 800,000 \text{ Hz or } 800 \text{ KHz or } .8 \text{ MHz}$$

(Review both Conversions and Scientific Notation in Appendix.)

THE ACCEPTOR CIRCUIT

First, consider the performance of a series RCL circuit at its resonant frequency. The diagram is drawn in Fig. 8-2. At resonance $X_L = X_c$ so the impedance of the circuit equals,

$$Z = \sqrt{R^2 + (X_L - X_c)^2} \text{ or } \sqrt{R^2} = R$$

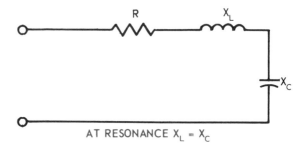

AT RESONANCE $X_L = X_C$

Fig. 8-2. The series resonant circuit is an acceptor circuit.

and only the ohmic resistance impedes the current flow in the circuit. <u>In a series resonant circuit the impedance is minimum.</u> At frequencies above or below the resonant frequency X_L is not equal to X_c and the reactive component <u>increases</u> the impedance of the circuit. The response of the series tuned circuit appears as a bell-shaped curve in Fig. 8-3A. Notice that the impedance of the circuit is minimum at resonance, and that maximum current flows at resonance and rapidly falls off on either side of resonance due to the increased

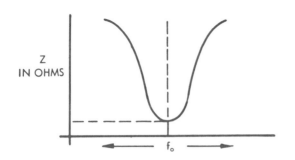

Fig. 8-3A. The curve shows the increase in impedance as the frequency is varied above or below resonance.

impedance, Fig. 8-3B. This described an ACCEPTOR CIRCUIT because it provides maximum response to currents at its resonant frequency.

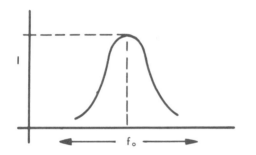

Fig. 8-3B. The response curve showing the current flow in the circuit above or below resonance.

Example: R = 10 ohms, L = 200 μ H, C = 200 pF. These components are connected across a 500 μ volt RF generator at 800 kilohertz. This circuit is diagrammed in Fig. 8-4. From the previous problem, the resonant frequency of this RCL circuit was 800 kilohertz, so $X_L = X_C$ at f_o.

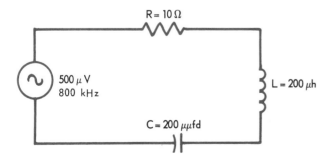

Fig. 8-4. The series-tuned acceptor circuit described in the text.

Then Z = 10 ohms and I $= \dfrac{E}{Z}$ or $\dfrac{500 \times 10^{-6}}{10} =$

50 μ amps

Also,

$$X_L = 2\ \pi \times 800 \times 10^3 \times 200 \times 10^{-6} = 1000\ \Omega$$
approx.

and

$$X_C = \dfrac{1}{2\ \pi \times 800 \times 10^3 \times 200 \times 10^{-12}} = 1000\ \Omega$$
approx.

The voltage drops around the circuit equal,

$$
\begin{aligned}
E_R &= 50\ \mu\ A \times 10 &&= 500\ \mu\ V \\
E_{X_L} &= 50\ \mu\ A \times 1000 &&= 50{,}000\ \mu\ V \\
E_{X_C} &= 50\ \mu\ A \times 1000 &&= 50{,}000\ \mu\ V
\end{aligned}
$$

It would appear that the sum of these voltages would be 100,500 μ V, but E_{X_L} and E_{X_C} are 180 deg. out of phase. Therefore a vector addition must be made,

$$E_{source} = \sqrt{(E_R)^2 + (E_{X_L} - E_{X_C})^2} = 500\ \mu V$$

It should be noted that, at resonance, the voltage drops across X_L and X_C are equal and also the voltage drop across R equals the source voltage. In a resonant circuit of this nature the interchange of energy between the inductance and the capacitance builds up a voltage which exceeds the supply voltage. In the first half cycle the magnetic field of the inductance stores the energy of the discharging capacitor. In the next half cycle the stored energy in the magnetic field of the inductance charges the capacitor. This action occurs back and forth, limited only by the series resistance between the two components. At resonance, the charging times for the inductance and capacitance must be the same, or they would have a cancelling effect.

Summarizing the series tuned circuit:

F	X_L OHMS	X_C OHMS	Z OHMS	I μA	E_{X_L} μV	E_{X_C} μV
200 KHz	250	4000	3750	.13	33	520
400 KHz	500	2000	1500	.33	165	660
600 KHz	750	1500	750	.66	495	990
800 KHz	1000	1000	10	5.0	50,000	50,000
1000 KHz	1250	800	450	1.1	1375	880
1200 KHz	1500	666	833	.6	900	400
1600 KHz	2000	500	1500	.33	660	165

Chart 8-1

Note: Above figures taken to nearest whole number for ease in understanding. Resistance of 10 ohms is insignificant, and is not included in computations except at resonance.

1. At reasonance, the impedance is minimum and the line current is maximum.
2. At resonance, the voltage drop, E_{X_L}, is equal to E_{X_C}, but is 180 deg. out of phase.
3. The vector sum of all voltage drops equals the applied voltage. Chart 8-1 shows the changing values as the circuit frequency is changed.

THE TANK CIRCUIT

The behavior of the circuit containing resistance, capacitance and inductance in parallel, differs from the series tuned acceptor circuit. The parallel tuned circuit is one of the more important circuits in the study of electronics and the student should become thoroughly acquainted with it.

A study of Fig. 8-5 will reveal the behavior of this circuit. In the first place, switch S_1 is closed, and capacitor C is charged to the supply voltage. Now S_1 is opened and C remains charged. When S_2 is closed, the capacitor discharges through L in the direction of the arrows. As the current flows through L, a magnetic field is built up around L, and this

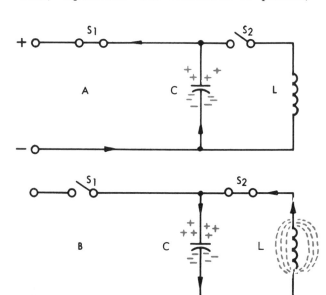

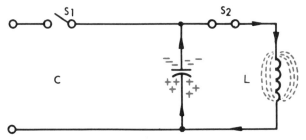

Fig. 8-5. The charge and discharge of C through inductance L is similar to flywheel action. It is called a tank circuit.

114

field remains as long as the current flows. However, when the charges on the plates of C become equalized, current ceases to flow and the magnetic field around L collapses, and the energy stored in this field is returned to the circuit. In the study of inductance in Chapter 6, we discovered that the induced emf opposes the current change.

This is also true in the parallel circuit. When the current drops to zero as a result of the discharged capacitor, a current is induced by the collapsing magnetic field which drives a charge onto the capacitor, but opposite to the original polarity. Now the capacitor discharges in the opposite direction. Once again the cycle of events occurs, and the capacitor again becomes charged as in its original state. This discharge-charge cycle repeats over and over and the current periodically changes direction in the circuit. The circuit is said to be OSCILLATING. The periodic reversals of current in the circuit may be described as FLY-WHEEL ACTION and the circuit is called a TANK CIRCUIT. If no energy were used during the cycles of oscillation, the circuit might oscillate indefinitely, but there is always some resistance due to the windings of the coil and the circuit connections. This resistance uses up the energy stored in the circuit and DAMPENS out the oscillation. The AMPLITUDE of each successive oscillation decreases due to the resistance.

The oscillators of a tank circuit may be likened to a child on a playground swing. The child may be swinging back and forth or oscillating, and if no one adds a little push to the swing, it will successively decrease in the amplitude of the swing until it finally comes to rest. If it were not for the friction or resistance, the swing might continue forever. Now if the swing was pushed every time it reached its maximum backward position, the added energy would replace the energy lost by friction, and the full swinging action would continue. The tank circuit, then, is similar. If pulses of energy are added to the oscillating tank circuit at the correct frequency, it will continue to oscillate.

Let us examine the meaning of correct frequency. During the discharge and charge of the capacitor C in the circuit, Fig. 8-5, a definite period of time must pass. In other words, during one cycle of oscillation a definite interval of time must pass. The number of cycles that occur in one second would be called the frequency of oscillation and is measured in cycles per second (cps) or hertz (Hz).

A thoughtful consideration of this circuit would disclose that if C or L were made larger so that it required a longer time to charge, the frequency of the circuit would decrease. This relationship may again be expressed mathematically by the resonance formula:

$$f_o = \frac{1}{2\pi\sqrt{LC}}$$

where,

f_o = resonant frequency in hertz

L = inductance in henrys

C = capacitance in farads

In electronic work, L usually has values in microhenrys (μH) and C in picofarads (pF). If this is the case, the equation may be simplified to read,

$$f_o = \frac{159}{\sqrt{LC}}$$

where,

f_o = resonant frequency in megahertz

L = inductance in microhenrys

C = capacitance in picofarads

THE REJECT CIRCUIT

If a parallel tuned circuit were connected across a generator of variable frequency, as in Fig. 8-6, you would discover that at the resonant frequency of the tuned circuit a

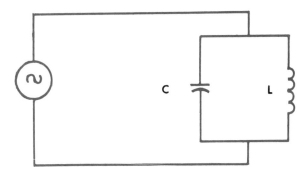

Fig. 8-6. A parallel LC circuit connected to a variable frequency generator.

minimum line current would flow. This is very useful information, because the resonant frequency may be determined by the minimum value or "dip" in the line current as measured by a current meter in the line. Radio transmitter operators always "dip the final" which means that they tune the final tank circuit to resonance which is indicated by a dip in current flow in the final circuit.

If minimum current flows in the circuit at resonance, then one may assume that at resonance a parallel-tuned circuit presents maximum impedance (Z). At frequencies other than resonance, the impedance is less. So, at resonance the circuit rejects currents at resonant frequency, and allows currents of frequencies other than resonance to pass. The characteristics are illustrated in the response curves in Fig. 8-7. There will be many applications of this principle in later studies of electronics.

It is difficult for the beginning student to understand why a parallel tuned tank presents maximum impedance at resonance. At resonance, $X_L = X_C$, and both have reactive values in parallel across the generator source. It would appear that these two reactive branches would combine to form a low reactive path for the line current. However, the current flowing in the X_L branch is lagging the applied voltage and the current in the X_C branch is leading the applied voltage. The currents, therefore, are 180 deg. out of phase and cancel each other. The total line current is the sum of the branch currents

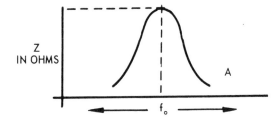

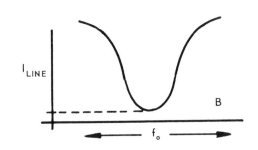

Fig. 8-7. A—The curve shows maximum impedance at resonant frequency. B—The circuit response shows minimum line current at resonance.

and therefore is zero, except for a small amount of current which will flow due to the resistance of the wire in the coil.

A typical problem will further clarify these circuit characteristics. Referring to Fig. 8-8, a 200 μH inductor and a 200 pF capacitor are connected in parallel across a generator source of 500 μV. The resistance R = 10 ohms represents the lumped resistance of the wire of the inductor. These same components were used in the study of the series resonant circuit and comparison should be made.

The resonant frequency of this tuned circuit is:

$$f_o = \frac{159}{\sqrt{LC}} \text{ or } \frac{159}{\sqrt{200 \times 200}} = \frac{159}{200} = .8 \text{ MHz (approx.)}$$

(Note use of convenient formula when L is in μH and C is in pF and f_o is in megahertz.

At resonance:

$$X_L = 2\pi fL \text{ or } 1000 \ \Omega$$

$$X_C = \frac{1}{2\pi fC} \text{ or } 1000 \ \Omega$$

The voltage across both branches of the parallel circuit is the same as the applied voltage or 500 μ volts. The current, therefore, in the X_L branch is:

$$I_{X_L} = \frac{500 \ \mu \text{ V}}{1000 \ \Omega} \quad \text{or} \quad \frac{500 \times 10^{-6}}{10^3} \quad \text{or}$$

$$500 \times 10^{-9} \text{ amps or } .5 \ \mu \text{ amps}$$

The current in the X_c branch is:

$$I_{X_c} = \frac{500 \ \mu \text{V}}{1000} = .5 \ \mu \text{ amps}$$

Since these two currents are 180 deg. out of phase, then:

$$I_{X_L} - I_{X_c} = .5 \ \mu \text{A} - .5 \ \mu \text{A} = 0$$

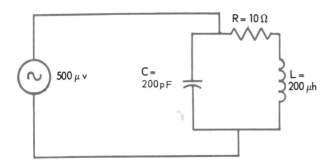

Fig. 8-8. The parallel resonant circuit used as an example in the text.

In this computation R has been neglected, because of its small value. Actually some current will flow as a result of R. For frequency above and below resistance, the performance of the circuit may be observed in Chart 8-2.

Note that the line current is the difference between the branch currents. This chart could be carried a step further to show the decreasing impedance due to frequencies other than resonance for,

$$Z = \frac{E}{I}$$

and as I increases, Z must decrease.

Q OF TUNED CIRCUITS

The Q of the circuit is described as the relationship between the inductive reactance and the resistance in the circuit, or

$$Q = \frac{X_L}{R}$$

In a series tuned circuit, for example, if the inductive reactance is 1000 ohms at resonance, and the resistance of the wire of the coil is 10 ohms, the FIGURE OF MERIT Q would be:

$$Q = \frac{1000}{10} = 100$$

f KHz	X_L OHMS	X_c OHMS	$I_{X_L} \ \mu\text{A}$	$I_{X_c} \ \mu\text{A}$	$I_{Line} \ \mu\text{A}$
200	250	4000	2.0	.125	1.875
400	500	2000	1.0	.25	1.75
600	750	1500	.66	.33	.33
800	1000	1000	.5	.5	0
1000	1250	800	.4	.625	.225
1200	1500	666	.33	.75	.42
1600	2000	500	.25	1.00	.75

Chart 8-2

The Q of a circuit is an indication of the sharpness of the reject or accept characteristics of the circuit. It is a quality factor. In the series acceptor circuit, an increase in resistance reduces the maximum current at resonance. See Fig. 8-9. The Q may also be used to determine the rise in voltage across L or C at resonance, for:

$$E_{X_L} = E_{X_C} = Q \times E_s$$

In the example on page 113, the supply voltage was 500 μ V and the Q is 100. Therefore, the voltage rise across X_L or X_C at resonance is equal to:

$$E_{X_L} = E_{X_C} = 100 \times 500 \ \mu V = 50,000 \ \mu V$$

If the circuit had a lower Q, the magnified voltage would be considerably less. For example, if R were increased to 20 ohms, then the Q would equal 50 and the voltage rise would equal:

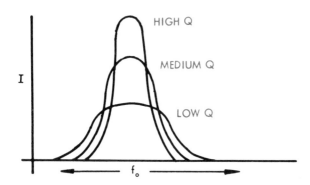

Fig. 8-9. As the Q of a circuit is lowered, the curve flattens out. Its selectivity decreases and its band pass increases.

$$E_{X_L} = E_{X_C} = 50 \times 500 \ \mu V = 25,000 \ \mu V$$

High Q circuits are very useful in selective electronic circuits and typical Qs range from 50 to 250. The higher value of Q produces a greater response of the circuit at resonance. Also a high Q circuit will have increased selectivity. Selectivity is determined by bandwidth which is the band of frequencies above

and below resonance, during which the circuit response does not fall below 70.7 percent (half power point) of the response at resonance.

The bandwidth of the tuned circuit may be found by the formula:

$$Bw = \frac{f_o}{Q}$$

Continuing the previous problem, the bandwith equals:

$$Bw = \frac{800,000 \ Hz}{100} = 8000 \ Hz$$

Since 800 KHz (the resonant frequency) is at the maximum response point, the bandwidth will extend 4000 Hz below resonance to 4000 Hz above resonance. The circuit may be considered, then, as passing all frequencies between 796 KHz and 804 KHz. Beyond either of these limits, the response will fall below the 70.7 percent value.

In the case of the parallel tuned circuit, the Q may be computed in like manner by using X_L at resonance and R which is the resistance of the coil L. The Q of a tank circuit may be used to determine the maximum impedance of the circuit at resonance for:

$$Z = Q \times X_L$$

where,

Z = impedance at resonance

$Q = \dfrac{X_L}{R}$ at resonance

X_L = reactance at resonance

In the example on page 116 the impedance at resonance would be:

$$Z = 100 \times 1000 \ ohms = 100,000 \ ohms$$

Another case which should be mentioned, is when a resistance is shunt across or in parallel with a tank circuit. In Fig. 8-10 this circuit is

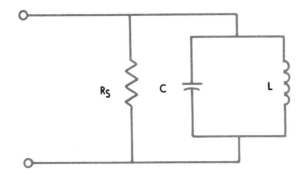

Fig. 8-10. The resistor R is a damping resistor and broadens the circuit response.

shown. R_S is called a damping resistor, and effectively broadens the frequency response of the circuit, because this resistor will carry a part of the line current which cannot be cancelled at resonance. Shunt damping lowers the Q of the circuit and makes it less selective.

LOADING THE TANK CIRCUIT

The parallel tuned circuit is used to couple energy from one circuit to another. Coupling transformers, used in radio and television sets, employ the tuned circuit to transfer signals from one stage to another. The radio transmitter uses a coupling device to a tank circuit to feed energy to the antenna system. If another coil is inductively coupled to the coil in the tank circuit, the varying magnetic field of the oscillating tank inductance will induce a current to flow in the coil. Fig. 8-11 shows the circuit of an intermediate frequency transformer used in a radio receiver.

Since both the primary and secondary of this transformer are tuned circuits, maximum transfer of radio frequency energy may be achieved, by carefully tuning the circuits to resonance. The intermediate frequency used in most radios is 455 KHz. The if transformer is tuned for maximum response at this frequency. A band pass of 5 KHz above and below the center frequency is maintained so that signals between 450 KHz and 460 KHz may pass without attenuation. The adjustment of if transformers is a part of radio servicing called ALIGNMENT.

FILTERING CIRCUITS

Inductance and capacitance have many useful applications in electronic circuitry. A circuit which is purposed to separate specified frequencies is known as a FILTER.

Filters may perform intended filtering action by separating the ac variations from a direct current. They may be designed to pass low frequencies and reject high frequency current or reject low frequency and pass high frequency currents or either pass or reject specified bands of frequencies. A filter derives its name from its function and is named:

1. Low Pass Filter.

2. High Pass Filter.

3. Band Pass Filter.

4. Band Reject Filter.

12-C1

Fig. 8-11. An intermediate frequency transformer used to couple energy from one tube to another in a radio receiver. (J. W. Miller Co.)

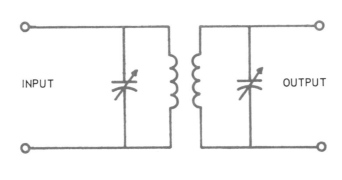

INPUT   OUTPUT

The output of a rectifier circuit is a pulsating direct current. For most purposes in electronics a pure direct current is needed with a low percentage of ripple. These filter circuits will be described in the section on power supplies.

In the study of filtering action these points should be reviewed:

1. A capacitor will block a direct current, but will pass an alternating current.

2. A conductor may carry a current which has both a dc component and an ac component.

Reference to Fig. 8-12A shows a graph representing a steady dc voltage of 10 volts and Fig. 8-12B represents an ac voltage with a peak value of 10 volts. The combination of these two voltages is shown in Fig. 8-12C. This wave

represents the algebraic sum of the two waves. Particular notice should be given to the new axis of the ac voltage, which is now varying above and below the 10 volt dc level. Because the axis has been raised by the dc voltage, the ac voltage no longer reverses polarity and the current is a varying direct current. It varies in amplitude between 0 and 20 volts. When this varying direct current is connected to a circuit, such as shown in Fig. 8-13, the capacitor C immediately charges up to the average dc level

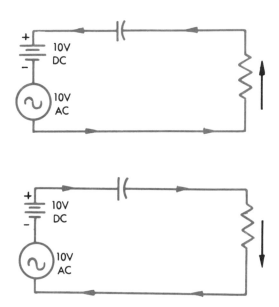

Fig. 8-13. The ac generator and dc source are connected in series to the RC circuit.

of voltage, which in this case would equal 10 volts. When the incoming voltage rises to 20 volts, current flows through R to charge capacitor C. In the next half cycle, the incoming voltage drops to zero, and capacitor C discharges to zero through R.

IMPORTANT: A voltage appears across R due to the charge and discharge of C, so the output of this circuit, taken from across R, represents only the ac component of the incoming signal voltage. See Fig. 8-14. The dc component is blocked by capacitor C. There is very little phase shift between the input and output waves, because the value of R is selected as ten times or more the value of the reactance

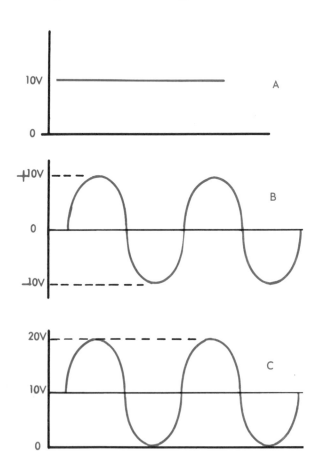

Fig. 8-12. A—A steady direct current. B—A 10V peak alternating current. C—The combined dc and ac.

of C at the frequency of the input voltage. When this ratio of resistance to reactance is maintained, the phase shift is negligible.

EXPERIMENT: Select a 1 mfd. capacitor from your stock of parts and connect circuit as in

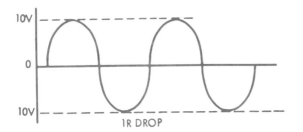

Fig. 8-14. The voltage across R is the result of the charge-discharge current and represents the ac component of the signal.

Fig. 8-15. As 60 Hz ac will be used in this experiment, the reactance of C should be computed:

$$X_c = \frac{.16}{fC} \text{ or } \frac{.16}{60 \times 1 \times 10^{-6}} = .0026 \times 10^6$$

or 2600 ohms

Now R should equal 10 times X_c or 10 x 2600 ohms or 26,000 ohms. A 25,000 ohm resistor will be close enough for this experiment. Connect the 6.3 volts (rms) terminals of a power supply in series with the 10 volt dc supply and connect to the circuit. Measure-

ments with dc voltmeter across C should read 10 volts. An ac meter across the same points should read zero. An ac meter across R should read 6.3 volts and a dc meter would read zero across R.

CONCLUSIONS: The dc component of the incoming signal appears across C. The ac component appears across R. This is a very important principle, and will be discussed in more detail in later chapters when it is necessary to couple a signal voltage (ac voltage) from one radio tube or stage to another. It will be called RC Coupling.

BYPASSING

In certain applications it is desirable to bypass an ac signal or voltage around a resistance, so that the voltage drop across the resistor will be the result of only the dc component of the signal voltage. These circuits are shown in Figs. 8-16A and 16-B. To aid in understanding the bypass capacitor, Fig. 8-16A consists of two

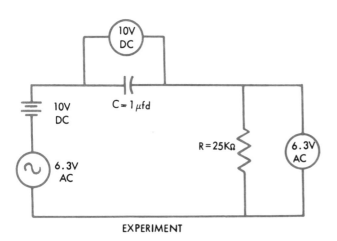

EXPERIMENT

Fig. 8-15. This experiment shows the blocking of the dc signal component and the passing of the ac component.

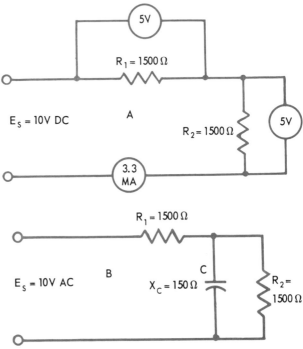

Fig. 8-16. A—A dc voltage produces equal drops across R_1 and R_2. B—The voltage drop across R_2 is held constant by bypassing the ac component.

resistors of 1500 ohms connected to a source of 10 volts dc. Total resistance is equal to $R_1 + R_2$ or 1500 ohms + 1500 ohms = 3000 ohms. The current in the circuit is:

$$I = \frac{E}{R} \text{ or } \frac{10V}{3000\,\Omega} = .0033 \text{ amps or } 3.3 \text{ mA}$$

The voltage drop across each resistor is:
R_1 = I x R or .0033 x 1500 = 5V
R_2 = I x R or .0033 x 1500 = 5V

When a 10 volt ac source is now connected, an ac voltage would appear across both resistors, but it is desired to bypass this ac component around R_2. To accomplish this, a capacitor of such a value as to have a low reactance to the ac voltage is connected in parallel with R_2. If the reactance of C is one tenth of resistance R, the greater part of the varying current will flow through C and not through R. For further clarification, assume values as indicated in Fig. 8-16B. C is chosen to have a reactance of 150 ohms to the frequency of the alternating current. This is one-tenth the value of R_2. For all practical purposes the impedance of CR_2 in parallel can be considered as 150 ohms, and the voltage distribution around the circuit may be measured or computed.

Total resistance equals $R_1 + R_2 X_C = 1650$ ohms. And $R_2 C$ represents one-eleventh of the total resistance to alternating current. As voltage drop is a function of resistance, then ten-elevenths of the voltage appears across R_1 and only one-eleventh across $R_2 C$ in parallel. The applied ac voltage is 10 volts so;

$$E_{R_1} = 9.1V \text{ and } E_{R_2} = .9V$$

If a 10V dc voltage is connected in series with the 10V ac voltage for the combined input voltage, the dc voltage divides equally between R_1 and R_2 with 5 volts each. C is an open circuit for a direct current. The ac voltage divides according to the above mentioned ratio, and most of this voltage appears across R_1. The voltage across R_2 remains relatively constant

due to the bypass capacitor action. In both audio and radio frequency circuits many applications are made of bypass capacitors. Summarizing, a capacitor may be selected which will form a low reactance path around a resistor or to ground for currents of selected frequencies.

LOW-PASS FILTERS

It is sometimes desirable to construct a filter which will pass low frequencies, yet attenuate high frequency currents. This filter is called a LOW-PASS FILTER. The circuit configuration always places a resistance or an inductor in series with the incoming signal voltage, and a capacitor in shunt or across the line, as in Fig. 8-17. As the frequency is increased, the reactance of L increases so that a larger amount of the voltage appears across L. Also as the frequency increases, the reactance of C will decrease, thus providing a bypass for the higher

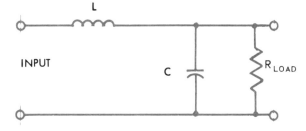

Fig. 8-17. A low-pass filter circuit.

frequency currents around the load resistance R. In both respects the higher frequencies appear in small amounts across the load, and low frequencies develop higher voltages across the load. To summarize: Low frequencies pass; high frequencies are rejected.

HIGH-PASS FILTER

The opposite to the low-pass filter is the HIGH-PASS FILTER, which passes selected high frequency currents, and rejects the low frequency currents. The circuit configuration always places a capacitor in series with the incoming signal voltage, and an inductance shunt across the line as in Fig. 8-18. As the

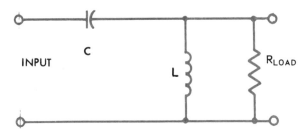

Fig. 8-18. The high-pass filter circuit.

frequency increases, X_L increases, and a higher voltage is developed across L and R in parallel. As frequency increases X_C decreases providing a low reactance path for high frequency signals. Low frequencies are shunted or bypassed around the load R by the low reactance of L at low frequencies.

To improve the filtering action of both low and high-pass filters, frequently two or more sections are connected together. They are named as suggested by the circuit configuration. In Fig. 8-19, the schematic drawings and the names of several types of filters are given.

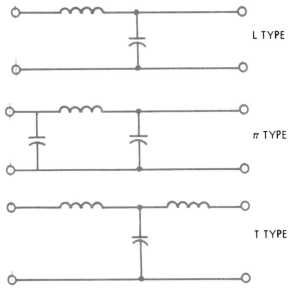

Fig. 8-19. Filter circuits may be named by the circuit configuration.

TUNED CIRCUIT FILTERS

Earlier in this chapter the series and parallel resonant circuits were described as accept or reject circuits respectively. These tuned circuits may be used conveniently and effectively as filters, because of their abilities to give maximum response at their tuned resonant frequency. In Fig. 8-20, the series resonant circuit

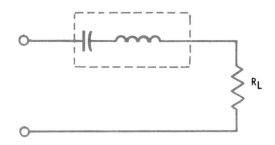

Fig. 8-20. The series tuned circuit used as a BAND PASS filter.

is used as a band pass filter and accepts only currents near its resonant frequency. In Fig. 8-21 the series resonant circuit is shunted across or parallel to the load. This provides a low impedance path, around or bypassing the load, for currents at resonant frequencies. In this case the filter would respond as a band reject filter.

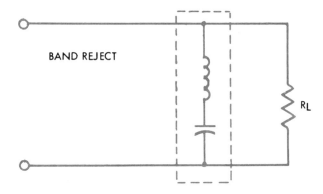

Fig. 8-21. The series tuned circuit arranged as a BAND REJECT filter.

The circuits using the parallel tuned circuit appear in Fig. 8-22. Its effect is the opposite to the series tuned circuit. When the tank circuit is in series with the incoming currents, it provides maximum impedance or rejects the frequencies at its resonant frequency. When the tank circuit is shunt across the load, it causes the maximum response across the load at resonance. Frequencies other than resonance are bypassed

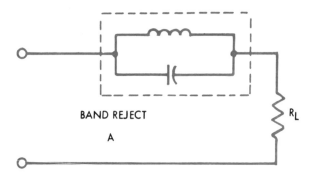

BAND REJECT

A

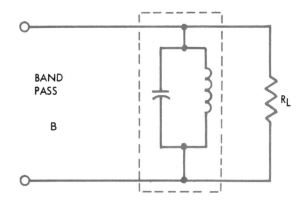

BAND
PASS

B

Fig. 8-22. A—The parallel tuned circuit as a BAND REJECT filter. B—The parallel tuned circuit as a BAND PASS filter.

Fig. 8-23. A high-pass filter used to remove interference to the television by other radio services.

THE NOMOGRAPH

The nomograph or alignment chart illustrated in Fig. 8-24 provides a rapid and convenient method of solving problems involving X_L, X_C and resonance. These examples will permit you to use and understand this chart.

1. What is the reactance of a .01 μ F capacitor at 10 KHz? Using upper chart, place a ruler or straight edge on 10 KHz on the frequency scale and .01 on the capacitance scale. The rule crosses the reactance scale at 1590 ohms.

2. What is the reactance of an 8 henry choke coil at 60 Hz? Use lower chart. Place rule on 60 Hz and 8 H and read 3014 ohms.

3. What is the resonant frequency of a tuned circuit when C = 250 pF and L = 200 mH? Use upper chart. Place rule on 200 mH and 250 pF and read 22.5 KHz on frequency scale.

because of the decreased impedance of the shunt tuned-tank circuit. Combinations of both series and parallel tuned circuits may be used to provide sharper cut-off and greater attenuation. A typical high-pass filter to remove interference from your TV is illustrated in Fig. 8-23. This filter attenuates all signals below Channel 2, yet all TV channels are passed through to the receiver.

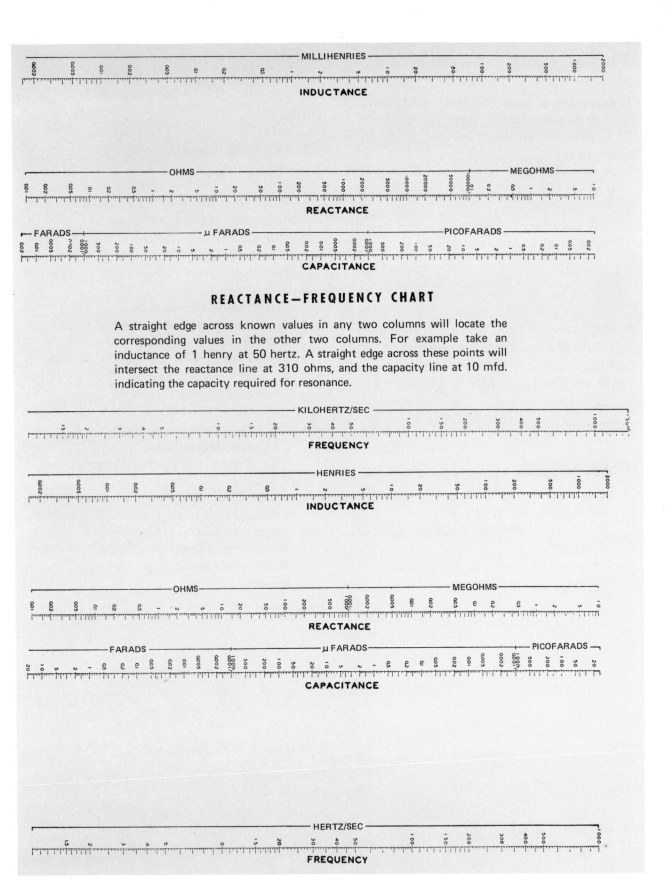

REACTANCE—FREQUENCY CHART

A straight edge across known values in any two columns will locate the corresponding values in the other two columns. For example take an inductance of 1 henry at 50 hertz. A straight edge across these points will intersect the reactance line at 310 ohms, and the capacity line at 10 mfd. indicating the capacity required for resonance.

Fig. 8-24. Nomograph or alignment chart.

FOR DISCUSSION

1. How can a series resonant circuit develop voltages higher than the applied voltage?
2. Explain fly-wheel action and relate it to frequency.
3. A simplified formula for resonant frequency is $f_o = \dfrac{159}{\sqrt{LC}}$ when f is in megahertz, L is in μH and C is in pF. Explain mathematically the derivation of this formula from

$$f_o = \frac{1}{2\pi\sqrt{LC}}$$

4. Why does the "dip" in the line current of a parallel LC circuit indicate resonance?
5. What is the relationship between the Q and the selectivity of a tuned circuit?

TEST YOUR KNOWLEDGE

1. A series resonant circuit is sometimes called an _____ circuit.
2. The formula for resonance is f_o = _____.
3. At resonance the impedance of the series LC circuit is _____.
4. Compute the resonant frequency of the following circuits:

L = 100 μH	C = 250 pF	f_o =
L = 200 μH	C = 130 pF	f_o =
L = 8 μH	C = 1 μF	f_o =

5 In a series resonant circuit E_R = _____. and the vector sum of E_L, E_C, and E_R = _____.

6. Refer to Chart 8-1. At frequencies above resonance the circuit appears _____. (inductive or capacitive)
What is the Q of this circuit? _____.
Why does circuit current decrease above or below resonance?
7. Frequency is measured in _____.
8. A parallel resonant circuit is sometimes called a _____ circuit.
9. In reference to a parallel LC circuit, at resonance, line current is _____. (minimum or maximum) Z is _____.
10. Refer to Chart 8-2. At frequencies above resonance the circuit appears _____. (inductive or capacitive)
Why does current increase above and below resonance?
11. What is the effect of a damping resistor across a parallel tuned circuit.
12. Name four general types of filters.
13. What is the purpose of a bypass capacitor?
14. Distinguish between a high-pass and a low-pass filter.

Chapter 9

ELECTRIC MOTORS

A philosopher once characterized electricity as the "invisible wonder of the world." He was correct in many respects, for scientists have yet to see electricity. However, the many useful effects of electricity have been discovered and harnessed, to make our lives richer and more pleasant. The conversion of electrical power into mechanical power represents one of the most useful and labor saving discoveries in electrical technology. Such a device is the now familiar electric motor. Motors are used to turn the wheels of industry. Motors are used in homes for refrigeration, air conditioning, mixing foods, cleaning rugs and on a myriad of other labor saving appliances. Motors are used on the farm for turning grinders, mixers, pumps and machinery. In the school shop the motor powers the bench saw, the lathe and numerous wood and metal machines. The motor is our willing and able servant.

PRINCIPLES OF MOTOR OPERATION

The operation of a motor depends upon the interaction of magnetic fields. A review of the Laws of Magnetism will reveal that:

LIKE POLES REPEL EACH OTHER

UNLIKE POLES ATTRACT EACH OTHER

or

A NORTH POLE REPELS A NORTH POLE

A SOUTH POLE REPELS A SOUTH POLE

but

A NORTH POLE ATTRACTS A SOUTH POLE

To understand the theory of the simple dc motor refer to the series of drawings, Figs. 1 to 9 inclusive, and their descriptive captions.

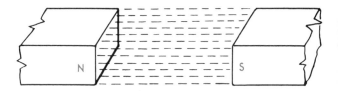

Fig. 9-1. A magnetic field exists between the North and South poles of a permanent magnet.

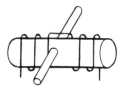

Fig. 9-2. An electromagnet is wound on an iron core and the core is placed on a shaft so it can rotate. This assembly is called the ARMATURE.

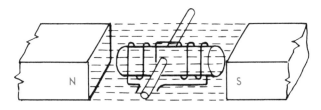

Fig. 9-3. The armature is placed in the permanent magnetic field.

In order to make the motor more powerful, the permanent field magnets may be replaced by electromagnets called FIELD WINDINGS.

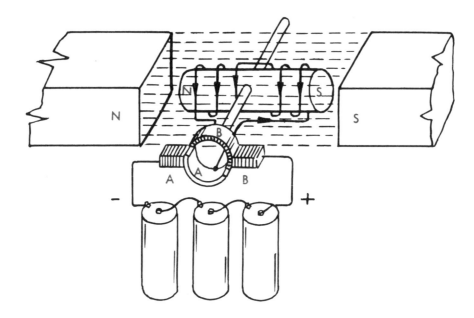

Fig. 9-4. The ends of the armature coil are connected to semicircular sections of metal called the COMMUTATOR. Brushes contact the rotating commutator sections and provide a means of energizing the armature coil from an external power source. NOTE: The polarity of the armature electromagnets depends upon the direction of current flow through the coil. Review Chapter 4. A battery is connected to the brushes and current flows into brush A to commutator section A, through the coil to section B and back to the battery through brush B, completing the circuit. The armature coil is magnetized as indicated on drawing.

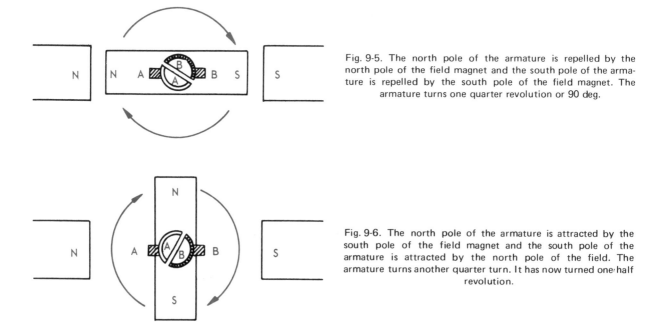

Fig. 9-5. The north pole of the armature is repelled by the north pole of the field magnet and the south pole of the armature is repelled by the south pole of the field magnet. The armature turns one quarter revolution or 90 deg.

Fig. 9-6. The north pole of the armature is attracted by the south pole of the field magnet and the south pole of the armature is attracted by the north pole of the field. The armature turns another quarter turn. It has now turned one half revolution.

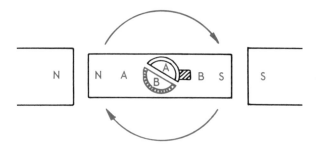

Fig. 9-7. As the commutator sections have turned with the armature, section B now contacts brush A and section A now contacts brush B. The current now flows into section B and out section A. The current has been reversed in the armature due to the commutator switching action. This reversal of current changes the polarity of the armature, so that unlike poles are next to each other.

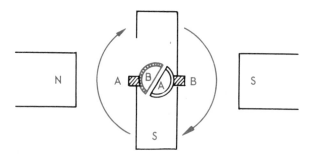

Fig. 9-8. Like poles repel each other and the armature turns another quarter turn.

Fig. 9-9. Unlike poles attract each other and the armature turns the last quarter turn, completing one revolution. The commutator and brushes are now lined up in their original positions, which causes the current to reverse in the armature again. The armature continues to rotate by repulsion and attraction. The current is reversed at each one-half revolution by the commutator.

These field windings may have an independent source of voltage connected to them, or they may be connected in series or parallel with the armature windings, to a single voltage source. All three types are diagramed in Fig. 9-10.

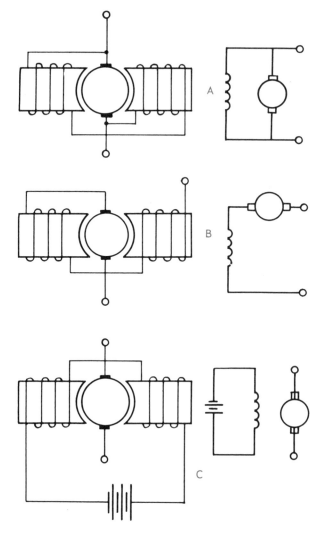

Fig. 9-10. Sketches and schematic diagrams of: A—Shunt wound motor. B—Series wound motor. C—Independently excited field motor.

A demonstration motor may now be constructed using the laboratory equipment, as illustrated in Fig. 9-11. Connect the motor as a series motor, then as a shunt motor, and compare the speed and power.

Commercially, motors are manufactured in a slightly different manner and obtain their rotational force by the interaction of the magnetic field around a current carrying conductor and the fixed magnetic field. A conductor carrying a current will have a magnetic field around it and the direction of the field depends upon the direction of the current. When this conductor is

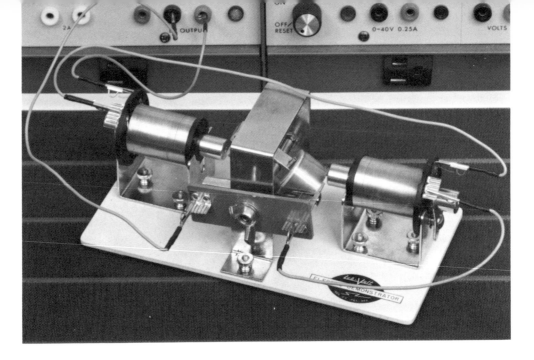

Fig. 9-11. The electrodemonstator may be set up to operate as a series or shunt motor.

placed in a fixed magnetic field, the interaction between the two fields will cause motion.

See Figs. 9-12 to 9-16 inclusive.

Fig. 9-12. A magnetic field exists between the poles of a permanent magnet. The arrows indicate the direction of field.

Fig. 9-13. A current carrying conductor has a magnetic field, its direction depends upon direction of current. USE LEFT HAND RULE, page 52 Chapter 4.

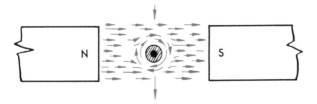

Fig. 9-14. In this illustration the field around the conductor reinforces the permanent field above the conductor, but opposes the permanent field below the conductor. The conductor will move toward the weakened field as indicated by the arrows.

Fig. 9-15. The current has been reversed in the conductor causing the conductor field to reverse. Now the field is reinforced below the conductor and weakened above the conductor. The conductor will move upward as indicated by arrow.

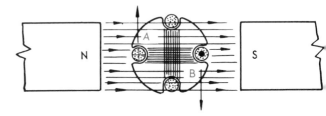

Fig. 9-16. The single conductor is replaced by a coil of conductors wound in the slots of an armature core. Notice how the interaction of the two fields will produce rotation. Coil side A moves up and coil side B moves down. The rotation is clockwise.

The armature coils are connected to commutator sections as in the experimental motor and the theory of operation is similar. A practical motor would have several armature coils wound in separate slots around the core. Each coil would have its respective commutator sections. Also, the number of field poles may be increased to give the motor greater power. A

130

four pole motor is sketched in Fig. 9-17. The current divides into four parts and the current flowing in windings under each field pole produces rotation and increases the turning power or TORQUE of the motor.

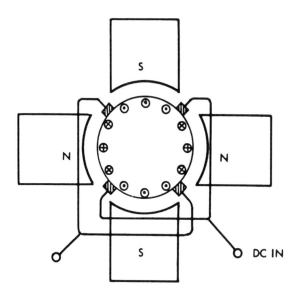

Fig. 9-17. The torque of the motor is increased by adding armature coils and field coils.

COUNTER EMF

In the study of generator action in Chapter 5, it was explained that if a conductor cut through a magnetic field, voltage would be induced in the moving conductor. Although a motor is intended to convert electrical energy into mechanical energy, it also becomes a generator when the armature begins to rotate. The induced electromotive force opposes the applied EMF, and is called COUNTER EMF as a result of the GENERATOR ACTION of the motor. The magnitude of the counter emf is directly proportional to the speed of rotation and the field strength. This may be represented mathematically by:

Counter emf = Speed x Field Strength x K

K equals other constants in any motor such as the number of windings. The actual effective voltage then applied to the windings in the armature must equal,

$$E_{source} - E_{counter} = E_{armature}$$

The current flowing in the armature windings at any given instant may be found by Ohm's Law, when the ohmic resistance of the windings is known:

$$I_{armature} = \frac{E_{armature}}{R_{armature}}$$

COMMUTATION AND INTERPOLES

As the motor armature rotates, the current in the armature windings is periodically reversed due to commutator action. Due to the self-inductance of the windings, the current does not immediately reverse. This results in sparking at the commutator brushes. Such arching may be prevented by moving the brushes slightly against the direction of rotation and taking advantage of the counter emf induced by the previous pole. This counter emf would oppose the self-induction caused by the decreasing current in the coil, thus eliminating the sparking. This is not a satisfactory method, because, as the load is varied on the motor, it would be necessary to manually change the position of the brushes.

Larger motors depend upon INTERPOLES or COMMUTATING POLES to reduce the

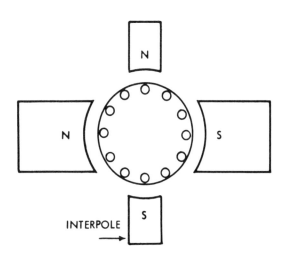

Fig. 9-18. Interpoles prevent "sparking" at the commutator.

sparking as a result of commutation. An inter-pole is another smaller field pole with the same polarity, following the main field pole in the direction of rotation. The counter emf developed as the armature passes the interpole overcomes the emf caused by self-induction in the armature windings. The windings of the interpole are connected in series with the armature and carry the armature current. As a result, the interpole field strength varies as the load varies, and provides automatic control of commutator sparking. For a sketch of a motor with interpoles, see Fig. 9-18.

SPEED REGULATION

Various motors are designed for special purposes. Some develop full power under load, whereas others must be brought up to speed before the load is applied. When the speed is determined for a motor on the job, it is desirable to have the motor maintain that speed under varying load conditions. A comparison of the speed under no-load, to the speed at full-load expressed as a percentage of the full-load speed, is called the percent of speed regulation. Thus:

Percent Speed Regulation =

$$\frac{\text{Speed no-load} - \text{Speed full-load}}{\text{Speed full-load}} \times 100$$

A low percent speed regulation would indicate that the motor would operate at a rather constant speed regardless of load applied.

THE SHUNT DC MOTOR

The shunt motor may be identified as having the field windings shunt across or in parallel to the armature. The schematic drawing of this motor may be seen in Fig. 9-19. The shunt motor is generally referred to as a <u>constant speed motor</u>. It finds applications in driving machine tools, or other machines requiring a relatively constant speed under varying loads.

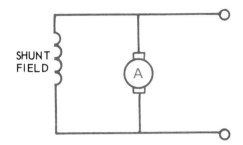

Fig. 9-19. Schematic of a shunt motor.

In the shunt motor, both the field and the armature are connected across the power line. Under no-load conditions the counter emf is almost equal to the line voltage, so very little armature current flows and very little torque is developed. When a load is applied and the armature decreases its speed, the counter emf also decreases, which increases the armature current and the torque. When the torque matches the load, the motor remains at constant speed. This action may be summarized as,

AT NO-LOAD

High counter emf
Low armature current
Low torque

AT FULL-LOAD

Decreased counter emf
Increased armature current
Increased torque

You should realize that the total current, used by this motor, is the sum of the field and armature currents and the input power may be computed by Watt's Law as:

Power = Applied E x Total Current

This, however, is not the output power, because the motor is not one hundred percent efficient.

THE SERIES DC MOTOR

In the series wound motor the field windings are connected in series with the armature windings. A diagram of this motor is given in Fig. 9-20. It is apparent that all the line current must flow through both the field and armature

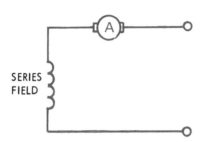

Fig. 9-20. Schematic of a series motor.

windings. Under loaded conditions the counter emf opposes the line voltage and limits the current to a safe value. If the load were suddenly removed, the armature would speed up and develop a higher cemf (counter emf). This reduces the current flowing through the field and reduces the field strength which causes the motor to increase its speed, because:

$$\text{Speed} = \frac{\text{cemf}}{\text{Field Strength x K}}$$

This action is cumulative, and the series motor would develop a speed which would cause the motor to literally fly apart, due to centrifugal force. A series motor is never operated without a load. It is usually directly connected to a machine or through gears. It is not safe to use a belt from motor to machine. If the belt should break or slip off, the motor would "run wild" and destroy itself.

One of the main advantages of the series motor is its ability to develop a high torque under load. Under load conditions, the armature speed is low and the cemf is low which results in a high armature current and increased torque. Series motors have heavy armature windings to carry these high currents. As the motor increases in speed the cemf builds up, the

line current decreases and the torque decreases. Series motors are used on electric trains, cranes and hoists, and other traction equipment.

COMPOUND DC MOTORS

The compound motor has both the series and the shunt field. The intention of the inventor, in this case, was to combine the desirable characteristics of each type of motor. The schematic drawings of this motor are shown in Fig. 9-21. As one might expect, the series winding which must also carry the armature current consists of several turns of heavy

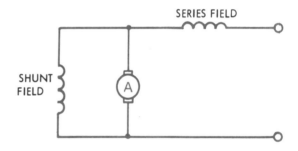

Fig. 9-21. Schematic of a compound motor.

wire; the shunt field winding consists of many turns of finer wire. Both windings are wound on the same field poles. Two methods may be used to connect these windings. If the magnetic field on the series winding aids or reinforces the magnetic field of the shunt winding, the motor is said to be a CUMULATIVE COMPOUND motor. If the two windings are so connected as to oppose each other magnetically, the motor is a DIFFERENTIAL COMPOUND MOTOR. A detailed study of compound motors is beyond the scope of this text; however, the characteristics of different types as described here should be noted.

The differential compound motor behaves in most respects like the shunt motor. The starting torque is low, and it has good speed regulation if loads do not vary too severely. However, it is not widely accepted.

The cumulative compound motor develops high starting torque. It is used where heavy loads are applied and some variance in speeds can be tolerated. The load may be safely removed from this motor.

MOTOR STARTING CIRCUITS

The armature resistance in most motors is quite low, and when the motor is stopped or at low speed, very little cemf is developed to oppose the line voltage. Consequently dangerously high currents will flow. In order to protect the motor, until it builds up its speed and cemf, a current limiting resistor may be placed in series with the armature to hold the current at a safe value. A simplified motor starting circuit is drawn in Fig. 9-22. The resistance is variable. Starting from a dead stop, maximum resistance is inserted in the armature circuit. As the speed builds up the resistance may gradually be decreased until the motor reaches full speed. The resistance may then be removed. An application of Ohm's Law will give meaning to this explanation. Assuming the armature resistance is .1 ohm and the applied voltage 100 volts, then the initial armature current would be:

$$I_{armature} = \frac{100V}{.1} \text{ or } 1000 \text{ amps}$$

This current would burn the insulation off from the wires and the motor would be destroyed. When the motor is up to speed and cemf has developed, then the voltage across the armature is equal to:

$$E_{line} - cemf = E_{armature}$$

If the motor develops a cemf of 95 volts as it approaches full speed, the armature current would be:

$$I_{armature} = \frac{100V - 95V}{.1} \text{ or } 50 \text{ amps}$$

This is a safe current for this motor armature to carry. Therefore, it is necessary to start the motor with the full series resistance in the armature circuit and gradually decrease the resistance until the motor can limit the current by its own cemf.

The adjustment of starting resistance may be done manually by a lever which decreases the resistance step-by-step until the motor reaches full speed. This circuit is diagramed in Fig. 9-23. In step one the maximum resistance is in series and with each successive step the resistance is decreased. At step 4, the lever arm is held magnetically by the holding coil which is also in series with the circuit. If the line voltage should fail, the lever arm would snap back to "off" position and the motor would need to be restarted. This is a protective device. Operators must use their good judgment when starting the motor. They must not move the lever to the next lower resistance position until the motor has gained sufficient speed.

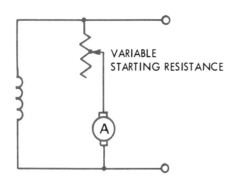

Fig. 9-22. Simplified motor starting circuit.

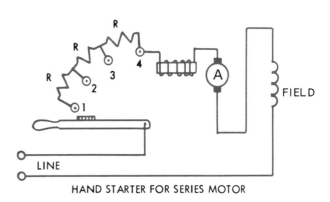

HAND STARTER FOR SERIES MOTOR

Fig. 9-23. Circuit of a typical step motor starter.

Automatic starters remove the human element of error and also permit remote starting of the motor. Several devices are used.

Fig. 9-24 shows one type of starting switch. When line switch S is closed, voltage is applied to the field which is across the line, and to the armature through R_1, R_2 and R_3 in series with the armature CONTACTORS C_1, C_2 and C_3 are a special type which will not close until the current has <u>decreased</u> to a predetermined value. The initial surge of current in the armature circuit flows through series coil of C_1 which holds it open. As the motor increases speed and

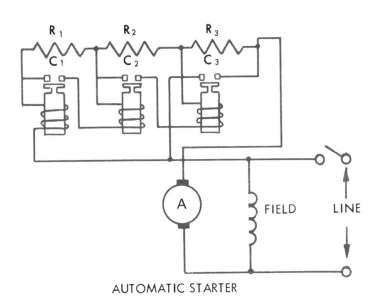

AUTOMATIC STARTER

Fig. 9-24. Circuit for an automatic motor starter.

the armature current decreases, C_1 contactor closes and cuts R_1 out of the circuit. Similar action occurs in C_2 and C_3. When the motor reaches full speed, all contactors will be closed. The protective resistance is no longer needed.

Before leaving motor starters, mention should be made of the PUSH BUTTON STARTER. This is a familiar device. The motors in your shop may be equipped with starting boxes containing RED and BLACK push buttons marked STOP and START. A push button starter is illustrated in Fig. 9-25. The circuit diagram is in Fig. 9-26. This starter is a relay

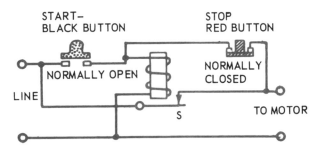

Fig. 9-26. Circuit of push-button control showing how relay is "locked out."

switch which operates on only a momentary pulse of current. Follow the action. When the black button is pushed, the contacts close and energize the relay coil which closes contacts S. The push button does not have to be held down, because once the relay switch is closed, it remains closed or "locked out." The closed switch S completes the circuit to the motor or device. The contacts of button "STOP" or RED are normally closed. To stop the motor, a momentary push on the RED button opens the circuit to the relay, and allows contacts S to open. For larger motors, the push button starter may be used to activate other starting devices.

THYRISTOR MOTOR CONTROLS

The thyristor is a name for a family of semiconductor devices that include the silicon controlled rectifier (SCR) and the triac. In the past few years these devices have been used in many circuits including motor controls. Fig.

9-27 shows the schematic symbols for each of these devices.

A half-wave SCR motor control circuit is shown in Fig. 9-28. This circuit will control up

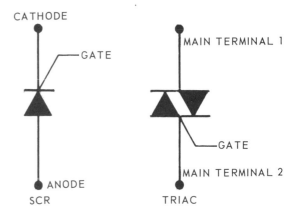

Fig. 9-27. Schematic symbols for SCR and TRIAC.

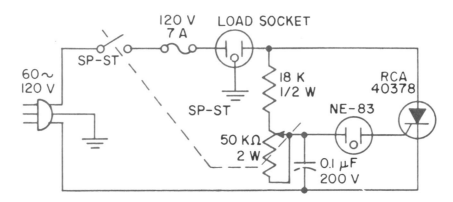

Fig. 9-28. Half-wave SCR motor control (RCA circuit).

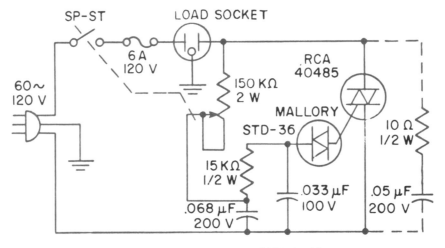

Fig. 9-29. Triac motor control (RCA circuit).

to seven amps in a universal motor such as found in a hand drill and hand saber saw.

The triac controls the full ac wave, while the SCR only controls half the wave. A typical full-wave triac circuit can be found in Fig. 9-29.

Fig. 9-30 shows a variety of SCR packages. The pencil points to an unencapsulated pellet. The two types in the foreground are capable of handling less than an amp of current while the larger types can control up to 2000 amps.

THE UNIVERSAL MOTOR

In the discussion of motor principles, the Laws of Magnetism were used to explain the operation of the dc motor. Will a dc motor operate on alternating current? Yes, to a limited extent. With an alternating current the poles of both the field and armature windings would be periodically reversing. However, two north poles repel each other as well as two south poles, so the motor action continues in the dc motor when ac is applied. For best results the series motor should be used. When the shunt motor is connected to ac, the inductance of the

Fig. 9-30. A variety of SCR packages.

Fig. 9-31. A fractional horsepower induction motor. Notice the stator coils and rotor.
(Delco Products, General Motors Corp.)

field windings will cause a phase displacement, and the motor action will be impaired. Experimentally, connect the laboratory constructed dc motor as a series motor, and apply about 6 to 8 volts ac to its terminals.

When this universal type motor is used commercially, the series wound type is preferred. These motors are not used for heavy duty purposes because of excessive sparking at the brushes. Commercial types are used for small fans, drills, and grinders.

INDUCTION MOTORS

One special class of motor, especially designed to operate on alternating current, is the INDUCTION MOTOR. The induction motor is constructed with the usual field poles and windings. Because the field poles and windings remain stationary, they are called the STATOR. Refer to Fig. 9-31. Since an alternating current is connected to these field coils, the polarity of the coils will change periodically according to the frequency of the current. Consider these windings as the primary of a transformer. If another coil of wire is closely coupled to this

primary, a current will be induced to flow in it. Review Chapter 6. According to Lenz's Law, the induced current will produce a field polarity which is opposite to the field which causes it, namely, the primary or stator windings. Therefore, there is an attractive force between the two coils. In the induction motor these secondary coils are mounted in slots in a laminated armature. The armature is called the ROTOR. See Fig. 9-32. Now this motor will not start rotation, it will only hum. If it is manually started, then it will continue rotation at a speed of 3600 rpm if used on 60 Hz ac. The 3600 is derived from 60 x 60 = 3600 or the number of

Fig. 9-32. The rotor and centrifugal switch assembly in an induction motor. (Delco Products, General Motors Corp.)

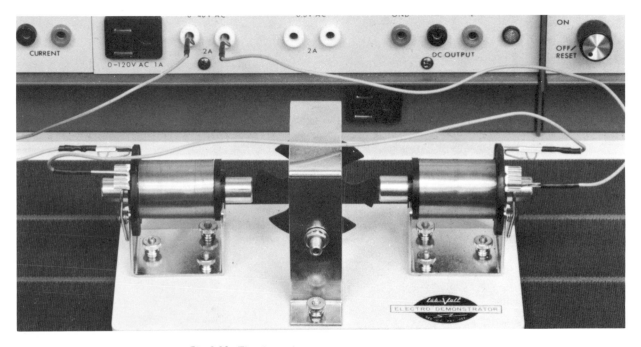

Fig. 9-33. The electrodemonstrator set up as an ac motor.

cycles per minute. A little thought on this motor action will disclose that the alternations of the ac are performing a function similar to the commutator in a dc motor. It is periodically reversing the current and polarity.

Experimentally, replace the armature on the laboratory constructed motor with the ac rotor. See Fig. 9-33. Connect the field windings to a 6 volt source of ac. The motor will not start by itself. Twirl the rotor with your fingers and notice that the motor continues to rotate. It appears that some method is needed to start the induction motor.

THREE-PHASE INDUCTION MOTOR

A three-phase alternating current represents three separate currents, 120 deg. out of phase with each other. This type of ac is useful in producing a rotating magnetic field in the stator. If the field rotates, then it will cause the rotor to follow the field around. A partial diagram of a stator with three poles is drawn in Fig. 9-34. Phase 1 current is connected to Pole A. Phase 2 is connected to Pole B. Phase 3 is connected to Pole C. Notice the sine waves showing the phase relationship on each line.

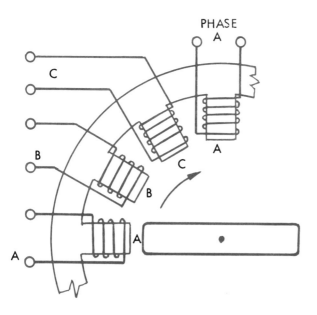

Fig. 9-34. One phase of the three-phase current is connected to each of the respective stator coils.

When phase 1 rises to maximum, the rotor is attracted to Pole A. As phase 1 starts to decrease, phase 2 rises to maximum and attracts the rotor to Pole B. As phase 2 starts to decrease, phase 3 rises to maximum and attracts the rotor to Pole C. Starting with Pole A and phase 1 again, the field rotates around the stator and attracts the rotor with it. The phase relationship between currents is shown in Fig. 9-35. The rotor consists of a laminated iron

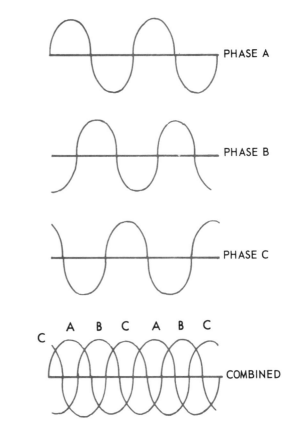

Fig. 9-35. The phase relationship between currents in a three-phase power source.

cylinder with heavy closed loops of copper wire imbedded in slots around the surface. Because the rotor appears like a cage, it is called a SQUIRREL CAGE rotor.

A polyphase motor is illustrated in Fig. 9-36. The rotor cannot follow the rotating field exactly. If both were in step, the rotor would not cut across any magnetic lines of force. No current would be induced in the rotor coils. A certain amount of SLIP exists between the

Fig. 9-36. A polyphase motor, (cut-away view).
(Delco Products, General Motors Corp.)

may be used for the purpose of phase splitting. A schematic diagram of a split-phase motor is shown in Fig. 9-37. The STARTING windings consist of many turns of relatively fine wire and the RUNNING windings have many turns of heavier wire. The windings are placed in slots around the inside of the stator. When electricity is applied to the motor terminals, current flows in both windings because they are in parallel. The main winding has many turns of wire and high inductance which causes the current to lag almost 90 deg. The starter windings of fine wire have considerably less inductance, and the current does not lag by as great an angle. The phase displacement creates a rotating field similar to the three-phase motor and the rotor starts to run. When the motor reaches speed, a centrifugal switch opens the starter winding

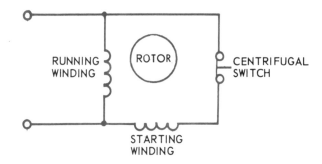

Fig. 9-37. Schematic diagram of a split-phase motor, showing starting windings.

rotating field and the rotor. Generally this is about 4 to 5 percent causing the rotor to turn at 1720 rpm when the field rotates at 1800 rpm. As the motor is loaded, the slippage increases and heavier currents are induced in the rotor.

SINGLE-PHASE INDUCTION MOTORS

In shops and industry, three-phase electricity is available for starting the large induction motors. At home, such currents are not readily available, so other means must be provided for starting the induction motor. If a means could be provided to divide a single-phase current into a polyphase current, the starting problem would be solved. This is called SPLITTING PHASES. Two methods are commonly used to produce the SPLIT-PHASE. In Chapter 6, it was learned that an inductance causes the current to lag in the circuit. If the single-phase current is caused to flow through two parallel paths that contain unequal amounts of inductance, then one current will lag behind the other and a phase displacement will exist between them. In effect, a two-phase current has been made of a single-phase current. The many turns of wire which make up the field windings of the induction motor are excellent inductances and

circuit and the motor runs as a single-phase inductor motor. The fine wire in the starting winding will not stand continuous operation and is cut out of the circuit after it has performed its function of starting.

The second method of phase splitting involves the use of a capacitor. Review Chapter 7. Capacitance causes the current to lead the applied voltage. If a capacitor is connected in series with the starting winding, a much larger phase displacement can be created. The capacitor is switched out of the circuit when the motor approaches full speed. Refer to diagram, Fig. 9-38. This motor is known as a CAPACITOR-START induction motor. These motors are used for many jobs around the shop and

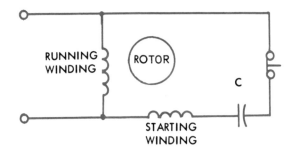

Fig. 9-38. Diagram of a capacitor-start split-phase motor.

home. They maintain a reasonably constant speed under varying loads. They do not develop a strong starting torque when compared to the three-phase squirrel cage motor. A capacitor-start motor is shown in Fig. 9-31. Note starting capacitor mounted on top.

REPULSION INDUCTION MOTOR

Another type of motor, strongly resembling the dc motor with commutator and brushes, is the repulsion-start motor. The major difference is that the brushes are connected together, rather than to the source of power. The brushes are so arranged that only selected coils of the armature are closed or complete circuits at a given time. The currents induced by transformer action in the rotor windings are displaced by the shorting brushes and poles are created which oppose the stator poles. Rotation is started by repulsion. When the motor reaches about 75 percent of its speed, a centrifugal shorting ring "shorts out" all the commutator

sections, and the motor runs as an induction motor. The repulsion induction motor has high starting torque, but does not readily come up to speed under load. It is being replaced by the cheaper capacitor – start motor.

THE SHADED POLE MOTOR

A simple two-pole motor involving an unusual method of starting rotation is sketched in Fig. 9-39. The rotor is the typical squirrel cage type and the stator resembles the usual dc motor, except that a slot is cut in the face of each pole and a single turn of heavy wire is wound in the slot. This is called SHADING THE POLE. The single turn of wire is the SHADING COIL. The action of the shading coil may be followed with reference to Fig. 9-40. As

SHADED POLE ACTION

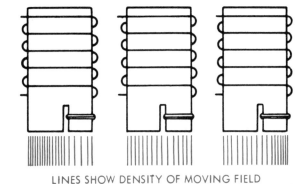

LINES SHOW DENSITY OF MOVING FIELD

Fig. 9-40. The magnetic field moves from left to right as the result of shading.

SHADED TWO POLE MOTOR

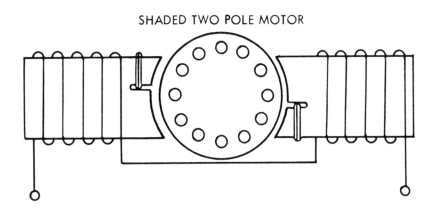

Fig. 9-39. Sketch of a shaded pole motor.

the current rises in the first quarter cycle of the ac wave, a magnetic field is produced in the field winding. However, the expanding magnetic field cuts across the shading coil which induces a current and polarity which opposes that of the field coil. This tends to subtract from and weaken the total field on the side of the shading coil. At the top of the ac wave very little change in current is taking place so the total magnetic field becomes equalized across the face of the pole. When the current starts to decrease in the next quarter cycle, the induced current in the shading coil is of such a polarity as to resist the decreasing current. So the pole has a strong magnetic field on the right side. Notice that during this half cycle, the magnetic field has moved from left to right. This movement of the field causes the rotor to start rotation. The starting torque of a motor of this type is relatively small. The motor has many applications such as small fans and electric clocks.

The studies of this chapter have been intended to give an exploratory experience in types of motors and their operation. Advanced design and theory of motors is detailed and complex and is beyond the scope of this text.

FOR DISCUSSION

1. Why does a motor increase in speed when its field strength is decreased?
2. Why does a motor get hot when overloaded?
3. If the load is removed from a series motor, it will destroy itself by centrifugal force. Explain.
4. Why are motor starters necessary on heavy duty motors?
5. Will your experimental dc motor run on 6 volts ac? Explain.
6. Compare an induction motor to a transformer.

TEST YOUR KNOWLEDGE

1. What is the purpose of the commutator in the dc motor?
2. A motor is a device which changes _____ energy into _____ energy.
3. Draw diagrams of the series motor and the shunt motor.
4. Draw a diagram and explain the motion of a current carrying conductor in a fixed magnetic field.
5. What is generator action in a motor?
6. What determines the current used by the motor?
7. Why are interpoles used on large dc motors?
8. Diagram a Push Button motor starting switch and explain its action.
9. If the starting windings were burned out in a split-phase motor, how would it behave when the power is turned on?
10. Name two common methods of splitting a single phase current to start a motor.

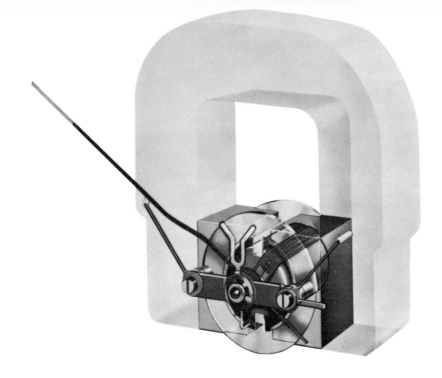

Fig. 10-1. A phantom view of the D'Arsonval meter movement.
(Daystrom, Weston Instruments Div.)

Chapter 10

INSTRUMENTS AND MEASUREMENTS

One of our American scientists recently remarked that "greater progress in scientific development depends upon accuracy of measurements."

The student in electricity and electronics must rely upon the accuracy of instruments to determine the performance and characteristics of a circuit. Skill in the use of instrumentation marks the level of technical competency. The technician must know what he or she is trying to measure and how to measure it. Experienced technicians and the correct measurements they make are valued assets in the design of electronic equipment as well as in the diagnosis and correction of circuit failures.

BASIC METER MOVEMENT

One of the more popular types of meter movements used to measure current and voltage is the D'Arsonval or stationary-magnet moving-coil meter. See Fig. 10-1. The movement consists of a permanent type magnet and a rotating coil in the magnetic field. To the rotating coil is attached an indicating needle.

See Fig. 10-2. When a current passes through the moving coil, a magnetic field is produced which reacts with the stationary field, and causes rotation or deflection of the needle. This deflection force is proportional to the intensity of the current flowing in the moving coil. The

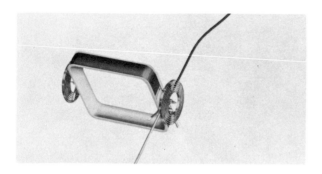

Fig. 10-2. Indicating needle is attached to rotating coil of meter.

moving coil is returned to its "at rest" position by hair springs which also act as connections to the coil. The deflecting force rotates the coil against the restraining force of these springs.

IMPORTANT: To change the direction of current in the moving coil would cause deflection of the needle in the opposite direction. Be sure that correct polarity of voltages and current are observed when connecting a meter to a live circuit.

An improvement on this simple meter movement includes a dampening device. A dampener prevents the indicating needle from overshooting and oscillating when making a measurement. To accomplish this, the moving coil is wound on an aluminum bobbin. As the bobbin tends to oscillate in the magnetic field, an emf is induced in it which opposes the change in motion. (Lenz's Law) As a result, any tendency for the coil to oscillate is quickly dampened out.

Meters are expensive instruments and should be used with skill and judgment. Precision types have jewel bearings like fine watches. The quality of the meter can be judged by its sensitivity and accuracy.

AMMETER

The instrument to be studied first is the AMMETER. This meter measures current flow in a circuit in amperes, milliamperes or microamperes, depending upon the scale selected. The moving coil of the meter is wound with many turns of very fine wire. If a large current were allowed to flow in the coil, it would quickly burn out. In order to measure larger currents with the basic movement, a SHUNT or alternate path is provided, through which the major amount of current will flow leaving only sufficient current to safely operate the moving coil. This shunt is in the form of a precision resistor connected in parallel with the meter coil.

In order to understand the use of shunts refer to Fig. 10-3 and follow this problem. The specification of a certain meter movement states that it requires .001 amps or one milliampere for full scale deflection of the needle

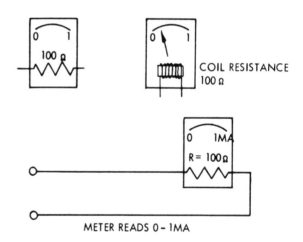

Fig. 10-3. Above. Sketch of basic ammeter. Below. Step 1. The voltage which causes the full scale deflection current is computed.

and the ohmic resistance of the moving coil is 100 ohms. Compute the shunt resistor values so that the meter will measure currents from 0-1 mA, 0-10 mA, 0-50 mA, and 0-100 mA. Each range involves using a different shunt.

Step 1. Calculate the voltage which, if applied to the coil, will cause one milliampere of current to flow and full scale deflection of needle.

$$E = I \times R \quad E = .001 \times 100 \text{ or } .1 \text{ volt}$$

where:

I = .001 amps for full scale deflection
R = resistance of meter coil

The meter will read from 0-1 mA without a shunt.

Step 2. To convert this same meter to read from 0-10 mA, a shunt must be connected which will carry nine-tenths of the current or 9 mA, leaving the one milliampere to operate the meter. Since .1 volt must be applied across the meter, and also across the shunt, the shunt resistance must be:

$$R_s = \frac{.1 V}{.009 \text{ amps}} = 11.1 \text{ ohms}$$

The meter, with 11.1 ohms shunt will measure from 0-10 mA.

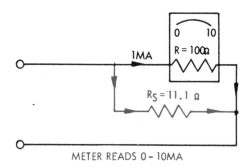

METER READS 0 – 10MA

Fig. 10-3. Step 2. The shunt carries 9/10 of the current.

Step 3. To convert this meter to read from 0-50 mA, a shunt must be used which will carry 49/50 of the current or 49 mA. The computation is the same as in Step 2:

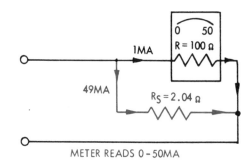

METER READS 0 - 50MA

Fig. 10-3. Step 3. The shunt carries 49/50 of the current.

$$R_s = \frac{.1 V}{.049 \text{ amps}} = 2.04 \text{ ohms}$$

Step 4. To convert the meter for the 0-100 mA scale, a shunt must be used which will carry 99/100 of the current or 99 mA.

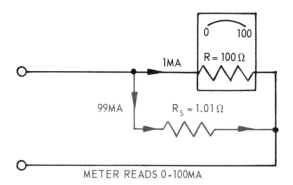

METER READS 0 -100MA

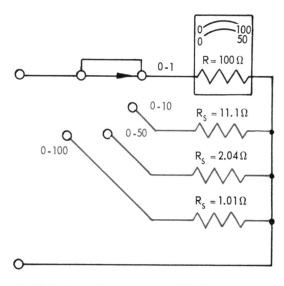

Fig. 10-3. Step 4. The shunt carries 99/100 of the current.

$$R_s = \frac{.1V}{.099 \text{ amps}} = 1.01 \text{ ohms}$$

The meter with the 1.01 ohms shunt will measure currents from 0-100 mA. Look again at the circuit in Fig. 10-3, and notice the switching device used to change the ranges of the meter. The correct scale on the dial must be used to correspond to the selected range.

IMPORTANT: An ammeter must always be connected in series with a circuit. Never across or parallel with it. To make a series connection requires breaking the circuit to connect the meter. A parallel connection may instantaneously destroy the meter. See Fig. 10-4.

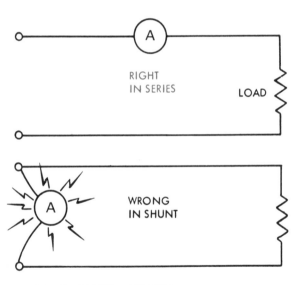

Fig. 10-4. The RIGHT and WRONG way to connect an ammeter.

VOLTMETER

The same basic movement used for the ammeter may also be used to measure voltage, providing the impressed voltage across the coil never exceeds .1 volt as computed for full scale deflection. To arrange the meter to measure higher voltages, MULTIPLIER RESISTORS are switched in series with the meter movement. Using the same meter as used for the ammeter, refer to Fig. 10-5 and follow the steps as the

multipliers are computed so that the meter will measure voltages from 0-1V, 0-10V, 0-100V and 0-500V.

Step 1. As no more than .1 volt is permitted across the meter coil at any time, a resistor must be placed in series with the meter which will cause a voltage drop of .9V, if the meter is used to measure one volt. Also the meter only allows .001 amperes for full scale deflection, which is the maximum allowable current in the circuit. The multiplier resistor must produce a .9 volts drop when .001 amperes flow through it.

$$R_M = \frac{.9V}{.001 \text{ amps}} = 900 \text{ ohms}$$

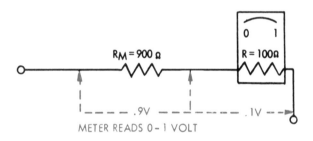

METER READS 0 - 1 VOLT

Fig. 10-5. Step 1. The multiplier causes an IR drop of .9V.

Step 2. To convert to the 0-10 volt range, a resistor must be selected to produce a 9.9 volt drop.

$$R_M = \frac{9.9V}{.001 \text{ amps}} = 9900 \text{ ohms}$$

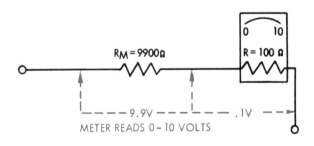

METER READS 0 - 10 VOLTS

Fig. 10-5. Step 2. The multiplier causes an IR drop of 9.9V.

Step 3. To convert to the 0-100 volt range, a resistor must be selected to produce a 99.9 volts drop.

$$R_M = \frac{99.9V}{.001 \text{ amps}} = 99,900 \text{ ohms}$$

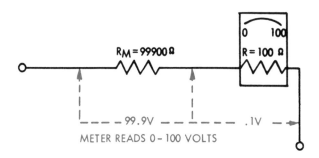

METER READS 0 – 100 VOLTS

Fig. 10-5. Step 3. The multiplier causes an IR drop of 99.9V.

Step 4. Finally, to use the 0-500 volt range, the resistor must cause a 499.9 volts drop.

$$R_M = \frac{499.9V}{.001 \text{ amps}} = 499,900 \text{ ohms}$$

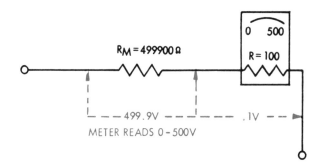

METER READS 0 – 500V

Fig. 10-5. Step 4. The multiplier causes an IR drop of 499.9V.

A switching device is again used to select the correct multiplier resistor for the range in use. Read the scale on the dial which corresponds to the range selected.

IMPORTANT: A voltmeter is always connected in parallel or across the circuit. Never in series. To measure voltage the circuit does not have to be broken. See Fig. 10-6. Added precautions: When measuring an unknown voltage, always

start with the meter on its highest range. Be sure that the leads are connected with the correct polarity. BLACK is negative. RED is positive.

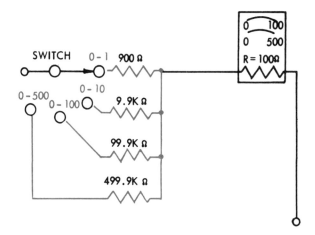

Fig. 10-5. Step 5. A switch selects the range.

VOLTMETER SENSITIVITY

The sensitivity of a meter is an indication of its quality. Ohms-per-volt for full scale deflection is the usual descriptive terminology. In Step 4 of the previous example, the total resistance of the meter and its multipler resistance is:

499,900 ohms in R_M

100 ohms for meter resistance
500,000 ohms total resistance

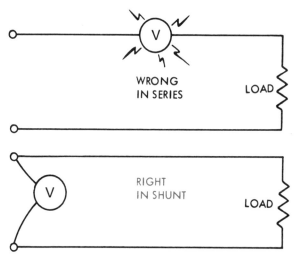

Fig. 10-6. The RIGHT and WRONG way to connect a voltmeter.

This amount of resistance is used in the 500 volt range which is equivalent to:

$$\frac{500,000 \text{ ohms}}{500V} \text{ or } 1000 \text{ ohms per volt}$$

By Ohm's Law, $I = \frac{E}{R}$ and the reciprocal of $I = \frac{R}{E}$ which is the same as sensitivity.

Therefore, sensitivity is equal to the reciprocal of the current required for full scale deflection. For the meter used in the above example:

$$\text{Sensitivity} = \frac{1}{I} \text{ or } \frac{1}{.001} \text{ or } 1000 \text{ ohms/volt}$$

This is an easy method of determining the sensitivity.

A good quality meter will have a sensitivity of 20,000 ohm/volt and precision laboratory meters are as high as 200,000 ohms/volt. Accuracy of the meter is usually expressed as a percentage such as 1 percent. The true value will be within one percent of the indicated scale reading.

OHMMETERS

A meter used to measure the value of an unknown resistance is called an OHMMETER. The same meter movement as used for the voltmeter and ammeter may be used for this purpose. A voltage source and a variable resistor are added to the circuit. The series type ohmmeter is diagramed in Fig. 10-7. A three-

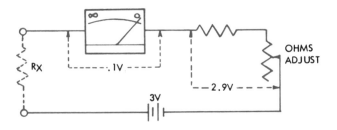

Fig. 10-7. The schematic diagram of the series ohmmeter.

volt battery is used as source, and is built into the meter case. As the meter movement permits only .1 volt for a current of .001 amps for full

scale deflection, a multiplier resistor is placed in series with the meter to drop the voltage:

$$R_M = \frac{2.9V}{.001 \text{ amps}} = 2900 \text{ ohms}$$

This resistance plus the meter resistance is equal to 3000 ohms. Part of this resistance is made variable to compensate for changes in battery voltage due to aging. To use the meter, the test terminals are shorted. This is equal to zero resistance and the meter deflects from left to right. The needle is adjusted to zero, by the variable resistor marked, "OHMS ADJUST." An unknown resistor, R_x, is now placed between the test terminals. The needle will deflect less than full scale depending upon the resistance value. The scale reads directly in ohms, or other ranges may specify that the scale value should be multiplied by 10, or by 1000 or by 100,000. It is interesting to note that with this circuit the meter scale reads right to left which is opposite to the voltmeter or ammeter scale.

A shunt ohmmeter may be connected as in Fig. 10-8. In this circuit the unknown resistance R_x is shunted across the meter. Low values of R_x cause lower currents through the meter. High values of R_x cause higher meter currents. With this arrangement, the indicating needle deflects from right to left in the customary manner. Zero resistance is on the left, and the scale increases from left to right.

MULTIMETER

Modern meters used by technicians, called multimeters, usually include the voltmeter,

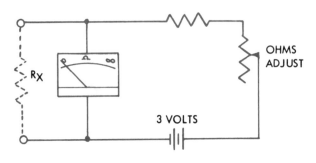

Fig. 10-8. The shunt ohmmeter.

ammeter and ohmmeter in one case. See Fig. 10-9. Adequate switching is provided to change function and range of the meter. Precaution: Before using the ohmmeter to test a circuit, be sure that the power is turned off the equipment under test. Better pull the plug to be sure. Also discharge all electrolytic capacitors by shorting terminals to ground with an insulated screwdriver. Multimeters are expensive and carelessness will ruin one.

WHEATSTONE BRIDGE

Many bridge circuits of various types are used in electrical measurements. To familiarize the student with this circuit the WHEATSTONE BRIDGE will be described to measure unknown resistance. Fig. 10-10 is the schematic diagram. R_1, R_2 and R_3 are precision variable resistors. R_x is the unknown. When voltage is applied to the circuit, current will flow through the branch $R_1 R_2$ and parallel branch $R_3 R_x$. Each resistor will cause a voltage drop. The bridge is balanced by connecting a meter between points A and B, and then adjusting the variable resistors until the meter reads zero. The balanced condition indicates that:

$$E_{R_1} = E_{R_3}$$

and

Fig. 10-9. The multimeter combines the voltmeter, ammeter and ohmmeter. (Simpson Electric Co.)

$$E_{R_2} = E_{R_x}$$

Since voltage is proportional to resistance,

$$\frac{R_1}{R_2} = \frac{R_3}{R_x} \quad \text{and} \quad R_x = \frac{R_2 \times R_3}{R_1}$$

The values of the precision variable resistors are indicated on an accurately calibrated dial.

IRON VANE METER MOVEMENT

The principle of operation of the iron vane meter movement may be understood by refer-

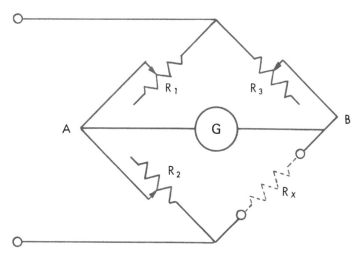

Fig. 10-10. The Wheatstone Bridge circuit.

ence to Fig. 10-11. If two pieces of iron are placed in the hollow core of a solenoid, they will both become magnetized with the same polarity, when a current passes through the solenoid. But, like poles repel each other. So the two pieces of metal are repelled. One piece of metal is fixed and the other pivoted so that it may turn away from the fixed metal. An indicating needle is attached to the moving vane, which also has hair springs so that the vane must move against the spring tension. An applied voltage causes a current to flow in the solenoid and produces the magnetic field. The moving vane is repelled against the spring in proportion to the strength of the magnetic field. The needle may indicate either voltage or current and is calibrated for the magnitude of the applied voltage or current.

When the iron vane movement is used for a voltmeter, the solenoid is usually wound with many turns of fine wire. Proper multipler resistances may be used to increase the range of the meter.

When used as an ammeter, however, the solenoid has relatively few turns of heavy wire, because the coil must be connected in series with the circuit and carry the circuit current.

Regardless of the polarity of the applied voltage or current, the iron vane meter movement always deflects in the same direction. (Two north poles repel as well as two south

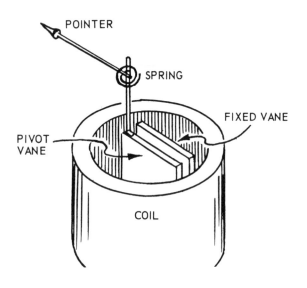

Fig. 10-11. A sketch which shows the operating principle of the iron vane meter movement.

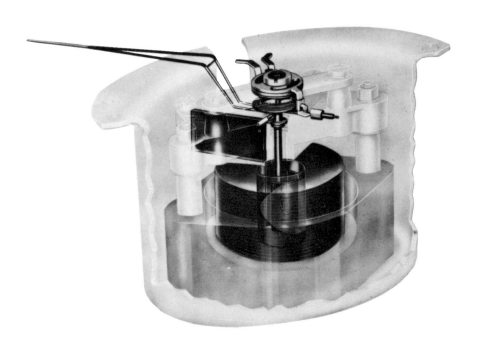

Fig. 10-12. The phantom view of a concentric type iron vane meter movement. (Daystrom, Weston Instrument Co.)

poles.) Either ac or dc may be measured by this instrument. Generally this type of meter is better suited for measurements in high power circuits. A phantom view of the interior construction of the concentric type iron vane meter is shown in Fig. 10-12.

THE WATTMETER

Power in an electric circuit is equal to the product of the voltage and the current. To devise a meter which may be read directly in watts, a movement similar to the D'Arsonval movement may be used, except the permanent magnetic field is replaced by the coils of an electromagnet. This type of meter is usually referred to as an electrodynamometer movement. The circuit diagram of a simple wattmeter is shown in Fig. 10-13. The moving

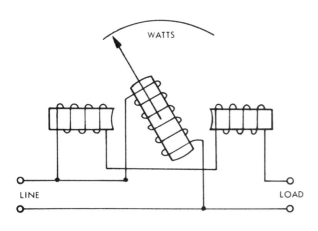

Fig. 10-13. Diagram of a wattmeter.

coil, with the proper multiplier resistance, is connected across the voltage in the circuit. The coils of the electromagnet are connected in series with the circuit under measurement. The interaction between the two fields is proportional to the product of the voltage and the current and deflection of the indicating needle is read on a calibrated scale in watts. A commercial type meter movement is illustrated in Fig. 10-14.

The wattmeter will measure the instantaneous power used in the circuit. To determine the amount of power used for a given time, an

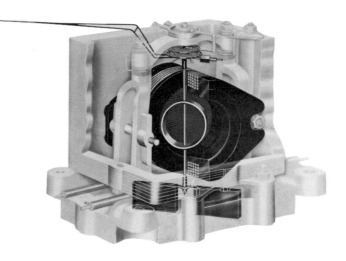

Fig. 10-14. Phantom view of the electrodynamometer movement used in a wattmeter. (Daystrom, Weston Instrument Co.)

entirely different device is used, called a watt-hour meter. This meter sums up all the powers used for an interval of time. Such a meter is installed by the power company on the outside of your home. As the watt-hour is a relatively small unit, the customary meter reads in kilo-watt-hours or 1000 watt-hours (KWH). Electric power consumed is paid for at the prevailing rates per KWH.

The watt-hour meter is a rather complicated type of induction motor using field coils in series with the line current and also field coils connected across the line voltage. An aluminum disc rotates within these fields, at a rate proportional to the power consumed. The disc is geared to an indicating dial, from which may be read the power used. This rotating disc may be observed in the meter on your home. To read the meter, observe the four dials in Fig. 10-15. Dial A, on the extreme right, reads from 0-10 units of kilowatt hours. For each revolu-

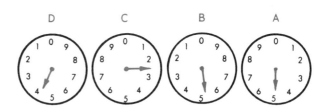

Fig. 10-15. The dials of the watt-hour meter.

tion of dial A, the next dial B reads one. For each revolution of dial B, dial C reads one. For each revolution of dial C, dial D reads one. Mathematically:

Dial A reads in units of one.
Dial B reads in units of ten.
Dial C reads in units of one hundred.
Dial D reads in units of one thousand.

To read the meter illustrated in Fig. 10-15:

Dial A points to 5.
Dial B points between 5 and 6.
Dial C points between 2 and 3.
Dial D points between 4 and 5.

The correct reading would be, 4255 KWH. Notice that when the indicating arrow is between numbers, the lowest number or the number just passed should be taken. To practice reading the meter, why not read the meter at your home each day for several days and compute the power used? The power consumed may be found by subtracting the previous reading from the present reading to obtain the difference.

AC METERS

The typical dc meter movement, such as the D'Arsonval movement, may also be used to measure alternating currents and voltages with certain changes in the circuit. Remember that the dc meter must always be connected with the correct polarity. Incorrect polarity, causing a reversal of current through the moving coil, would cause deflection to the left below zero and possible damage to the meter. If an ac voltage were connected to the meter, the periodic changes in direction of the current would cause the needle to vibrate at zero. In order to measure this ac voltage a rectifier must be used. Detailed discussion of rectifiers will be included in the chapter on Power Supplies. For the present, however, a rectifier may be described as a device which permits current to flow in only one direction. When used in a

meter, the rectifier allows one half cycle of the ac current to pass, but effectively cuts off the current when it reverse itself in the second half cycle. A simple circuit with a rectifier is shown in Fig. 10-16.

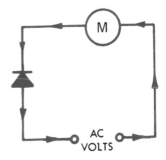

Fig. 10-16. Basic circuit for ac meter using a single rectifier.

Notice the symbol used for the rectifier. Its output is a pulsating direct current which will cause the meter to deflect in one direction only. An improved circuit, which serves the same purpose, is the full wave bridge rectifier shown in Fig. 10-17. By using four rectifiers, both half

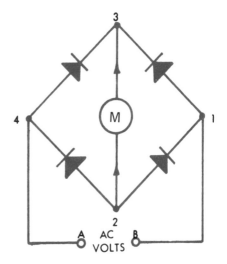

Fig. 10-17. A bridge rectifier circuit for an ac meter.

cycles of an ac current may be applied to the meter. When ac voltage is applied to terminals AB, and A is positive, electron flow is from B to 1 to 2 to 3 to 4 to A. When B is positive, electron flow is from A to 4 to 2 to 3 to 1 to B. Notice that, regardless of the polarity of the ac voltage, the current flow from point 2 to 3 through the meter is always in the same direction. The deflection of the meter needle usually is calibrated to indicate effective or rms voltage. Effective value of an ac voltage is equal to .707 x peak value. This is the most useful value of an ac voltage for the technician to measure.

LOADING A CIRCUIT

When a voltmeter is connected across a circuit to measure a potential difference, it is in parallel with the load in the circuit, and may introduce errors in voltage measurement. This is especially true when a meter with a low sensitivity is used. In the circuit of Fig. 10-18, two 10,000 ohm resistors form a voltage divider across a ten volt source. The voltage drops across R_1 = 5 volts and R_2 = 5 volts. If a meter with a sensitivity of 1000 ohms/volt on the ten volt range is used to measure the voltage across R_1, the meter resistance will be in parallel with R_1 and the combined resistance will equal to:

$$\frac{10,000\ \Omega}{2\ \Omega} = 5000\ \Omega$$

With the meter connected, the total circuit resistance becomes,

$$10,000\ \Omega + 5000\ \Omega = 15,000\ \Omega$$

and the current is .00066 amps.

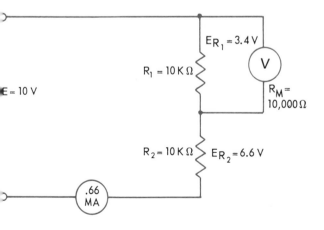

Fig. 10-18. The meter loads the circuit and introduces an error in the voltage reading.

By Ohm's Law, E_{R_1} = 3.4V and E_{R_2} = 6.6V the meter has caused an error of more than one volt due to its shunting effect. To avoid excessive errors resulting from this shunting effect, a more sensitive meter should be used.

In circuit of Fig. 10-19, a 5000 ohms/volt meter is used, and the combined resistance of meter and R_1 equals 8333 ohms and the total circuit resistance equals 18,300 ohms. By Ohm's Law then, I = .00054 amps and E_{R_1} = 4.6 volts and E_{R_2} = 5.4 volts. An error of .4 volts still exists, but the increased sensitivity of the meter has reduced this error. More expensive meters with a sensitivity of 20,000 ohms/ volt would reduce the error to a negligible amount.

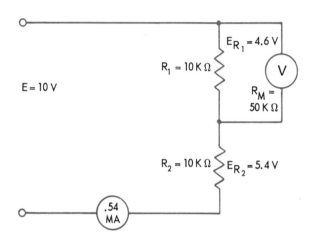

Fig. 10-19. A more sensitive meter gives more accurate readings.

HOW TO USE A METER

1. A meter is a delicate instrument. It should be handled with care and respect. Jarring, pounding, dropping and otherwise rough treatment may destroy a valuable meter.

2. When measuring voltage, the meter must be connected across the circuit (in parallel or shunt). Start on the <u>highest</u> range when measuring an unknown voltage and then to lower range for increased accuracy. Observe correct polarity! <u>Red or positive lead to the</u>

positive side of the circuit. Black or negative lead to the negative or grounded side of the circuit.

3. A meter has its greatest accuracy at about two thirds deflection on the scale. Use the range which will give as close to this deflection as possible.

4. When measuring current, the meter must be connected in SERIES with the circuit. A wire must be disconnected or unsoldered to insert the meter. It is wise to make a rough current calculation by Ohm's Law to determine proper current range on the meter. Observe correct polarity. Red or positive to positive side of circuit and black to the negative side.

5. When measuring resistance, be certain that no power is applied to the circuit and that capacitors are discharged. It is not necessary to observe polarity. Always zero and adjust meter on proper range before measurements are made. A closed circuit has zero resistance; an open circuit has infinite resistance.

ON ALL METER MEASUREMENTS, THE WISE TECHNICIAN WILL MAKE A "FLASH CHECK" BEFORE PERMANENTLY CONNECTING THE METER IN THE CIRCUIT. WHAT DOES THIS MEAN? MAKE YOUR DECISION ON HOW THE METER SHOULD BE CONNECTED IN THE CIRCUIT BUT LEAVE ONE METER LEAD DISCONNECTED. NOW VERY QUICKLY TOUCH AND REMOVE THIS UNCONNECTED METER LEAD, WHILE OBSERVING THE METER. DOES THE NEEDLE MOVE IN THE WRONG DIRECTION? CHANGE POLARITY. DOES THE NEEDLE MOVE TOO VIOLENTLY? CHANGE TO A HIGHER RANGE. THE "FLASH CHECK" WILL SAVE YOU MANY DOLLARS IN METER REPAIRS AND WASTED TIME.

6. High voltages are frequently measured in electronic circuits. Make sure your meter connections do not short or touch other

parts of the circuit. Use a well insulated test prod to reach difficult and obstructed test points in the circuit.

LIQUID CRYSTAL DISPLAYS

With the increased use of digital instruments today, the liquid crystal display is one means of showing the circuit condition. Liquid crystal displays are used in clocks, multimeters, timers, surveying equipment, and medical electronics. They have a life expectancy of over 50,000 hours. See Fig. 10-20 for a four digit liquid crystal display unit.

Fig. 10-20. A liquid crystal display chip.
(Liquid Xtal Displays Inc.)

APPLAUSE METER

No school should be without an applause meter. Why not build one and use it at school assembly programs and popularity contests? This is a real worthwhile experience in constructing electronic circuitry. Noise is converted to electricity by the speaker. It is amplified by the transistors, rectified by the diodes and measured on the meter. The louder the applause, the greater deflection of the meter.

Follow the circuit diagram, but design your own case, Fig. 10-21.

You have become acquainted with the loudspeaker as a device which produces music from your radio or hi-fi system. In this application the speaker is used in reverse to pick up the sound from the applauding audience. In later studies you will discover how this is possible. Potentiometer R_1 is a sensitivity control, and should be set for the correct level and not changed. The output of the three-stage grounded emitter transistor amplifier is rectified by the bridge circuit consisting of CR_1, CR_2, CR_3 and CR_4. The dc value is measured on the milliammeter.

In Fig. 10-21 the speaker has been omitted. It will plug into the jack on the front panel at the right. When in use the speaker should face the audience to pick up the sound.

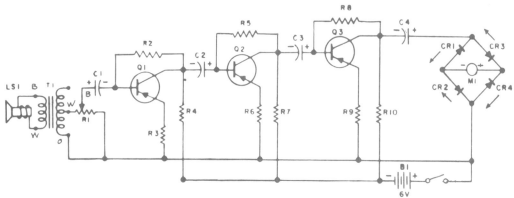

Fig. 10-21. An Applause Meter.

APPLAUSE METER
PARTS LIST

B_1 – 6-volt battery
C_1, C_2, C_3, C_4 – 10 mfd electrolytic, 25-volt
CR_1, CR_2, CR_3, CR_4 – Sylvania 1N64 diode or RCA SK 3091
LS_1 – 3-6 ohm loudspeaker
M_1 – 1 milliampere full-scale meter
Q_1, Q_2, Q_3 – Sylvania 2N1265 transistor or RCA SK 3003
R_1 – 5 K, 1/2W potentiometer
R_2, R_5, R_8 – 150 K, 1/2W
R_3, R_6, R_9 – 100 Ω 1/2W
R_4, R_7, R_{10} – 2.2 K, 1/2W
T_1 – Stancor TA-33 or equivalent

This is another way by which electronics serves you. Devices of this nature are used in industry and medical science. One interesting study has been made to determine the effect of noise upon human fatigue. Noise levels are measured by electronic sensing devices.

FOR DISCUSSION

1. Discuss the safety rules and precautions pertaining to meters.
2. Why is a rectifier necessary in measuring an ac current or voltage?
3. What is meant by "loading a circuit" with a meter?

4. Name some applications of the applause or noise meter which you may have heard about.
5. How could electronic circuitry be used to test human hearing?
6. Who was Alexander Graham Bell?

TEST YOUR KNOWLEDGE

1. Name three types of meter movements.
2. A meter movement has a moving coil resistance of 50 ohms and requires .001 amperes for full scale deflection. Compute the shunt values for meter to read in the following ranges:

   0-1 mA
   0-10 mA
   0-50 mA

3. Compute multipliers for the meter in questions 2 to use on the following ranges:

   0-1   volt
   0-10   volts
   0-100 volts

4. Draw the circuit of a series ohmmeter and explain its operation.
5. Ask your instructor for six different resistors. Measure and record their value and compare measured value to color code value.

	Measured	Coded Value
1.		
2.		
3.		
4.		
5.		
6.		

6. A voltmeter must always be connected in _____ with the circuit or component. The ammeter must always be connected in _____ with the circuit.
7. Draw a circuit to measure ac using a full wave bridge rectifier.
8. A Wheatstone bridge is used to measure an unknown resistance. The following values will balance the bridge.

   R_1 = 500 ohms, R_2 = 100 ohms,

   R_3 = 300 ohms, R_x = ?

9. What is the major difference between an iron vane current meter and an iron vane voltmeter?
10. Read and record the electric meter on the outside of your home. The reading represents _____ . Read and record again in one week. How much power has been consumed? At five cents per kilowatt hour, how much will this power cost?

Chapter 11

VACUUM TUBES

AND SEMICONDUCTORS

Thomas Alva Edison is best remembered for his invention of the incandescent lamp. The dramatic exhibition of the world's first electric lamp at Menlo Park in 1879 eclipses many of the other scientific discoveries made by Mr. Edison. An obscure experiment with the incandescent lamp led to the invention of the vacuum tube by which the "wonders of electronics" contribute daily to our pleasure and standard of living. Mr Edison placed a metal plate in the vacuum bulb containing his light. By connecting a battery and a meter in series between the light filament and the plate, he discovered that an electric current would flow through the light, if the POSITIVE terminal of the battery were connected to the plate. If the negative battery terminal were connected to the plate, NO CURRENT WOULD FLOW. This experiment is diagramed in Fig. 11-1. The electron theory was yet to be discovered and the scientific explanation of the current flow through the light was quite mysterious.

Today the electron theory provides a satisfactory explanation. Certain metals and oxides of metals give up free electrons when heated. The heat supplies sufficient energy to the electrons so that they break away from the forces holding them in orbit and become "free electrons." This is described as the THERMIONIC EMISSION. In Mr. Edison's experiment, a cloud of these free electrons was emitted from the filament of his light bulb. When a positive plate was placed in the bulb, the free electrons were attracted to the plate. (Unlike charges attract.) A flow of electrons constitutes a current, so the meter in the plate circuit indicated that electrons were passing from filament to plate. The modern vacuum tube, although improved by more recent inventions, still uses the principles discovered in the laboratory of Thomas Edison.

THERMIONIC EMITTERS

Better emitting materials have been developed in the years since Edison's carbon filament. Many materials, when heated to the point of emission, will melt. Tungsten seemed to be the most satisfactory material for many years, but it required considerable heat for satisfactory emission. Yet, it was strong and durable. Tungsten is still used in large high-power vacuum tubes.

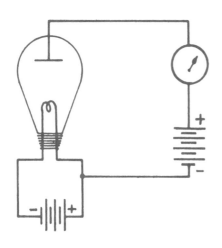

Fig. 11-1. Circuit diagram of the Edison discovery.

To lower the operating temperatures and power consumption the thoriated-tungsten emitter was developed. An extremely thin layer of thorium is deposited on the tungsten emitter. THORIUM is one of the heavier metallic elements (symbol TH, atomic weight 232.2.) It is mined in India. This type of emitter produces satisfactory emission at much lower temperatures than pure tungsten.

The most efficient of emitters is the oxide-coated type. Usually the emitter is metal such as nickel on which is formed a thin layer of barium or strontium oxide. Because of its low power consumption and high emission, this type is widely used in vacuum tubes for radios, television and other electronic devices.

CATHODE

The emitter in the vacuum tube is called the CATHODE. Heat may be supplied either directly or indirectly. Both methods have certain advantages. Schematic diagrams of both types are shown in Fig. 11-2. For portable

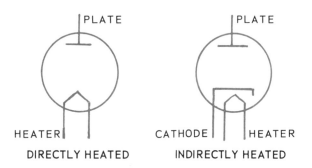

Fig. 11-2. Schematic symbols showing directly and indirectly heat cathodes.

equipment that employs batteries as a source of power, the directly heated type is more advantageous. There are less heat losses and the filament can be so designed that only a small amount of power is consumed during use.

When a source of alternating current is available, the indirectly heated cathode is more practical. The power loss is insignificant. The indirect method provides isolation between the

heater voltage source and the cathode, so that HUM may not be introduced into the circuit.

Research and development in electron tube technology has led to the use of many specialized materials in the construction of the modern tube. An appreciation of engineering materials may be gained by studying the illustration in Fig. 11-3. Note the several special alloys and combinations of materials used in the manufacturing of the tube.

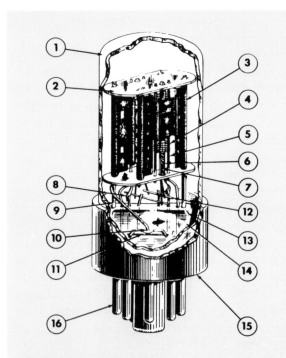

1. ENVELOPE - Lime glass
2. SPACER - Mica sprayed with magnesium oxide
3. PLATE - Carbonized nickel or nickel-plated steel
4. GRID WIRES - Manganese-nickel or molybdenum
5. GRID SIDE-RODS - Chrome copper, nickel, or nickel-plated iron
6. CATHODE - Nickel coated with barium-calcium-strontium carbonates
7. HEATER - Tungsten or tungsten-molybdenum alloy with insulating coating of alundum
8. CATHODE TAB - Nickel
9. MOUNT SUPPORT - Nickel or nickel-plated iron
10. GETTER SUPPORT AND LOOP - Nickel or nickel-plated iron
11. GETTER - Barium-magnesium alloys
12. HEATER CONNECTOR - Nickel or nickel-plated iron
13. STEM LEAD-IN WIRES - Nickel, dumet, copper
14. PRESSED STEM - Lead glass
15. BASE - Bakelite
16. BASE PINS - Nickel-plated brass

Fig. 11-3. Typical electron tube. (RCA)

THE DIODE

A tube which contains a cathode and a plate is called a DIODE. The prefix "Di" signifies two elements in the tube. The cathode may be either directly or indirectly heated. In either case the heater is not considered when counting the tube elements. The plate is usually a cylindrical piece of metal surrounding all the elements in the tube. The plate acts as the collector of electrons, emitted from the cathode. A circuit employing a diode is drawn in Fig. 11-4. This is very similar to the original

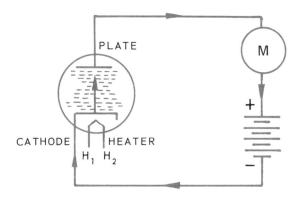

Fig. 11-4. Diode circuit showing direction of electron flow.

circuit used by Thomas Edison. The heater connections are shown as H_1 and H_2 and may be connected to an alternating current source. The first number in a tube designation expresses the approximate voltage which should be applied to the heaters. For example, a 6H6 tube requires 6.3 volts, a 12AX7 requires 12.6 volts and a 25Z6 needs 25 volts. This is important to remember.

When the tube in Fig. 11-4 is turned on, the heaters indirectly heat the cathode and cause thermionic emission of many electrons. If the plate of the diode is connected to the positive terminal of the battery in the circuit, ELECTRONS WILL FLOW IN THE CIRCUIT, from cathode to plate. If the connections are reversed and the negative battery terminal is connected to the plate, NO ELECTRONS WILL FLOW. This electron tube acts as a one-way value

permitting electron flow in one direction only. In fact, the term "valve" is still used in England to describe the tube.

At a given operating temperature of the diode the cathode emits the greatest possible number of electrons. These form a space charge around the cathode. Some of these electrons are attracted back to the cathode which has been made slightly positive as a result of emitting electrons. When the plate is made positive, many of the electrons are attracted to it which constitutes electron flow. If the plate is made MORE POSITIVE BY APPLYING A HIGHER VOLTAGE, a greater number of electrons is attracted and consequently there is an increase in current. The chart in Fig. 11-5 graphically illustrates the increase in current as a result of an increase in plate voltage. A point may be reached by increasing the voltage, at which all the emitted electrons are attracted to the plate and any further increase in voltage will not increase the current. This is called the SATURATION POINT of the electron tube.

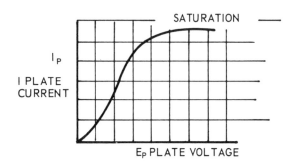

Fig. 11-5. This graph indicates that as plate voltage is increased, plate current increases.

Diodes find wide application as rectifiers, voltage regulators and detectors. Each of these uses will be discussed in later chapters in this text.

SEMICONDUCTORS (TRANSISTORS)

On July 1, 1948, an announcement of a new scientific discovery was made in the New York Herald Tribune:

"A tiny device that serves nearly all of the functions of a conventional vacuum tube, and holds wide promise for radio, telephone and electronics was demonstrated yesterday by the Bell Telephone Laboratories scientists who developed it. Known as the transistor . . ."

The tiny transistor has truly revolutionized the electronics industry. The transistor provides immediate circuit operation. No warm-up time is needed, as in the vacuum tube circuits. Power is not required for heaters. The transistor is characterized by small size, ruggedness, low power requirements, long life, reliability and lightweight.

In the study of transistors or semiconductors some entirely new concepts in electronic theory must be understood. Briefly, what do we know about conductors? We learned that a copper wire used for electrical connections is a good conductor. We learned also, that certain materials such as glass and rubber, are classified as insulators. Somewhere between the insulator and the conductor are many materials classified as semiconductors. The transistor falls into this class; it is neither a good conductor nor a good insulator.

ATOMIC CHARACTERISTICS

A few minutes spent in review of Chapter 1 will materially aid your understanding of the atomic structure of materials. In Fig. 11-6, the atomic diagram of the elements of GERMANIUM (Ge) and SILICON (Si) are drawn. Note that the atomic number of Ge is 32 and Si is 14 which means that these elements have electrons in orbits in outer rings of 32 and 14 respectively. Note also that each has four electrons in its outer ring.

In the crystalline structure of pure germanium shown in Fig. 11-7, the four electrons in the outer rings of each atom tend to join in what is called a covalent bond. These electrons seem to have a binding force which joins them together, yet each electron is bound to its own

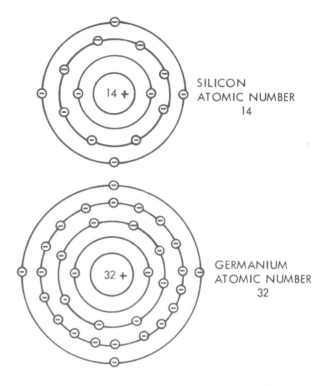

SILICON
ATOMIC NUMBER
14

GERMANIUM
ATOMIC NUMBER
32

Fig. 11-6. The atomic structure of germanium and silicon.

nucleus. This overlapping or sharing effect, resulting from the covalent bonds, presents a picture of germanium called the LATTICE CRYSTALLINE STRUCTURE. Only the electrons in the outer orbit are illustrated. In this form the germanium is a pure insulator and will not conduct electricity.

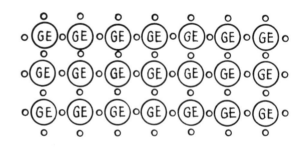

Fig. 11-7. The crystalline structure of germanium showing covalency between electrons in outer rings.

CONDUCTION OF ELECTRICITY

At this point in our studies, it has been well established that current flow in a conductor is a result of the transfer of energy by electrons.

For emphasis, a copper conductor is sketched in Fig. 11-8. As an electron is added to one end of the conductor, an electron leaves the opposite end. This chain reaction explains the conduction of electricity by ELECTRONS.

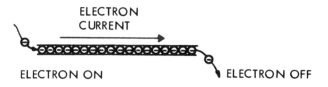

ELECTRON CURRENT

ELECTRON ON ELECTRON OFF

Fig. 11-8. Electricity is conducted in a copper wire by the transfer of energy by electron movement.

Assume now, in Fig. 11-9 that an electron is removed from one end of the conductor, leaving a vacancy or hole in which there should be an electron, but there is none. This hole has a strong attraction for an electron and can be considered as positive. So the next electron fills up the hole and leaves a positive hole in its place. The positive holes are successively filled until the vacant hole appears at the other end of the conductor. Conduction has taken place by the movement of HOLES through the material. In the study of transistors we are concerned with both electrons and holes as conductors.

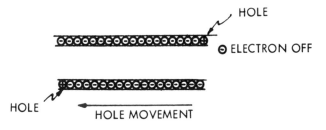

HOLE

ELECTRON OFF

HOLE

HOLE MOVEMENT

Fig. 11-9. Conduction through a material as the result of holes.

SEMICONDUCTORS

Returning now to the pure germanium crystal, which is a good insulator, a minute amount of an impurity may be added which will change its conduction characteristics. Impurities commonly used in transistors may be classified as:

TRIVALENTS: Aluminum, Gallium, Indium, Boron
PENTAVALENTS: Arsenic, Bismuth, Antimony, Phosphorus

The prefixes "TRI" and "PENTA" refer to the number of electrons in the outer ring of the atom. When a pentavalent, such as arsenic, is added to the crystalline germanium, four of the electrons in the outer ring of the arsenic atom join in covalent bond with four electrons in the outer ring of the germanium, leaving one electron from the arsenic FREE. This is illustrated in Fig. 11-10. These free electrons change the characteristics of the germanium from an insulator to a semiconductor. Conduction through the material will be by means of FREE-ELECTRONS. Since the pentavalent impurity contributed free electrons to the crystal, it is called a DONOR IMPURITY. Since conduction will be by negative electrons, the crystal now becomes an N TYPE crystal.

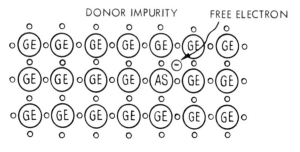

DONOR IMPURITY FREE ELECTRON

Fig. 11-10. A pentavalent impurity leaves free electrons. It is called a DONOR impurity.

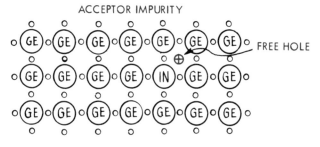

ACCEPTOR IMPURITY

FREE HOLE

Fig. 11-11. A trivalent impurity leaves positive "holes." It is called an ACCEPTOR impurity.

Consider now the addition of a trivalent impurity which has only three electrons to join in covalent bond. One more electron is needed to complete the lattice structure, so therefore a HOLE is left, as illustrated in Fig. 11-11. Since the impurity has created a hole or vacancy, it is called an ACCEPTOR IMPURITY. Conduction through this crystal will be the result of POSITIVE HOLES and is named a P TYPE crystal. The simple circuit in Fig. 11-12 shows

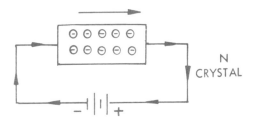

Fig. 11-12. Conduction through an N type crystal by electrons.

conduction through an N type crystal by means of free electrons. In Fig. 11-13 conduction is shown through a P type crystal. Notice that the hole movement is in the opposite direction to electron flow.

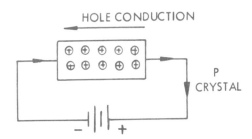

Fig. 11-13. Conduction through a P type crystal by holes.

SEMICONDUCTOR DIODES

The semiconductor diode results from the fusion of a small N type crystal to a P type crystal. See Fig. 11-14. At the junction between the two crystals, the carriers, (electrons and

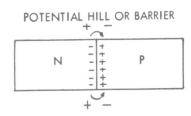

Fig. 11-14. A crystal diode made of N and P type crystals.

holes) tend to diffuse. Some of the electrons move across the barrier to join the holes; some of the holes move across the barrier to join the electrons. This action is explained by the LAW OF CHARGES. Unlike charges attract each other. Due to the diffusion, the region around the junction in the P type crystal becomes negative, having collected electrons from the N crystal. Also, the region around the junction in the N type crystal becomes positive, having lost electrons and gained holes. Therefore, a small voltage or potential exists between the regions close to the junction. This is called a POTENTIAL HILL or POTENTIAL BARRIER. The barrier prevents the other electrons and holes in the crystal from combining.

A voltage or potential is connected across the diode in Fig. 11-15. In this case the positive terminal of the source is connected to the P crystal and the negative source to the N crystal.

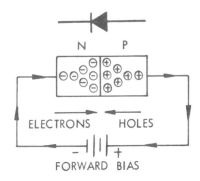

Fig. 11-15. Current conduction through the crystal diode when biased in the forward direction.

The negative electrons in the N crystal move toward the barrier and the positive holes in the P crystal move toward the barrier. The source voltage opposes the potential hill and reduces its barrier effect and allows the electrons and holes to join at the barrier. Therefore, current flows in the circuit. In the P crystal, by holes; in the N crystal, by electrons. The diode is BIASED IN A FORWARD DIRECTION.

A comparison to Fig. 11-16 shows the same junction diode connected with REVERSE BIAS, that is, the positive source to the N crystal and the negative source to the P crystal.

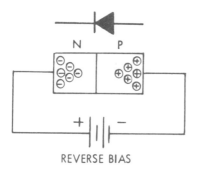

Fig. 11-16. Conduction through the diode when biased in a reverse direction.

In this case the source voltage aids the potential hill and limits the electron-hole combinations at the junction. The electrons in the N crystal are attracted toward the positive terminal of the source. Very little current will flow in the circuit. The reverse voltage may be increased to a point where the diode will break down. This is similar to the inverse peak voltage rating of a vacuum tube.

CRYSTAL DIODE

The crystal diode has many applications in electronic circuits for detection and rectification. The point-contact diode consists of a small piece of N type germanium crystal, against which is pressed a fine phosphor bronze wire. During manufacture, a high current is run through the diode from wire to crystal which forms a P type region around the contact point in the germanium crystal. The point-contact diode, therefore, has both the P and N type

Fig. 11-17. A typical crystal diode. (Sylvania)

crystals and its operation is similar to the junction diode. A crystal diode used in a radio detector is shown in Fig. 11-17, together with its schematic symbol. This diode is used in several projects in this text.

SILICON RECTIFIERS

Because diodes conduct current more easily in one direction than in the other, this process is called RECTIFICATION. One of the more important semiconductor diodes is the SILICON RECTIFIER. These devices are used in power rectification circuits that will be discussed in the next chapter on Power Supplies.

Silicon rectifiers come in a variety of shapes and sizes. Fig. 11-18 shows some common diode outlines.

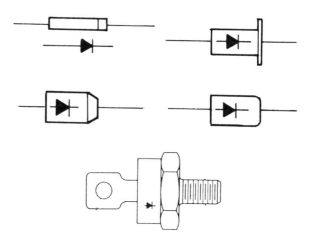

Fig. 11-18. Some widely used silicon rectifier outlines.

Because of the high forward-to-reverse current ratios of silicon rectifiers, they can achieve rectification efficiencies of greater than 99 percent. These rectifiers are very small and light. They can be made very resistant to shock and physical abuse. Silicon rectifiers have excellent life characteristics. They do not deteriorate with age like a vacuum tube rectifier.

DIODE CHARACTERISTICS AND RATINGS

Diodes and rectifiers are usually rated according to their current and voltage capabilities as well as their PEAK REVERSE VOLTAGE (PRV). This peak reverse voltage rating is used by the manufacturer to define the maximum allowable reverse voltage that can be applied across the rectifier without damaging it.

Many diodes are labeled by a "1N" in their identification. Examples of this are 1N34 or 1N4004. Some manufacturers attach their own special labels, such as HEP 320 or SK 3051, to help identify the diodes.

SERIES AND PARALLEL RECTIFIER ARRANGEMENTS

Diodes or rectifiers can be connected in series or parallel to improve the voltage or current capabilities of a single rectifier. By connecting diodes in series, the voltage rating may be increased over the value of a single rectifier. One may connect diodes in parallel to improve the current handling ability of the combination over that of only one diode.

FOR DISCUSSION

1. Why does a vacuum tube have to have a vacuum inside?
2. Explain why electrons normally do not flow from the plate to the cathode of a vacuum tube diode.
3. What is "saturation point?"
4. What are some good basic semiconductor materials that have four electrons in the outer shell?
5. A conductor has _____ (low or high) resistance to current flow.
6. A space missing an electron in a semiconductor is called a _____.
7. What is the primary difference between trivalent and pentavalent materials?
8. Discuss the difference between forward bias and reverse bias.
9. Draw the symbol for a diode and show the cathode and anode connections.
10. Define rectification.
11. Why are diodes connected in series?

TEST YOUR KNOWLEDGE

1. What is meant by covalent bonds in a lattice crystalline structure?
2. Name three donor impurities. Why are they called donor impurities?
3. Name three acceptor impurities. Why are they called acceptors?
4. How is the potential barrier formed at the junction of a PN crystal?
5. Draw a block diagram of the crystal diode showing forward bias connections.
6. Draw the same diagram showing reverse bias.
7. A N type crystal conducts by _____.
   A P type crystal conducts by _____.
8. What materials are used as thermionic emitters?
9. Draw the schematic for:
   Diode (directly heated cathode).
   Diode (indirectly heated cathode).
10. What is meant by the manufacturer's rating called "Peak Reverse Voltage?"

Chapter 12

POWER SUPPLIES

For the proper operation of electronic equipment, a variety of source voltages must be available. Alternating current must be supplied for the heaters of the tubes. High direct current voltages are needed for the B+ or plate and screens of the electron tubes. Negative voltages must be supplied for the C supply or grid bias. Although batteries may be used for all these requirements, the common source of electric power is that furnished by the local power company and is 115 volt alternating current at a frequency of 60 hertz.

The complete power supply circuit may be used to perform these circuit functions:

1. Step-up or step-down by transformer action the ac line voltage to required voltages.

2. Change the ac voltage to a pulsating dc voltage by either half-wave or full-wave rectification.

3. Filter the pulsating dc voltage so that a relatively pure dc is available for equipment use.

4. Provide some method of voltage division to meet the needs of equipment.

5. Frequently, the power supply includes circuitry and components to regulate its output in proportion to the applied load. See Figs. 12-1 and 12-2 for some typical commercial power supplies.

Fig. 12-2. Low voltage dc power supply. (Dynascan Corp.)

THE POWER TRANSFORMER

The first component in the Power Supply is the transformer. Its purpose is to step-up or to step-down the alternating source voltage to values needed for the radio, TV or electronic circuit. It would be wise at this time to review transformer action as discussed in Chapter 6.

Most transformers do not have any electrical connection between the secondary windings and the primary windings. This gives the transformer a feature called ISOLATION. Since the secondary is isolated from the primary, a person has to have their body or hands connected across both of the secondary connections in order to become shocked, which is a safety factor. This is not true in the primary where commercial alternating current (ac) provided by

Fig. 12-1. Typical "Bench" style power supply used by experimenters. (Kepco, Inc.)

the power company is used. One of the connections is HOT and the other is GROUNDED or NEUTRAL. This means that if a person were standing on the ground and touched the hot connection, they would be shocked. If they touched the ground connection they would <u>not</u> be shocked. See Fig. 12-3.

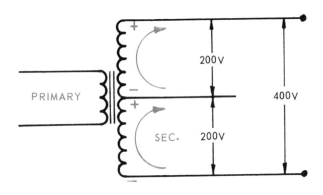

Fig. 12-3. Isolation in a transformer.

Secondary windings can be tapped to provide different voltages. The tapped voltage are 180 deg. out of phase. See Fig. 12-4.

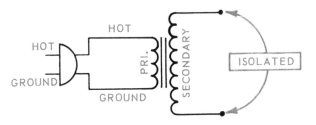

Fig. 12-4. A center-tapped transformer.

If the tap is placed half the way from one end of secondary winding to the other end, it is called a CENTER TAP. Many power supplies use center tapped secondary transformer winding, where two diodes provide a full-wave output.

Since transformers of a variety of types and sizes may be found in almost all electronic devices, the student should be quite familiar with the basic theory and purposes of this valuable component.

LESSON IN SAFETY: Transformers produce high voltages which may be extremely dangerous. Proper respect and extreme caution must be practiced at all times when working with or measuring high voltages. Do not, under any circumstances, attempt to work with these high voltages until your instructor has given you approval to go ahead.

HALF-WAVE RECTIFICATION

The process of changing an alternating current to a pulsating direct current is called RECTIFICATION. In the circuit of Fig. 12-5, the output of a transformer is connected to a

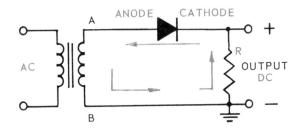

Fig. 12-5. Schematic of basic diode rectifier.

diode and a load resistor, in series. The input voltage to the transformer appears as a sine wave, which periodically reverses polarity at the frequency of the applied voltage. The output voltage of the transformer secondary also appears as a sine wave, the magnitude of which depends upon the turns ratio, and it is 180 deg. out of phase with the primary. The top of the transformer (point A) is connected to the diode anode. When point A is positive during the first half cycle, the diode conducts, producing a voltage drop across R equal to IR. During the second half-cycle, point A is negative and the diode anode is negative. No conduction takes place and no IR drop appears across R. An oscilloscope connected across R would produce the wave form shown in Fig. 12-6. It is important to understand that the output of this circuit consists of pulses of current flowing in <u>only one direction</u> and at the <u>same frequency as</u>

Fig. 12-6. Wave forms of input and output of diode rectifier.

the input voltage. The output is a pulsating direct current. Since only one-half of the alternating current input wave is used to produce a useful output voltage, it is therefore named a HALF-WAVE RECTIFIER. The polarity of the output voltage should be observed. One end of resistor R is connected to ground. The electron or current flow is from ground to the cathode which makes the cathode end of R positive as indicated.

A negative rectifier may be made by reversing the diode in the circuit. See Fig. 12-7. The diode conducts when the cathode becomes

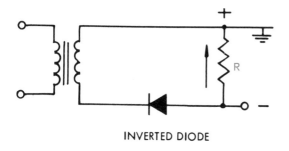

INVERTED DIODE

Fig. 12-7. Inverted diode produces a negative voltage.

negative, which is the same as making the anode positive. The current through R would be from anode to ground, making the anode end of R

negative and the ground end of R more positive. Voltages taken from across R, the output, would be negative in respect to ground. This circuit is called an inverted diode, and is used when a negative supply voltage is required.

Vacuum tube diodes, Fig. 12-8, may be used in power supplies as rectifiers as well as semiconductor diodes. When a positive potential is applied to the plate and the cathode is less

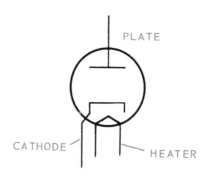

Fig. 12-8. Vacuum tube diode symbol.

positive or negative with respect to the plate, a current will flow through the diode. No current will pass through the diode when the plate is negative in respect to the cathode. Thus the diode will pass current flow easily in one direction and block the current when the voltage polarity is changed. Fig. 12-9 shows a vacuum tube diode as a half-wave rectifier.

FULL-WAVE RECTIFIER

The pulsating direct voltage from the half-wave rectifier is difficult to filter to a pure dc voltage. The half-wave rectifier uses only one

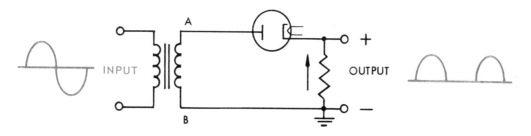

Fig. 12-9. Schematic of basic diode rectifier.

half of the input alternating current wave. By employing two diodes, as diagramed in Figs. 12-10 and 12-11, both half cycles of the input wave may be rectified. In order to do this, a center tap is made on the secondary winding which is attached to ground. In Fig. 12-10, point A is positive and diode anode of D_1 is

Fig. 12-12. The wave forms of input and output of full-wave diode rectifier.

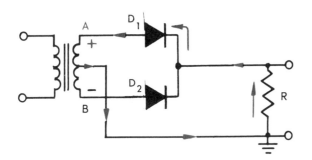

Fig. 12-10. Arrows show current in full-wave rectifier during first half-cycle.

positive. Current flows as indicated by arrows. During the second half of the input cycle, point B is positive, diode anode of D_2 is positive, and current flows as indicated in Fig. 12-11. Notice that, regardless of which diode is conducting, the current through load resistor R is always in the same direction. Both positive and negative half-cycles of the input voltage cause current

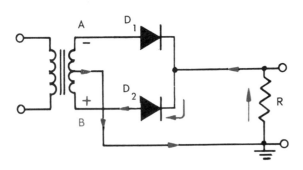

Fig. 12-11. Direction of current during second half-cycle.

through R in the same direction. Output voltage of this FULL-WAVE RECTIFIER is taken from across R and consists of direct current pulses at a frequency of TWICE the input voltage. See Fig. 12-12. In order to produce full-wave rectification in this circuit, it was necessary to cut the secondary voltage in half by means of the center tap.

A variation of the full-wave rectifier circuit is drawn in Fig. 12-13. In this diagram a duo-diode (two diodes in one envelope) tube is used and the directly heated cathode is supplied by a separate 5 volt secondary winding. The load resistor R is replaced by two resistors and the tap between R_1 and R_2 is grounded. The

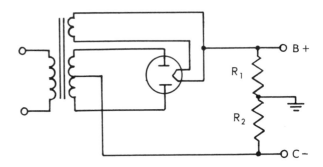

Fig. 12-13. Full-wave rectifier using a duo-diode tube which produces both negative and positive voltages in respect to ground.

transformer center tap is connected to the bottom of R_2. This circuit operates exactly like the previous circuits. The reference point for the output voltage which is ground has been moved. Now point B is positive in respect to ground but point C is negative with respect to ground. Both negative and positive voltages may be supplied by this circuit.

THE BRIDGE RECTIFIER

It is not always necessary to use a center-tapped transformer for full-wave rectification. The full secondary voltage can be rectified by using four diodes in a BRIDGE RECTIFIER CIRCUIT. See Figs. 12-14 and 12-15. Two circuit drawings are illustrated so that current flow may be observed on each half cycle. In Fig. 12-14 point A of transformer secondary is positive and current flows in direction of

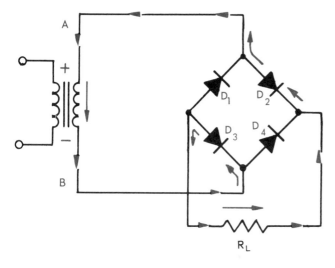

Fig. 12-14. Current in bridge rectifier during first half-cycle.

arrows. When point B is positive, current flows as shown in Fig. 12-15. Again, notice that the current through R is always in one direction. Both halves of the input voltage are rectified and the full voltage of the transformer is used.

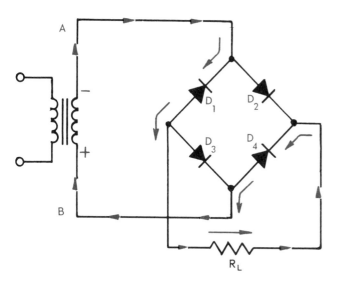

Fig. 12-15. Current in bridge rectifier during second half-cycle.

FILTERS

The resulting output of either the half-wave or full-wave rectifier is a pulsating voltage. Before it can be applied to the other circuits in a radio or television receiver, the pulsations must be removed to a negligible amount. Pure

direct current is desirable. This is accomplished by means of a FILTER NETWORK. Referring to Fig. 12-16, the output waveform of a full-wave rectifier is shown. The average voltage

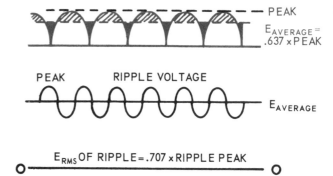

Fig. 12-16. Average value of full-wave rectifier output.

of this pulsating wave is designated by the line. E_{av} and is equal to .637 x the peak voltage. The shaded portion of the wave above the average line is equal in area to the shaded portion below the line. The fluctuation above and below the average voltage is called the ac RIPPLE. It is the ripple that requires the filtering. The PERCENTAGE OF RIPPLE to the average voltage must be limited to small value. Mathematically, the ripple percentage may be found by:

$$\% \text{ Ripple} = \frac{E_{rms} \text{ of Ripple Voltage}}{E_{average}} \times 100$$

A capacitor connected across the rectifier output, as shown in Fig. 12-17, provides some filtering action. The capacitor has the ability to store electrons. When the rectifier tube is conducting, the capacitor charges rapidly to approximately the peak voltage of the wave, limited only by the resistance of the rectifier tube and the reactance of the transformer

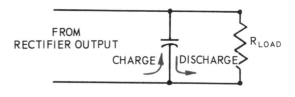

Fig. 12-17. Filtering action of a capacitor.

windings. As the voltage from the rectifier drops, between the pulsations in the wave, the capacitor then discharges through the resistance of the load. The capacitor, in effect, is a storage chamber for electrons which accepts electrons at peak voltage and supplies them to the load when the rectifier output is low. A comparison of input and output wave forms may be seen in Fig. 12-18. Capacitors used for this purpose are the electrolytic types, because large capac-

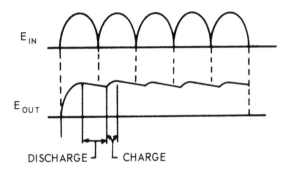

Fig. 12-18. Input and output of capacitor filter showing change in wave form.

itances are needed in a limited space. Customary values range from 4 to 50 microfarads or more with working voltages in excess of the peak voltage from the rectifier.

The filtering action may be improved by the addition of a choke in series with the load. This filter circuit appears as in Fig. 12-19. The filter choke consists of many turns of wire wound on a laminated iron core. In our earlier studies of magnetism and inductive reactance, it was learned that inductance was that property of a circuit which resisted a change in current. A rise in current induced a counter emf which opposed the rise; a decrease in current induced

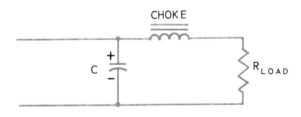

Fig. 12-19. Further filtering is produced by the choke in series with the load.

a counter emf which opposed the decrease. So, as a result, the choke continually opposes any fluctuation in current, yet offers very little opposition to a direct current. Chokes used in radios have values from 8 to 30 henrys with current ratings from 50 to 200 milliamperes. Larger chokes may be used in transmitters and other electronic devices. The additional filtering action as a result of the filter choke may be observed in Fig. 12-20.

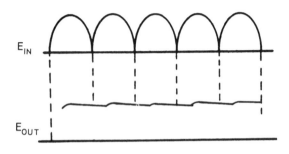

Fig. 12-20. Wave forms show filtering action of capacitor and choke.

A second capacitor may be used in the filter section after the choke, to provide additional filter action. See Fig. 12-21. The action of this capacitor is similar to the first capacitor. The circuit configuration appears as the Greek letter π and the filter is called a pi-section filter.

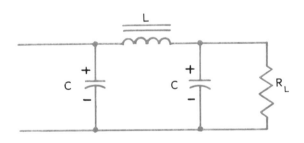

Fig. 12-21. Pi (π) section filter configuration.

Since the first filtering component is a capacitor, it is referred to as a capacitor input filter. In Fig. 12-22 the choke is the first filtering component, so it is called a choke-input filter. The circuit configuration now appears as an inverted L and is therefore called an L section filter. Several of these filter sections may be used in series to provide additional filtering.

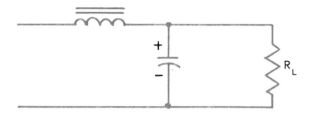

Fig. 12-22. Choke input L filter.

In the capacitor input filter, the capacitor charges to the peak voltage of the rectified wave. However, in the choke input, the charging current for the capacitor is limited by the choke and the capacitor does not charge to the peak voltage. As a result, the output voltage of the power supply using the capacitor input filter is higher than one using the choke input filter.

VOLTAGE REGULATION

The output voltage of a power supply will usually decrease when a load is applied. This is not a desirable effect and should be eliminated. The voltage decrease under load, to the power supply voltage with no load, when expressed as a percentage, is called the PERCENTAGE OF VOLTAGE REGULATION. It is a factor considered in the quality of a power supply. Expressed mathematically,

$$\% \text{ Voltage Regulation } = \frac{E_{nl} - E_{fl}}{E_{fl}} \times 100$$

where,

E_{nl} = voltage with no load

E_{fl} = voltage with full load

Example: If a power supply has a no load voltage of 30 volts and this voltage drops to 25 volts when a load is applied, what is its percentage of regulation?

$$\% \text{ Regulation} = \frac{30V - 25V}{25V} \times 100$$

$$= \frac{5}{25} \times 100 = 20\%$$

LOAD RESISTOR

To complete the basic power supply circuit a load resistor is connected across the supply as in Fig. 12-23. This resistor serves a threefold purpose.

1. As a BLEEDER. During operation of the power supply, peak voltages are stored in the capacitors of the filter sections. These capacitors remain charged after the equipment is turned off, and can be dangerous if accidentally touched by the technician. The load resistor serves as a "bleeder" and allows these capacitors to discharge when not in use. The wise technician will always take the added precaution and short-out capacitors to ground with an insulated screwdriver.

2. To Improve Regulation. The load resistor acts as a preload on the power supply and causes a voltage drop. When equipment is attached to the supply, the additional drop is relatively small and the regulation is improved. An example will clarify this point. Assume the terminal voltage of the power supply is 30 volts with no load resistor and

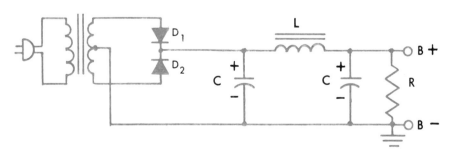

Fig. 12-23. Complete power supply circuit with load resistor.

no equipment connected to it. When the equipment is connected and turned on, the voltage drops to 25 volts and the regulation is 20 percent. (See previous example under voltage regulation.) If the resistor across the power supply produces an initial drop to 26 volts, then the output voltage is considered as 26 volts. If the equipment now connected to the supply causes the voltage to drop to 25 volts, then the power supply regulation may be considered as,

$$\% \text{ Regulation} = \frac{26V - 25V}{25V} \times 100$$

$$= \frac{1V}{25V} \times 100 = 4\%$$

The usable voltage of the supply has only varied four percent. A further advantage of preloading the supply is an improvement in the filtering action of the choke. The resistor allows a current to flow in the supply at all times. A choke has better filtering action under this condition of current than when it varies between a low value and zero.

3. As a Voltage Divider. The load resistor provides a convenient method of obtaining several voltages from the supply. By replacing the single load resistor by separate resistors in series as in Fig. 12-24, several fixed dc voltages are available. A sliding tap resistor may be used, if intermittent adjustments of voltage are required. This is called a

VOLTAGE DIVIDER and takes advantage of Ohm's Law which states that the voltage drop across a resistor equals current times resistance, (E = I x R). In Fig. 12-24, the three resistors are equal. The voltage drop divides equally between them and their sum equals the applied voltage. Other values of resistors may be used, if other voltages are required.

The beginning student sometimes neglects the effect of connecting a load to one of the taps on a voltage divider. This load is in parallel to the voltage divider resistor, decreases the total resistance, and therefore changes the voltage at that particular tap. This is illustrated in Fig. 12-25. The divider is made up of three 5 K resistors. The 30 volts of the supply divides

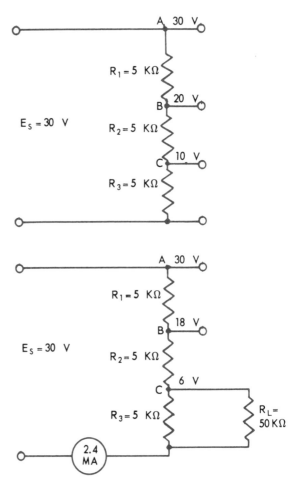

Fig. 12-25. Diagrams show change in resistance in a voltage divider when a load is attached.

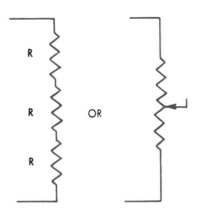

Fig. 12-24. Voltage divider across power supply output.

to 10 volts, 20 volts and 30 volts at terminals C, B and A respectively. If a load of 5 K Ω is now connected to terminal C as shown, it is parallel with R_3 and resistance becomes:

$$R_T = \frac{R_3 R_L}{R_3 + R_L} = \frac{5000 \times 5000}{5000 + 5000} = 2500 \ \Omega$$

Now the total series resistance across the power supply is only, 5 K Ω + 5 K Ω + 2500 Ω = 12,500 Ω and the current through the divider is:

$$I = \frac{E}{R} = \frac{30}{12,500 \ \Omega} = .0024 \text{ amps}$$

and the voltage at point C is:

$$E = I \times R = .0024 \times 2500 \ \Omega = 6V$$

and the voltage at point B is 18 volts. If another load were connected to point B, a further change of voltage division would result.

An electronic device that may be used as a voltage regulator is the ZENER DIODE. A schematic symbol is shown in Fig. 12-26.

CATHODE ——————————— ANODE
(K)   (A)

Fig. 12-26. Zener diode symbol.

Fig. 12-27 shows a characteristic curve for a zener diode. When the diode is forward biased, it acts like an ordinary diode or a closed switch.

When the applied voltage is increased the forward current increases. When the diode is reverse biased, a small reverse current flows until a point where the diode reaches the zener breakdown region, V_2. At this point the zener diode has the ability of maintaining a relatively

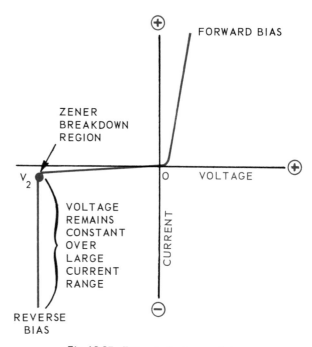

Fig. 12-27. Zener diode characteristics.

constant voltage as the current varies over a certain range. Because of this condition, the diode provides excellent voltage regulation.

Fig. 12-28 shows a zener diode being used as a simple shunt regulator.

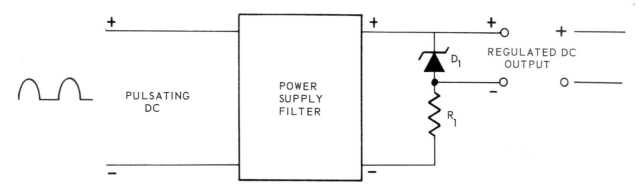

Fig. 12-28. Zener diode as shunt voltage regulator.

VOLTAGE REGULATORS

It is apparent that some means should be provided to assure a constant voltage output at the power supply under varying load conditions. The theory of this circuit takes advantage of the fact that a voltage drop across a resistor is equal to the product of current and resistance. The fundamental circuit is diagramed in Fig. 12-29. The total output of the supply is applied to terminals A and B, and the voltage is divided between the load and variable resistor R in correct proportion. If the voltage of the supply increases, the voltage across the load would tend to increase. If the resistance R is manually increased, the voltage drop across R would increase and offset the increase of voltage from the supply, leaving a constant voltage for the load. Manually, this is impossible, so sensitive electronic circuits, or voltage regulators, have been devised as a means of maintaining a constant voltage output at the power supply under varying load conditions.

VOLTAGE DOUBLERS

Up to this point in the study of power supplies it has been assumed that the convenient source of power was 115 volt ac found in homes and schools. The transformer was used to step-up or step-down the voltages required for the electronic circuits. Because transformers are heavy and expensive, voltage multiplying circuits have been devised to raise voltages without the use of transformers. A study of Figs. 12-30A, and 12-30B, will explain the

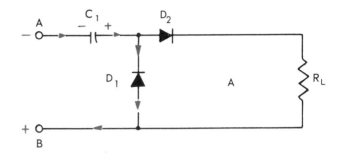

Fig. 12-30A. During first half-cycle C_1 charges through conduction of rectifier D_1.

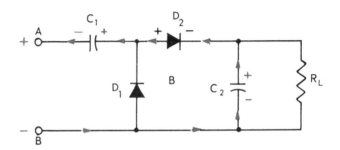

Fig. 12-30B. During second half-cycle applied line voltage is in series with charge on C and current flows through D_2.

action in a half-wave voltage doubler circuit. In Fig. 30A, the input ac voltage is on the negative half-cycle, making point A negative and current flows from point A, through the rectifier D_1 and charges capacitor C_1 to the polarity shown. During the positive half-cycle, point A is positive and the applied voltage is in series with the charged capacitor C_1. In the series connection the voltages add together so the output from the doubler is the applied voltage plus the voltage of C_1. Current cannot flow through the rectifier D_1 due to its unilateral conduction. The output wave form shows half-wave rectification with an amplitude of about twice the input voltage. Rectifier D_2 permits current to flow in only one direction to the load.

A full-wave voltage doubler is drawn in Figs. 12-31A and 12-31B. During the positive peak of the ac input, point A is positive and current flow is from point B, charging C_1 in the polarity shown, through D_1 to point A. (Follow arrows) During the negative cycle of the input point A is negative and current flows through D_2 to C_2 and charges it to indicated polarity, to

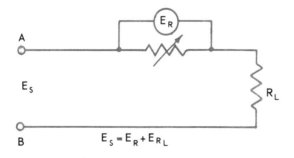

Fig. 12-29. Basic circuit shows theory of voltage regulation.

$$E_S = E_R + E_{R_L}$$

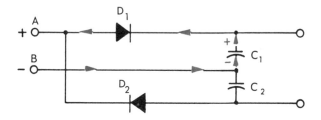

Fig. 12-31A. C_1 charges during first half-cycle.

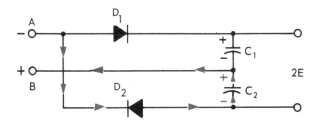

Fig. 12-31B. C_2 charges during second half-cycle. Output voltage equals $C_1 + C_2$ in series.

point B. Notice that during one cycle of ac input, capacitor C_1 and C_2 have been charged so that the voltages across C_1 and C_2 are in series. The output is taken from across these capacitors in series and the output voltage is the sum of both voltages or twice the input voltage.

Voltage doubler circuits are useful in producing high voltages for circuits requiring low currents. As the output voltage depends upon charged capacitors the voltage regulation is poor. Conventional filter circuits are added to smooth out the ripple voltage as in the transformer rectifier circuits.

AC-DC SUPPLY

In Fig. 12-32 a common type of the ac-dc power supply is illustrated. It is used in many small radios and calculators. When used with ac, the output is the pulsating dc which must be filtered. The capacitor input filter is used to secure the higher voltage. Generally in a radio using this supply, the negative of B— is grounded to the chassis. If the ac plug is inserted the wrong way in the wall receptacle, it is possible to get a severe shock if you are grounded and accidently touch the chassis.

When this supply is used on dc, the positive side of the dc line must be connected to the anode of the diode. The set will not operate otherwise. Current will only flow from cathode to plate.

LESSON IN SAFETY: The ac-dc radio described in Fig. 12-32 is very dangerous if the ac plus is connected the wrong way. One wire of your house wiring system is grounded to a water pipe, which will ground your radio only if the plug is inserted correctly. If the plug is reversed, an accidental contact between the radio and a water faucet would give a severe shock across your body. If the radio came in contact with any grounded table or appliance, sparks would fly with the possible danger of fire.

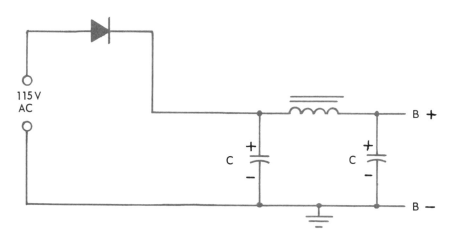

Fig. 12-32. An ac-dc power supply circuit.

FLOATING GROUND

To avoid some of the hazards of the ac-dc supply the radio chassis may be isolated from ground by a buffer capacitor. The chassis is not used as a common ground. The negative terminals are wired together in a "floating ground" system. See Fig. 12-33.

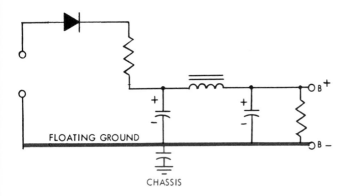

Fig. 12-33. An ac-dc supply with floating ground.

12 VOLT AUTOMOBILE BATTERY TRICKLE CHARGER

Many times the electronics experimenter needs a battery charger for an automobile. Notice the one shown in Fig. 12-34. It is called a "trickle" charger since it provides a low amount of charging current as compared to faster service station chargers. However, this

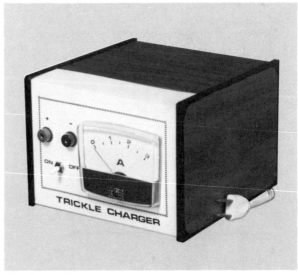

Fig. 12-34. Trickle charger used for recharging batteries.

battery charger is excellent to use in recharging an automobile battery overnight, or for 12 to 24 hours. The ammeter indicates output current while charging. A minimum amount of output current means that the battery is reaching a maximum charge. Fig. 12-35 shows a schematic for a power supply that can be used as a battery charger. All wiring is uncomplicated and this project is fun to construct. The parts list is also shown.

This power supply can also be used to operate such equipment as 12 volt automotive tape players and radios by adding C_1, shown in the schematic.

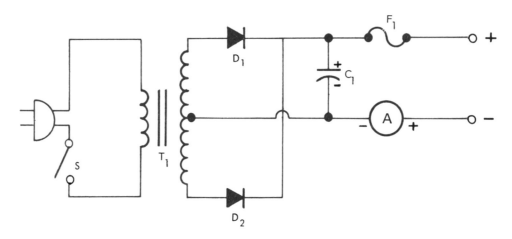

Fig. 12-35. 12 volt power supply.

PARTS LIST FOR 12 VOLT POWER SUPPLY

S_1 SPST switch

T_1 Step-down transformer, 117V primary, 12 VCT secondary   @3A

D_1, D_2 Diodes, 50 PIV @5A (RCA SK 3586)

*C_1 1000-5000 MFD, 50 volt capacitor (electrolytic)

F_1 3 amp fuse with holder

A   0-3 amp, ammeter

Also needed: Line cord, chassis output leads with alligator clips.

*C_1 is only needed when operating 12 volt automotive equipment such as a radio or a tape player. It is not needed when using this power supply as a battery charger.

0-15 VOLT DC LOW AMPERAGE POWER SUPPLY

This simple 0-15V dc power can be used to power transistor radios or provide a low voltage substitute for many dc operated electronic devices. The maximum output current is 100 mA. The circuit used in this power supply was provided by Graymark Enterprises. It may be purchased in their Model 803 power supply. See Fig. 12-36.

See Fig. 12-37 for the schematic and parts list.

Fig. 12-36. Solid state, low voltage power supply.
(Graymark Enterprises, Inc.)

SCHEMATIC DIAGRAM

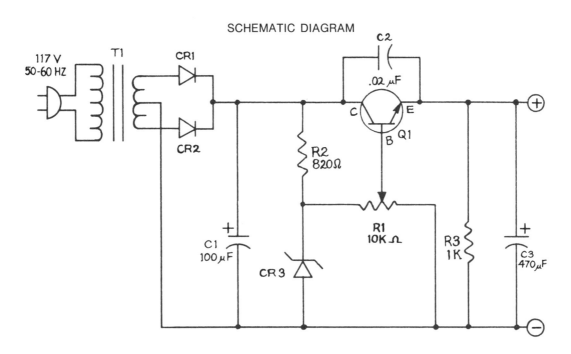

Fig. 12-37. Schematic diagram for 0-15V dc power supply.

PARTS LIST MODEL 803 POWER SUPPLY

R_1 – (1) Potentiometer, 10K ohms, 2W (62110)

R_2 – (1) Resistor, solid, 820 ohms, 1/2W, 10% (61821)

R_3 – (1) Resistor, solid, 1 K ohms, 1/2W, 10% (61377)

C_1 – (1) Capacitor, electrolytic, 100 μ F, 35WV (62100)

C_2 – (1) Capacitor, ceramic disc., .02 μ F, 50V (61231)

C_3 – (1) Capacitor, electrolytic, 470 μ F, 16WV (62101)

CR_1, CR_2 – (2) Diode, silicon, power, 50 PIV, 1A (62102)

CR_3 – (1) Diode, zener, 16V, .5W RCA SK 3142 (62109)

Q_1 – (1) Transistor, power, with 2-insulative washers* (62103)

T_1 – (1) Transformer, power, P = 117V, S = 40V, at 135 mA* (62104)

* Attached to panel

This power supply may be ordered as Model 803 from: Graymark Enterprises, Inc., P.O. Box 54343, Los Angeles, California 90054. Individual parts may be ordered by quoting the part number to Graymark.

FOR DISCUSSION

1. What is the purpose of a power supply?
2. Do all power supplies need filters?
3. How does a transformer provide isolation?
4. What is the purpose of a center tap on the secondary winding of a transformer?
5. How does a zener diode provide voltage regulation?
6. Explain how a full-wave voltage doubler operates.

TEST YOUR KNOWLEDGE

1. State four functions of the power supply.
2. State three purposes of the load resistor across the power supply.
3. The terminal voltage of a power supply is 30 volts. When a load is applied, the voltage drops to 28 volts. What is the percentage of regulation?
4. The average dc value of the output of a power supply is 25 volts. the ripple voltage has a peak value of .2 volts. What is the percentage of ripple?
5. Why does the output wave form of the capacitor filter approximate a saw tooth wave?
6. Why is the output from a capacitor-input filter higher than a choke-input filter?
7. Explain the use of a buffer capacitor in a "floating ground" system.
8. A power supply has a 10K Ω load resistor. Its terminal voltage is 300 volts. What wattage rating should this resistor have?
9. A voltage divider is needed in place of the load resistor in problem 8 to give us 75 volts, and 150 volts. What values of resistors should be used?
10. Draw a bridge rectifier circuit and place arrows showing current direction during each half-cycle.

Chapter 13

ELECTRON AMPLIFIERS

Electronic amplifiers have been in existence since the early part of the twentieth century. AMPLIFICATION is the ability of a circuit to control a large force by a smaller one. In electronics amplification may be accomplished by vacuum tubes or semiconductors in circuits. GAIN is often referred to as the ratio between the output power, current or voltage to the input power, current, or voltage. A very simple block diagram of an amplifier is shown in Fig. 13-1.

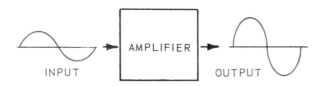

Fig. 13-1. Block diagram of amplifier.

ELECTRON TUBE AMPLIFIERS

THE TRIODE

One of the greater contributions to the scientific age was made by Dr. Lee DeForest and his experimentation with a third element, called a GRID, added to the electron tube. This three element tube is called the TRIODE. By placing a mesh of fine wire between the cathode and the plate of the tube, the scientist was able to control the flow of electrons through the electron tube. The grid usually is cylindrical in shape and surrounds the cathode. Sufficient space is allowed between the grid wires to permit electrons to pass through to the plate. Because this grid controls the electron flow, it is commonly referred to as the CONTROL GRID. Probably no other single discovery has stimulated the development of radio, television and electronic equipment to its present state than the GRID in the triode. Upon this discovery is based the ability of an electron tube to AMPLIFY.

In the previous study of the diode it was learned that the electron flow was controlled by variations in plate voltage. In the triode, the grid also has a pronounced effect upon this electron flow. If the grid is made negative in respect to the cathode, many of the electrons emitted from the cathode are repelled by the negative grid back to the cathode, thus limiting the number of electrons passing on to the plate. As the grid is made more and more negative a point may be reached at which no electrons will flow to the plate. This is described as the CUT-OFF point of the tube. It represents the negative voltage applied to the control grid which cuts off the electron flow through the tube. The voltage applied to the control grid is the BIAS VOLTAGE and at cut off it would be called the CUT-OFF-BIAS.

The circuit in Fig. 13-2 shows a triode with both plate and grid voltages. Notice that the grid bias battery has its negative terminal connected to the grid. In electronic work, one should remember, these voltages are referred to as:

A voltage — for the heaters in the tube.
B voltage — for the plate of the tube.
C voltage — for the grid of the tube.

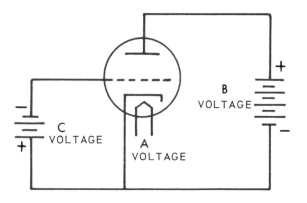

Fig. 13-2. This triode circuit shows the connections for plate voltage and grid bias voltage.

Portable radios and equipment use dry batteries called A and B batteries for these voltage sources. In the familiar radio operating on ac power these voltages are obtained from the power supply. Earlier portable radios used the C battery also, but now improvements have permitted the C battery to be discarded. The C bias voltage may be obtained by other methods which will be studied later.

The graph in Fig. 13-3 shows the current flow through an electron tube as the grid bias is changed. The plate voltage is held at a constant value. The curve in this graph is plotted by measuring the value of current at each change of grid voltage. At a grid bias of negative 2 volts the current is 8 mA. At negative 6 volts the current drops to 3 mA.

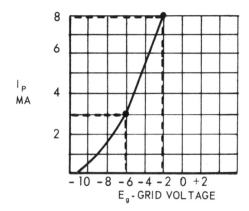

Fig. 13-3. The graph shows the change in plate current I_p as a result of a change in grid voltage E_g. The plate voltage is held at a constant value.

ELECTRON TUBE CHARACTERISTICS

The performance of any electron tube with different applied voltages to the grid, plate and cathode may be plotted on a series of graph curves which represent the STATIC characteristics of the tube. These characteristic curves differ somewhat from curves produced under actual operation conditions. Curves of characteristics under operating conditions would be termed DYNAMIC CHARACTERISTICS. The ability of a tube to amplify was previously mentioned. What does amplification mean? If you were unfortunate to have a flat tire on your car and wished to change to the spare, it would be necessary to jack up the car to make the change. A jack is a mechanical device to amplify your power to lift. In fact, only a few pounds of pressure on the jack handle will lift hundreds of pounds of automobile. In the case of the electron tube, a small change of voltage on the grid of the tube may produce a relatively large change of voltage at the plate. The ability of a tube to amplify a voltage or signal is known as its AMPLIFICATION FACTOR or MU. It represents the ratio between the change in plate voltage to the change in grid voltage, which will produce an equal change in plate current. It is expressed as:

$$\mu \ (MU) = \frac{\Delta E_p}{\Delta E_g} \ \text{(with } I_p \text{ held constant)}$$

where:

μ or MU equals the amplification factor.

ΔE_p means a change in plate voltage.

ΔE_g means a change in grid voltage.

I_p means plate current or electron flow through tube.

To further clarify the μ of a tube, refer to the typical circuit in Fig. 13-4 used for the determination of tube characterisctics. Notice the voltmeter for measuring the plate voltage, E_p, the voltmeter to measure grid voltage, E_g, and the ammeter to measure plate current, I_p. The plate voltage and grid voltage supplies are variable. Assuming the plate voltage is set at 120 volts and the grid voltage at negative 2 volts, the current through the tube, I_p, is 8 mA. If the grid voltage is changed to negative 4 volts (a change of 2 volts) the current will drop to 3 mA. Now increase the plate voltage until the current rises to the original 8 mA. If the plate voltage is increased to 160 volts (a change from 120 volts to 160 volts equals 40 volts) the current rises to 8 mA. From this explanation then, a 40 volt change in plate voltage was required to overcome a 2 volt change in grid voltage, while the current was held at 8 mA. The μ then equals,

$$\mu \text{ (MU)} = \frac{\Delta E_p}{\Delta E_g} = \frac{40 \text{ volts}}{2 \text{ volts}} = 20 \text{ (amplification factor)}$$

Plate current, I_p is held at 8 mA

The above example applies to a 6J5 tube and values are taken from the average plate characteristics chart in the Tube Manual.

A second characteristic of an electron tube is the PLATE RESISTANCE, designated as r_p. It may be determined by holding the grid at a constant voltage and measuring the change in plate current resulting from a change in plate voltage. It is expressed in ohms and the mathematical relationship appears as:

$$r_p = \frac{\Delta E_p}{\Delta I_p}$$

Grid voltage, E_g, is held constant

Continuing the previous experiment with reference to the 6J5 tube and the circuit in Fig. 13-4, set the grid voltage at negative 4 volts. At a plate voltage of 120 volts the plate current reads approximately 3 mA. When the plate voltage is increased to 160 volts (a change of 40

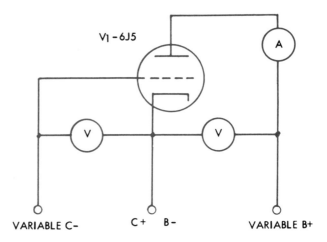

Fig. 13-4. The typical circuit used to determine the static characteristics of a triode.

volts) the plate current increases to 8 mA (a change of 5 mA). Substituting these results in the equation, then:

$$r_p = \frac{\Delta E_p}{\Delta I_p} = \frac{40 \text{ volts}}{5 \text{ mA}} = \frac{40 \text{ volts}}{.005 \text{ amps}} = 8000 \text{ ohms}$$

A curve resulting from this experiment is shown in Fig. 13-5. The characteristics of AC PLATE RESISTANCE of the tube may be determined from this curve.

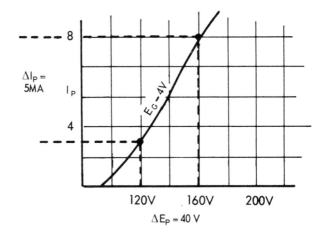

Fig. 13-5. This curve shows graphically the determination of the plate resistance of a tube.

The third characteristic of the electron tube is the TRANSCONDUCTANCE, g_m, sometimes called the control grid-plate transconductance or MUTUAL CONDUCTANCE. This characteristic may be expressed as a ratio between the change in plate current resulting from a change in grid voltage, while the plate voltage is held at a constant value. This ratio is expressed in "mhos" (ohms spelled backwards). Because a mho is a relatively large unit, the trans-conductance is converted to "micromhos" or μ mhos. The mathematical expression of this ratio becomes:

$$g_m = \frac{\Delta I_p}{\Delta E_g} \text{ (with } E_p \text{ constant)}$$

To further clarify the meaning of trans-conductance the previous experiment is continued. Refer to Fig. 13-6. At a fixed plate voltage of 120 volts and a grid voltage of negative two volts, the plate current reads approximately 8 mA.

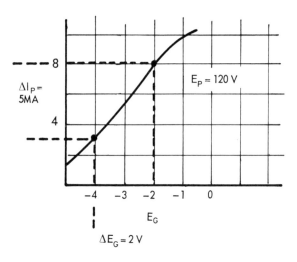

Fig. 13-6. The determination of the transconductance of an electron tube by using its characteristic curves.

When the grid voltage is changed to negative four volts (a change of two volts), the plate current decreases to 3 mA (a change of 5 mA). Then,

$$g_m = \frac{I_p}{E_g} = \frac{5 \text{ mA}}{2 \text{ volts}} = \frac{.005 \text{ amps}}{2 \text{ volts}} = .0025 \text{ mhos}$$

Converting, .0025 mhos = 2500 micromhos

The three electron tube characteristics now are summarized in these three formulas:

Amplification Factor, $\mu = \dfrac{E_p}{E_g}$ I_p constant

AC Plate Resistance, $r_p = \dfrac{E_p}{I_p}$ E_g constant

Transconductance, $g_m = \dfrac{I_p}{E_g}$ E_p constant

An interrelationship between these three characteristics exists and may be expressed as:

$$\mu = g_m r_p \text{ or } g_m = \frac{\mu}{r_p} \text{ or } r_p = \frac{\mu}{g_m}$$

when g_m is in mhos and r_p is in ohms.

A set of characteristics curves for the 6J5 tube is illustrated in Fig. 13-7. A family of average plate characteristics curves for a tube may be found in a Tube Manual. Practice in applying the curves leads to a better understanding of the electron tube. The engineer uses the curves when designing electronic circuitry.

INTERELECTRODE CAPACITANCE

In our studies of capacitance in Chapter 7, we discovered that by definition a capacitor was two plates of metal separated by a dielectric or insulating material. Also the action of the capacitor was described as blocking a direct current, yet effectively passing an alternating current. An examination of the triode electron tube discloses that the cathode, grid and plate are metallic parts separated from each other by a vacuum, and by definition are capacitors. The capacitance resulting from these electrodes is termed INTERELECTRODE CAPACITANCE. Actually three capacitances are involved. These are shown schematically in Fig. 13-8.

C_{gp} = capacitance between grid and plate

C_{gk} = capacitance between grid and cathode

C_{pk} = capacitance between cathode and plate

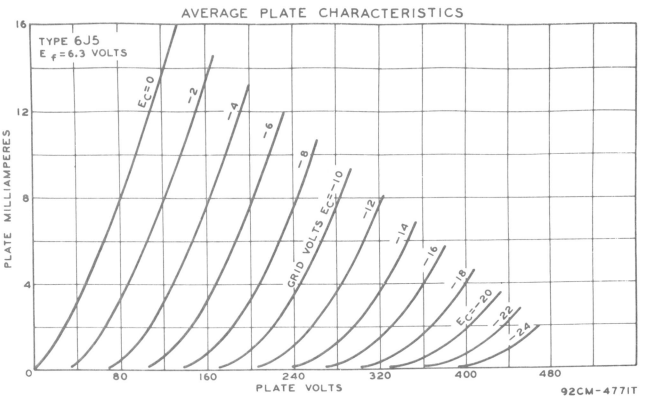

Fig. 13-7. Average plate characteristics of the 6J5 tube.
(RCA)

Interelectrode capacitance can be undesirable or desirable depending upon the application of the electron tube. If the tube is used at low frequencies, the high capacitive reactance prevents little interaction between the tube elements. However, as the frequency is increased the reactance decreases, and causes a shunting effect between the elements. This shunting effect may be counteracted by feeding a small part of the plate voltage back to the grid of the tube by means of a capacitor. This feedback voltage will be 180 deg. out of phase with the internal feedback caused by the interelectrode capacitance and therefore cancels it. This is called NEUTRALIZATION and is used extensively in rf amplifier circuits. Feedback and neutralization will be discussed in later chapters.

In some electronic circuits, such as oscillators, interelectrode capacitance is purposely used to transfer energy from the plate circuit to the grid circuit in order to sustain oscillation.

THE TETRODE

Because of the shunting effect of the electrode capacitance at high frequencies, the triode is limited as an amplifier, without neutralization circuits. An improvement upon the triode is made by inserting another grid, called the SCREEN GRID, between the control grid and

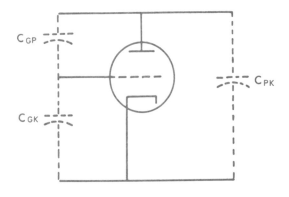

Fig. 13-8. Dotted symbols show capacitance between the elements in a tube.

the plate. This electron tube now becomes a four element tube (cathode, control grid, screen grid and plate) called a TETRODE. It is represented schematically in Fig. 13-9. The screen grid is bypassed to ground, externally, through a capacitor making the grid an effective screen between the control grid and plate and

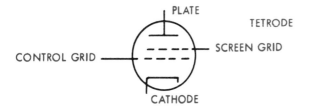

Fig. 13-9. Symbol for a TETRODE.

eliminating the grid-plate capacitance, (C_{gp}). A dc voltage of slightly less value than the plate voltage is usually applied to the screen grid and the screen grid acts as an accelerator for the electrons passing between the cathode and plate. Some electrons are attracted to the screen grid and will cause a current to flow in the screen circuit. However, most electrons pass through the screen to the plate. High amplification is possible in the tetrode because the control grid is placed quite close to the cathode. A decrease or increase in plate voltage has comparatively little effect on the plate current due to the isolation effect of the screen grid. The transconductance of the tetrode is relatively high.

THE PENTODE

In the tetrode, the accelerated electrons strike the plate with considerable velocity. When this occurs, loosely held electrons on the plate are knocked off into free space and form a space charge around the plate. This is called SECONDARY EMISSION. Some of these electrons are attracted to the screen and subtract from the useful plate current through the tube. This effect is more pronounced at times when the plate voltage is below the screen voltage. To overcome the undesirable effects of

the secondary emission, a third grid is placed in the tube between the screen grid and plate. The tube now has five elements (cathode, control grid, screen grid, suppressor grid and plate) and is called a PENTODE. This third grid is named the SUPPRESSOR GRID. It is usually internally connected to the cathode and is at cathode potential. This grid repels the free electrons resulting from secondary emission and drives them back to the plate. The pentode is diagramed in Fig. 13-10. These tubes have high

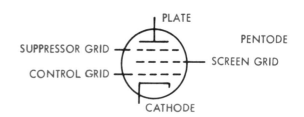

Fig. 13-10. Symbol for a PENTODE.

amplification factors, high plate resistance and transconductance. Interelectrode capacitance is at a minimum. Pentodes have wide application as rf amplifiers and also as audio power amplifiers. An exploded view of a miniature pentode is shown in Fig. 13-11.

BEAM POWER TUBES

A special development in electron tubes is the BEAM POWER TUBE which was designed for use in high powered amplifiers. It differs from the usual tetrode or pentode by the introduction of special plates which form the electron stream into a concentrated beam. The grid wires also are lined up so that they will cause only a minimum interference to the electron stream. Since the screen grid is in the shadow, so to speak, of the control grid very few electrons actually hit the screen grid and the screen current is low in the beam power tube. In comparison to the conventional tetrode or pentode the beam power tube provides a greater power output for the similar plate and screen voltages. The construction of a beam power pentode is illustrated in Fig. 13-12.

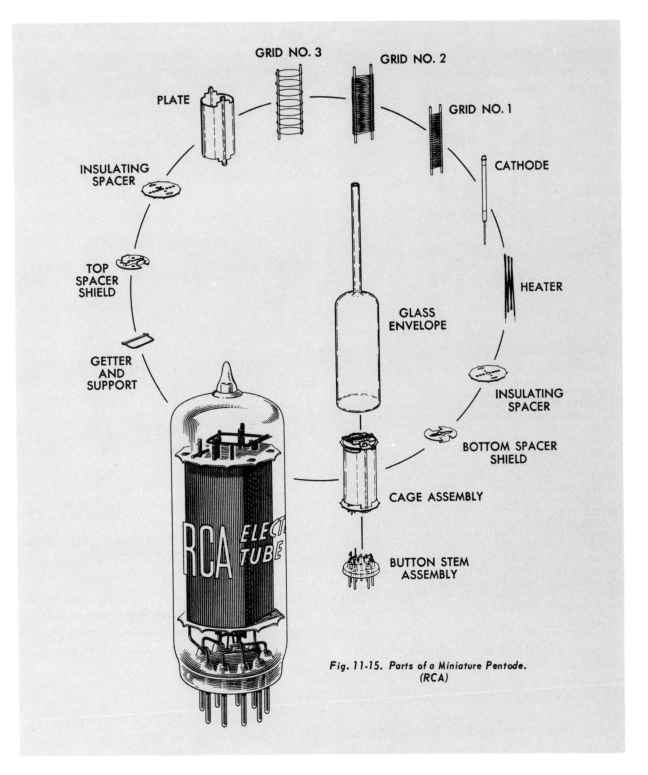

GRID NO. 3

GRID NO. 2

GRID NO. 1

PLATE

CATHODE

INSULATING SPACER

HEATER

TOP SPACER SHIELD

GLASS ENVELOPE

GETTER AND SUPPORT

INSULATING SPACER

BOTTOM SPACER SHIELD

CAGE ASSEMBLY

BUTTON STEM ASSEMBLY

Fig. 11-15. Parts of a Miniature Pentode.
(RCA)

Fig. 13-11. Parts of a minature pentode.
(RCA)

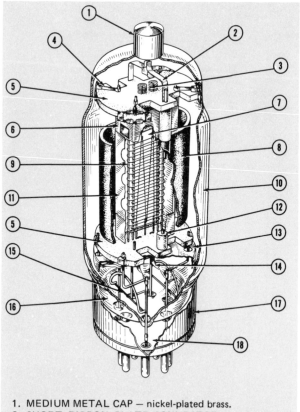

1. MEDIUM METAL CAP — nickel-plated brass.
2. SHORT RIBBON PLATE CONNECTOR — molybdenum.
3. FILAMENT SUPPORT SPRINGS — tungsten.
4. MOUNT SPACER — nickel-chromium strip.
5. MOUNT SUPPORT — ceramic.
6. TOP SHIELD — nickel.
7. HEAVY-DUTY FILAMENT — thoriated tungsten.
8. PLATE — zirconium-coated graphite.
9. ALIGNED-TURN CONTROL GRID (GRID No. 1) AND SCREEN GRID (GRID No. 2) — molybdenum.
10. BULB OR ENVELOPE — hard glass.
11. BEAM-FORMING ELECTRODE — nickel.
12. PLATE-SUPPORT SPACER — ceramic.
13. BOTTOM SHIELD DISK — nickel.
14. FILAMENT CONNECTOR — nickel-plated steel.
15. DIRECTIVE-TYPE GETTER.
16. MOLDED-FLARE STEM — hard glass.
17. GIANT BASE — nickel-plated brass with ceramic insert.
18. TUNGSTEN-TO-GLASS SEAL.

Fig. 13-12. Construction of RCA-813 beam power tube.

MULTI-ELEMENT TUBES

In addition to the electron tubes previously discussed, many special purpose tubes with more elements have been developed. One such tube is the PENTAGRID converter used in the superheterodyne radio. Extra grids allow the addition of signals of other frequencies. The schematic diagrams in Fig. 13-13 show a few of the many dual purpose and special purpose tubes. Note that in some cases, two tubes are enclosed in one envelope. The name of each tube suggests its application.

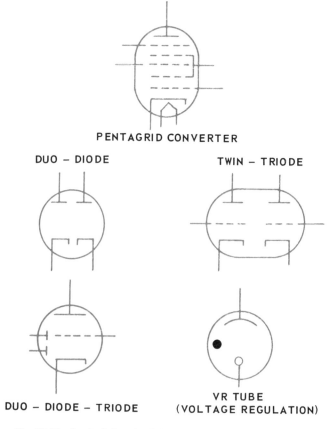

Fig. 13-13. Symbol for the PENTAGRID CONVERTER and special combination tubes.

It is also possible for a vacuum tube to have three or more functions in the same envelope. Fig. 13-14 shows the outline drawing for a 6AF11 which has two triodes and a pentode in the same case.

Fig. 13-14. A three element tube — 6AF11.

HOW THE ELECTRON TUBE AMPLIFIES

The ability of an electron tube to amplify a voltage or signal is called its AMPLIFICATION FACTOR or MU and represent the ratio between the change in plate voltage to a change in grid voltage when the plate current is held at a constant value. The amplification factor of any tube may be found by consulting a Tube Manual.

The following diagrams and examples explain the theory of vacuum tube amplification. See Figs. 13-15 to 13-20.

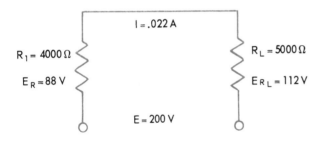

R_T = 10,000 Ω

I_T = .02 amps

E_{R_1} = 100V and E_{R_L} = 100V

Fig. 13-15.

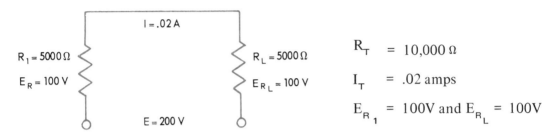

R$_1$ is reduced to 4000 Ω

R_T = 9000 Ω

I_T = .022 amps

E_{R_1} = 88V and E_{R_L} = 112V

Fig. 13-16.

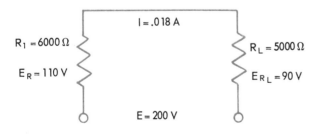

R$_1$ is increased to 6000 Ω

R_T = 11,000 Ω

I_T = .018 amps

E_{R_1} = 110V and E_{R_L} = 90V

Fig. 13-17.

The three simple series resistor circuits demonstrate a principle. In all cases R_L is held constant. As R_1 is varied above and below 5000, the voltage drops across R_1 and R_L vary. When R_1 is increased, a larger proportion of the total voltage, 200V, appears across R_1. When R_1 is decreased, a smaller proportion appears across R_1. Consider the case of R_1 with infinite resistance:

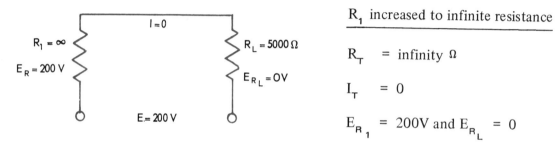

R$_1$ increased to infinite resistance

R_T = infinity Ω

I_T = 0

E_{R_1} = 200V and E_{R_L} = 0

Fig. 13-18.

Since no current is flowing in the circuit, the voltage drop across E_{R_L} = I x R$_L$ = 0 x 5000 = 0.

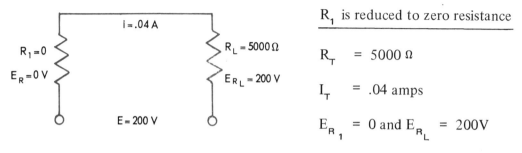

R$_1$ is reduced to zero resistance

R_T = 5000 Ω

I_T = .04 amps

E_{R_1} = 0 and E_{R_L} = 200V

Fig. 13-19.

A vacuum tube may be considered as a variable resistance and its resistance depends upon the bias voltage applied to its grid. Refer back to Fig. 13-7 showing the characteristic curves of the 6J5 tube. When the plate voltage is 200 volts and the negative bias is −6 volts, the plate current is 7.5 mA. If the bias is changed to −8 volts, the current decreases to 3.5 mA. The tube appears as more resistance when the grid is made more negative.

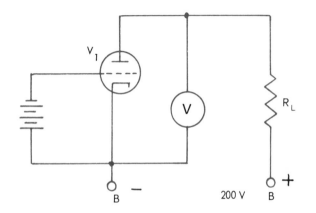

Tube V$_1$ is substituted for R$_1$

If V$_1$ is made more resistive by making the grid more negative, then E_p increases and E_{R_L} decreases.

If V$_1$ is made less resistive by making grid less negative, then E_p decreases and E_{R_L} increases.

Fig. 13-20.

Special cases are:

When V_1 is biased to CUT-OFF, it has infinite resistance, and E_p = B+ voltage of 200 volts.

When V_1 is biased positively to SATURATION, it has minimum resistance and E_p = approximately 0.

A varying voltage, therefore, applied to the grid of the tube, produces varying voltages E_p and E_{R_L} . This characteristic of a tube explains its ability to amplify an input voltage. For example: If a one volt change in grid bias produced a 20 volt change in plate voltage, E_p, the tube has amplified the input, and a GAIN of 20 times results.

$$\text{(Gain) } A = \frac{\text{Output E}}{\text{Input E}} = \frac{20}{1} = 20$$

The GAIN of the tube is a result of both the amplification factor of the tube and the resistance R_L selected. They are related in the following formula:

$$A = \mu \ \text{x} \ \frac{R_L}{R_L + R_p}$$

where,

μ = amplification factor

R_L = Load resistance

R_p = Plate resistance

Notice that as R_L is made larger, the gain A approaches the amplification factor, but it can never exceed it. The overall gain of several stages will equal the product of individual gains resulting from tubes and other devices, including transformers and tuned circuits.

The basic amplifier circuit is drawn in Fig. 13-21. For this example the grid bias voltage is fixed at negative four volts, and a two peak ac signal is also applied to the grid. The instantaneous grid bias voltage is equal to the algebraic sum of the fixed voltage and the signal voltage.

When signal is maximum positive then:

$(+2)V + (-4)V = -2$ volts

When signal is maximum negative then:

$(-2)V + (-4)V = -6$ volts

The grid voltage is varying between the limits of -2 and -6 volts. The output voltage wave form is also shown in Fig. 13-21 as the plate voltage of the tube. Observe that the output signal has been inverted or is 180 deg. out of phase with the input signal. This will be explained by a careful review of the previous discussion on "How a tube amplifies." The inverted amplified output voltage may be coupled to another tube or circuit for further amplification.

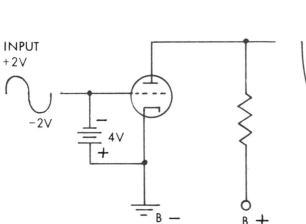

Fig. 13-21. Basic amplifier circuit with input and output signals.

GRID BIAS VOLTAGE

The term "grid bias voltage" has been defined as the voltage applied between the grid of a tube (usually negative) and its cathode. Three common methods are used to supply this voltage and each should be thoroughly understood:

1. FIXED BIAS. This bias is supplied by a C battery or negative power supply. Circuits used for this purpose were described in Chapter 12. The fixed supply has wide application in radio transmitters and other equipment. Refer to Fig. 13-21 in which a C battery is used.

2. THE GRID LEAK. In Fig. 13-22 a capacitor has been inserted in series with the control grid. When the incoming signal is positive, the grid of the tube is driven positive. Some of the electrons emitted from the cathode will be attracted to this positive grid. When

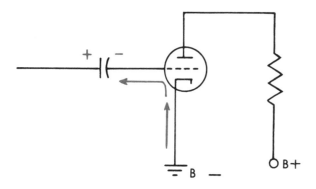

Fig. 13-22. Capacitor is connected in series with grid to establish a bias voltage.

the incoming signal is negative, the grid becomes negative, but the grid is a cold element in the tube and cannot give up its electrons. As a consequence, more electrons are collected by the grid with each positive swing of the signal, until the grid becomes so negative that the tube will become CUT-OFF. To overcome this objectionable behavior, a resistor, called a "leak" is placed in parallel to the capacitor as in Fig. 13-23.

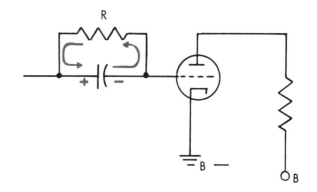

Fig. 13-23. Resistor in parallel with C provides an electron "leak" and permits C to discharge.

This leak resistor allows the capacitor to partially discharge, and maintain a level or charge for satisfactory operation. The function of the leak may be better understood by referring to Fig. 13-24. In this diagram, the bucket containing water may be considered the capacitor C. The hole or leak in the bucket is R and the supply pipe is the source of water or the grid. If the leak is too large, the level of water will decrease; if the leak is too small, the level will increase. Proper selection of values of R and C will permit the correct voltage level or bias to be maintained on the grid. It is important to note that no bias will be developed for the tube, unless there is an incoming signal to drive the grid positive. Circuits similar to these were used as the grid-leak detector of early radios. They now find application in limiting circuits and oscillators.

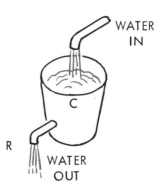

Fig. 13-24. Leak action is compared to maintaining water level in a bucket.

3. CATHODE OR SELF-BIAS. One of the more common methods of supplying the bias for proper operation of a vacuum tube is the cathode resistor, as diagramed in Fig. 13-25. When the tube conducts, a voltage drop develops across R_K and the cathode end of R_K becomes more positive than ground. But the grid is connected to ground, so the

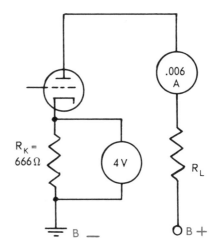

Fig. 13-26. Resistor of 666 ohms creates a −4 bias.

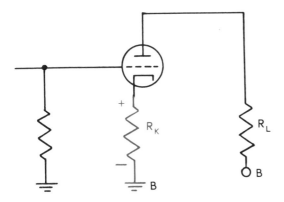

Fig. 13-25. Resistor R_K in cathode circuit causes cathode end of R_K to be more positive than ground.

cathode is more positive than the grid. Is this not the same as applying a negative voltage to the grid? It means exactly the same. Either the grid may be more negative than the cathode, or the cathode may be more positive than the grid. The voltage drop E_R determines the bias on the tube.

PROBLEM: A vacuum tube is designed to operate with a plate current of 6 mA with a bias voltage of negative 4 volts. What value cathode resistor should be used?

SOLUTION:

$$R = \frac{E}{I} \qquad R = \frac{4V}{.006} = 666\ \Omega$$

See diagram in Fig. 13-26.

PROBLEM: In the diagram in Fig. 13-27, the resistance and current values are given. Find the B+ voltage, the plate voltage and the grid bias.

SOLUTION: Since the total current is 20 mA and 10 mA is flowing through the tube, then

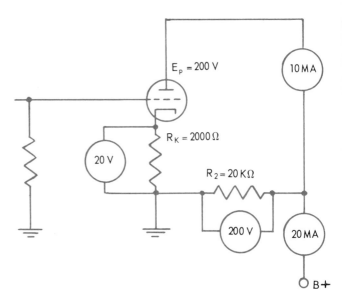

Fig. 13-27. Schematic diagram of problem in text.

10 mA must be flowing through R_2. So the voltage drop across R_2 is:

$$E_{R_2} = I \times R =$$
$$.01 \times 20,000\ \Omega = 200 \text{ volts, B+ supply}$$

Since 10 mA is flowing through the tube and R_K is in series, the voltage drop across R_K is:

$$E_{R_K} = .01 \times 2000\ \Omega = 20 \text{ volts}$$

The cathode is 20 volts more positive than ground to which the grid is connected, and the tube has a <u>negative 20 volt bias</u>. The plate voltage E_p must equal <u>200V</u>, measured between plate and ground.

DEGENERATION BY CATHODE BIAS

It is apparent from previous studies that the current flow in a vacuum tube is controlled by the incoming signal and the bias voltage on the grid. As the signal swings positive, the bias on the tube is less negative and the current through the tube increases. But the self-biasing cathode resistor produces a voltage drop in direct proportion to the current. As a result a positive signal increases the voltage drop and thus makes the grid more negative in respect to the cathode. These two effects are opposing each other and result in a decrease in gain in the tube. If part of the output of the tube is fed back to the input grid circuit, so that it opposes the input, (180 deg. out of phase) it is called DEGENERATION. This is undesirable in many circuits, so a method must be employed to maintain the voltage drop across the cathode resistor. A capacitor of the correct value, called a cathode bypass capacitor, is connected in parallel with R_K in Fig. 13-28. The capacitor

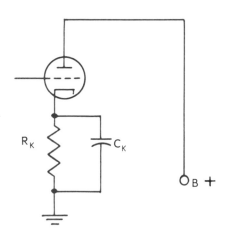

Fig. 13-28. Cathode by capacitor is used to limit degeneration.

provides a low reactance path for a varying or ac current around the resistor, thus leaving a relatively constant voltage drop across R_K due to the direct current. The value of C depends

upon the frequency of the signals being amplified and should have a reactance of one-tenth or less of the value of the cathode resistor.

PROBLEM: A tube used to amplify audio frequencies down to 100 hertz has a cathode resistor of 1200 ohms. What value bypass capacitor should be selected?

SOLUTION: Since R_K equals 1200 ohms, C should have a reactance of one-tenth or 120 ohms or less at 100 hertz. Now:

$$X_c = \frac{1}{2 \pi fC} \quad \text{or } C = \frac{1}{2 \pi fX_c}$$

Then:

$$C = \frac{1}{6.28 \times 100 \times 120}$$

$$= \frac{1}{75,360} = .000013 \, f = 13 \, \mu f$$

Capacitors are manufactured in 10 and 16 μ f sizes, so the 16 μ f would be selected for this circuit and X_c would be less than 120 Ω . A capacitor of this value would effectively short circuit the variation in voltage at a frequency of 100 Hz and when used as a cathode bypass, it would reduce degeneration to an allowable minimum.

If the tube were used to amplify radio frequency signals at 1000 KHz for example, then the bypass capacitor value would be:

$$C = \frac{1}{6.28 \times 1000 \times 10^3 \times 120} \quad \text{or } .0013 \, \mu f$$

A .001 μ f capacitor would be used.

CLASS A AMPLIFIERS

One convenient method of classifying vacuum tube amplifiers is by the operating point as determined by the fixed grid bias. Referring to the characteristic curve of a common vacuum tube in Fig. 13-29, the

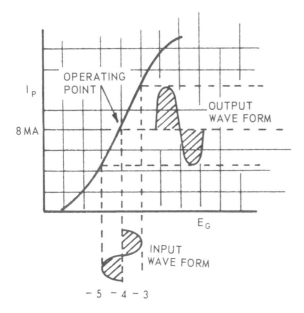

Fig. 13-29. Tube is operating around center of linear portion of characteristic curve as Class A.

operating point is set at negative four volts by a fixed voltage source. The curve shows that a plate current of 8 mA will flow with a plate voltage of 160 volts. A one volt peak signal applied to the grid causes the grid voltage to vary between −3 and −5 volts and the plate current to vary between 5 mA and 11 mA. Both the input and output wave forms are shown in Fig. 13-29. A tube operated in this manner is classified as a CLASS A amplifier and may be identified by:

1. The operating point is near the center of the straight line portion of the curve or it is operating on the linear portion of the curve.

2. The output wave form is a faithful reproduction of the input wave form except that it is amplified.

3. Plate current is always flowing in the circuit, whether an input signal is applied or not.

4. The average plate current is the same whether an input signal is applied or not.

5. Considering one cycle of the input signal as 360 deg., then current flows in the plate circuit during 360 deg. of the input cycle.

Characteristics of a Class A amplifier may be summarized as:

1. Low distortion, faithful reproduction. Such an amplifier enjoys wide usage in audio amplifiers when high fidelity reproduction is required.

2. Low power output and low efficiency. The tube is conducting at all times, thus allowing no time for cooling.

3. High power amplification.

It should be noted that the grid in the above mentioned amplifier is not driven positive by the input signal. This is usually the case. In special appplications, the signal does drive the grid positive and the grid collects electrons which causes a grid current to flow. Amplifiers have a further designation by a subscript number such as:

Class A$_1$ – grid is not driven positive.

Class A$_2$ – grid is driven positive and grid current flows. If no subscript is used, it is assumed to be Class A$_1$.

CLASS B AMPLIFIERS

On the characteristic curve in Fig. 13-30, a tube is biased at its cut-off point of negative 8 volts. A peak input signal of 4 volts causes the grid voltage to vary between −4 volts and −12 volts. The tube conducts only between −8 volts and −4 volts. During the negative half-cycle of the input signal the tube does not conduct and plate current is zero. The tube acts as a half-wave rectifier and produces plate current only during the positive half of the input signal. A tube operated in this manner, is called a CLASS B amplifier and may be identified by:

1. The tube is biased at approximately cut-off, which is its operating point.

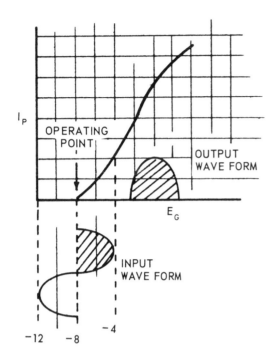

Fig. 13-30. Tube biased at cutoff is operating Class B.

2. Plate current flows only during one half the input signal or 180 deg. of the cycle.

3. Considerable distortion results, because one half of the input wave form has been removed.

The amplifier may be further characterized by:

1. Medium power output and efficiency. The tube is resting (not conducting) one half of the time.

2. Medium power amplification.

Again, if the input signal drives the grid positive, the subscript $_2$ is used, as Class B_2.

PUSH-PULL AMPLIFIERS

The medium power and efficiency characteristics of the Class B amplifier may be used to advantage to produce a high fidelity output by a push-pull circuit. Two tubes are used. Each tube amplifies faithfully one half of the input signal. Fig. 13-31 shows the circuit configuration. The input signal is applied to the grids of V_1 and V_2 by coupling transformer T_1. During the positive half of input cycle, the upper end of the secondary of T_1 is positive, making grid V_1 more positive and driving it into conduction. As a result, both halves of the input signal combine in the output as a relatively true reproduction of the input wave form.

It should be noted that the design of a push-pull amplifier requires the division or separation of the input signal into positive and negative half-cycles. These half signals are fed respectively to the grids of the tubes. This was accomplished in the previous example by means of an input or driver transformer.

Transformers which will give a satisfactory frequency response in a circuit of this nature are expensive. Other circuits have been designed which are more economical. In Fig. 13-32 is a diagram of a triode amplifier using equal load resistors in both the cathode and plate circuits. The varying current in the tube causes an equal voltage drop across both R_1 and R_2. However, the signal taken from the plate of the tube is 180 deg. out of phase with the input signal and the signal taken from across the cathode resistor is in phase with the input signal. These two signals are coupled by means of capacitors to the respective grids of the push-pull amplifier. This circuit is called a PHASE SPLITTER.

In order to provide greater gain as well as phase splitting or inversion, two triodes may be used as diagramed in Fig. 13-33. A duo-triode may be used. The tube used in this schematic is a 12AX7. Both triodes are enclosed in one glass envelope. To trace the action of this circuit, it must be remembered that both triodes have equal amplification factors. Assuming a voltage gain of 10, a one volt signal will produce a ten volt change at the plate of V_{1_A}. This signal is fed to the grid on one push-pull amplifier tube. The load resistor in the plate circuit of V_{1_A} consists of two resistors in series. One of these

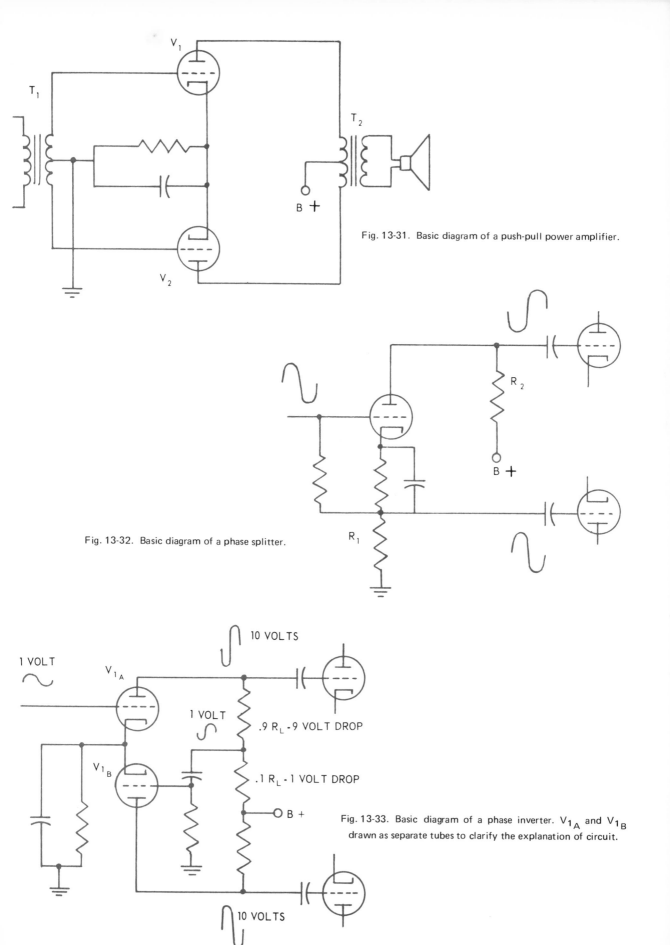

Fig. 13-31. Basic diagram of a push-pull power amplifier.

Fig. 13-32. Basic diagram of a phase splitter.

Fig. 13-33. Basic diagram of a phase inverter. V_{1_A} and V_{1_B} drawn as separate tubes to clarify the explanation of circuit.

10 VOLTS

1 VOLT

1 VOLT

.9 R_L -9 VOLT DROP

.1 R_L -1 VOLT DROP

B +

10 VOLTS

resistors is valued at one tenth and the other nine tenths of the total load resistance. The voltage drop across the load can then be divided to one volt and nine volts respectively. The one volt signal is fed to the grid of V_{1_B} and amplified to produce a ten volt change at the plate of V_{1_B}. This signal is coupled to the grid of the other push-pull amplifier tube. The phase of each signal in the circuit should be carefully studied as indicated on the schematic.

SUMMARIZING: A positive signal at the grid of V_{1_A} produces a negative signal or 180 deg. phase shift at the plate of V_{1_A}. One tenth of this negative signal is fed to the grid of V_{1_B} which produces a positive signal at the plate of V_{1_B}. This circuit is frequently called a PHASE INVERTER, or a PARAPHASE AMPLIFIER.

SEMICONDUCTOR AMPLIFIERS

THE JUNCTION TRANSISTOR

The ability of a transistor to amplify is the key to its popularity in electronics. The JUNCTION TRANSISTOR consists of three layers of the impure germanium crystals. There are two junctions. Transistors are illustrated by blocks and schematic symbols in Fig. 13-34. In the first case a thin layer of N type crystals is sandwiched between two P type crystals and is designated as a PNP transistor. In the second case a thin layer of P type crystal separates two N type crystals and is called an NPN transistor. In either case, the first crystal is called the

EMITTER; the center section, the BASE; and the third crystal, the COLLECTOR. In the schematic symbols, notice the direction of the arrowhead which signifies whether it is a PNP or NPN transistor. Popular transistors used in many of your projects are illustrated in Fig. 13-35. Can you identify them?

Fig. 13-35. Typical transistors used in projects in this text. (Sylvania)

The theory of operation of an NPN transistor is illustrated in Fig. 13-36. For ease in understanding, two batteries are used. The negative terminal of the battery is connected to the N emitter and the positive terminal of the same battery is connected to the P type base. Therefore, the emitter-base circuit is forward

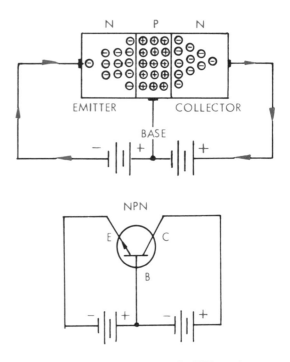

Fig. 13-36. The current flow in the NPN transistor.

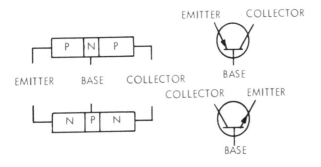

Fig. 13-34. Block diagrams and symbols for NPN and PNP transistors.

biased. In the collector circuit, the N collector is connected to the positive battery terminal and the P base to the negative terminal. The collector-base circuit is reverse biased.

Electrons enter the emitter from the negative battery source and flow toward the junction. The forward bias has reduced the potential hill of the first junction so the electrons combine with the hole carriers in the base to complete the emitter-base circuit. However, the base is a very thin section, approximately .001 in. and most of the electrons flow on through to the collector. This electron flow is aided by the low potential hill of the second PN junction. Approximately 95 to 98 percent of the current through the transistor is from emitter to collector and about 2 to 5 percent between emitter and base. A small change in emitter bias voltage will cause a relatively large change in emitter-collector current, while the emitter-base current change is quite small. Although the current gain is less than one, the voltage gain can be many times. The current gain in a junction transistor is designed as alpha ($\propto$). In the previous discussion, $\propto$ = .95 to .98.

The ability of the transistor to amplify is the result of the ratio between input and output resistance. Since the emitter-base is biased in a forward direction, its resistance is very low, 200 to 500 Ω. The collector-base circuit is biased in a reverse direction and has a very high resistance, 100K Ω – 1 meg Ω. The voltage gain may be computed as:

$$A = \propto \frac{R_{output}}{R_{input}}$$

Example: A transistor has an alpha characteristic of .98. Its input resistance is 500 Ω and its output resistance is 500K Ω. What voltage gain may be expected?

$$A = .98 \times \frac{500,000}{500} = 980$$

The previous discussion has used a NPN transistor as an example. The theory of operation would be the same for a PNP, except that

battery voltages must be reversed. In any case, the emitter is always biased in a forward direction and the collector in a reverse direction.

Now that we have inquired into the theory of transistor operation, a comparison to the familiar vacuum tube is desirable. They are alike in many respects. In Fig. 13-37, the emitter may be compared to the cathode which emits electrons. The plate and collector receive the

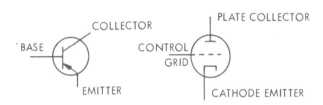

Fig. 13-37. A comparison between the transistor and the triode vacuum tube.

electrons. In the case of the vacuum tube a negative bias on the grid limits the electron flow through the tube. In most cases the grid is not driven positive. Therefore, the tube has a high input impedance. In the transistor, due to foward bias, current does flow from emitter to base and this circuit has a low impedance. The vacuum tube is considered as a VOLTAGE OPERATED device, whereas the transistor is CURRENT OPERATED.

TRANSISTOR CIRCUIT CONFIGURATIONS

The transistor may be connected in three circuit configurations, depending on which element is common to both input and output circuits. Usually the common element is at ground potential. In Fig. 13-38 the COMMON BASE amplifier circuit is drawn and it is compared to the grounded-grid vacuum tube circuit. The ac input signal between the emitter and base will vary the forward bias by alternately adding to and subtracting from the fixed bias. The increase and decrease in current in the emitter-collector circuit produces the amplified voltage across R_L. The input signal and the output signal are in phase. There is no signal inversion.

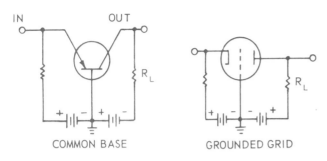

Fig. 13-38. The common base or grounded base configuration and its vacuum tube counterpart.

In Fig. 13-39, the COMMON EMITTER circuit is drawn. This is by far the more common transistor circuit and will be studied in more detail later in this chapter. Again the comparative vacuum tube circuit is drawn. Notice that this circuit conforms very closely to the conventional vacuum tube amplifier. In the common emitter circuit a positive input signal would make the base more positive or less

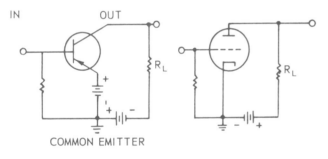

Fig. 13-39. The common emitter circuit compared to the triode vacuum tube.

negative with respect to the emitter, thus reducing the foward bias of the circuit. A negative input signal would make the base more negative and increase the forward bias. As a result, the voltage across R_L in the output circuit is 180 deg. out of phase. The signal has been inverted in the same manner that signals are inverted in a vacuum tube circuit.

In the common emitter circuit the gain is computed as the ratio between the change in collector current and the change in base current. The current gain is specified by the Greek letter β (beta) and may be mathematically stated as:

$$\beta = \frac{\Delta I_c}{\Delta I_b} \qquad V_c \text{ Constant}$$

where:

β (beta) equals current gain in common emitter circuit

ΔI_c equals the change in collector current

ΔI_b equals the change in base current

while holding V_c (collector voltage) constant.

Note: In transistor circuits, voltage is represented by the capital letter V instead of E as used in vacuum tube circuits and electricity.

Fig. 13-40 represents a common collector circuit and its vacuum tube comparison. The signal is applied between the base and collector. The output is taken from the collector-emitter circuit. This circuit compares to the conven-

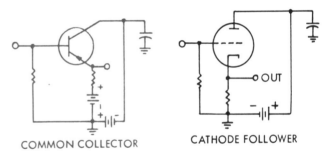

Fig. 13-40. The common collector circuit may be compared to the cathode follower vacuum tube circuit.

tional CATHODE FOLLOWER vacuum tube circuit. It will have a very high input impedance and a low output impedance. It is useful as an impedance matching circuit. The input and output signals are in phase. The voltage gain in this circuit must always be less than one.

THE TRANSISTOR AMPLIFIER

In Fig. 13-41, the common emitter amplifier circuit is drawn and its vacuum tube comparison. Only one battery is used and its positive terminal is connected to ground for the PNP transistor. Follow the circuit and notice that the emitter-base circuit is forward biased and the collector-base circuit is reverse biased.

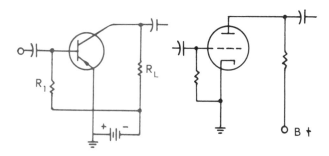

Fig. 13-41. A single stage transistor amplifier with common emitter configuration compared to the vacuum tube amplifier.

The resistor R_1 limits the emitter-base current and its value is selected for the desired operating point from the transistor specifications. R_L acts as the load resistor for the transistor and the amplified signal voltage appears across this resistor.

Transistors differ from resistors in respect to temperature effects. As the operating temperature increases, there is greater thermal activity in the crystalline structure and there is a decrease in resistance. The cumulative effect of decreasing resistance and increasing current may destroy the transistor by THERMAL RUN-AWAY. In Fig. 13-42, the common emitter amplifier is drawn with a stablizing resistor R_3. Any increase in current will make the emitter

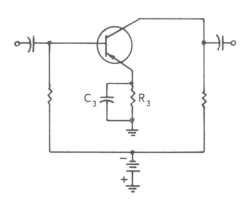

Fig. 13-42. Common emitter stage with stabilizing resistor and bypass capacitor.

end of R_3 more negative in respect to ground and thereby decrease the forward bias of the emitter-base circuit. This acts as a limiter to current increased resulting from thermal runaway. C_3 is a bypass capacitor and serves the same function as its counterpart in a vacuum tube amplifier. It prevents degeneration and loss of gain.

CASCADED AMPLIFIERS

As in vacuum tube circuits, several transistor stages may be connected in CASCADE to produce the desired amplification. There is a drawback which adds some expense to transistor amplifiers. This is the problem of matching the high impedance output of one transistor to the low input impedance of the following transistor without severe loss in gain. Specifically transistor stages may be cascaded by three primary methods:

1. Transformer coupling.

2. RC coupling.

3. Direct coupling.

These methods will be explained next.

TRANSFORMER COUPLING

The circuit in Fig. 13-43 shows two stages of transistor amplifiers coupled with a transformer. Assuming that transistor Q_1 has an output impedance of 20K Ω and Q_2 has an input impedance of 1K Ω, a severe mismatch and loss of gain would result.

The transformer is a convenient device to match these impedances. A step-down transformer is required. A low secondary voltage means a higher secondary current, which is just fine for transistors because they are current operated. Many special sub-ouncer and sub-sub-ouncer transformers have been developed for this purpose. Some of these are illustrated in Fig. 13-44.

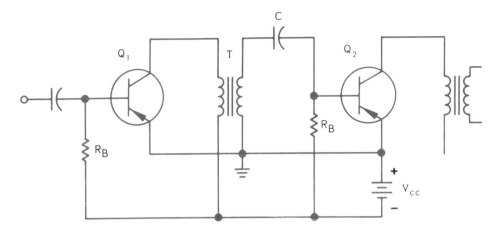

Fig. 13-43. A transformer coupled amplifier circuit.

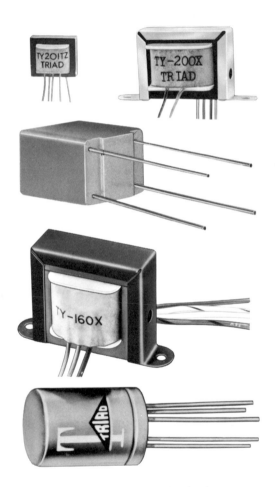

Fig. 13-44. These tiny sub-ouncer and sub-sub-ouncer transformers are used in transistor circuitry. (Triad)

The purpose of C in the transformer coupling circuit of Fig. 13-43 is to block the dc bias voltage of the transistor from ground. Notice that with C omitted, the transistor base would be grounded directly through the transformer secondary.

The major disadvantage of transformer coupling, besides cost, is the poor frequency response of transformers. They tend to saturate at high audio frequencies. At radio frequencies, the inductance and winding capacitance will present problems.

A variation of transformer coupling you will see in many circuits uses a tapped transformer. These taps can be at medium, low and high impedance points. With this version of a transformer, good impedance matching can be attained as well as good coupling and gain. Study the circuit in Fig. 13-45. For radio frequency amplifier circuits, both the transformer primary and secondary windings can be tuned by C for frequency selectivity.

One other point about the transformer should be called to your attention. The primary impedance of the transformer acts as a collector load for the transistor. Since this impedance only appears under signal conditions, the load is X_L of primary. From the dc point of view, the only load is the ohmic resistance of the wire used to wind the transformer primary. You will use this information later in the design of a power amplifier which will require an understanding of both dc and ac load lines.

RC COUPLING

An expensive and simple method of coupling transistor amplifier stages is by using resistor-capacitor, RC, coupling. A step-by-step explanation of how this is done is shown in Figs. 13-46 through 13-49.

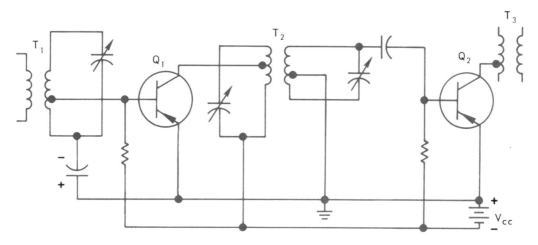

Fig. 13-45. In this circuit the taps on the transformer primary and secondary windings provide a convenient matching point. The transformer can be designed for good overall gain.

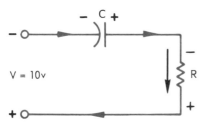

Fig. 13-46.

In Fig. 13-46, capacitor C will charge to V equals 10 volts. Only during the charging of C will a current cause a voltage to appear across R. After C is charged, $V_C = 10V$ and $V_R = 0$.

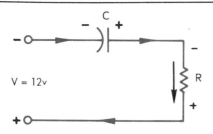

Fig. 13-47.

If the source voltage is changed to 12 volts. C will increase its charge to 12 volts also. The charging current will produce a momentary two volt pulse of voltage across R in the polarity shown.

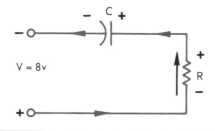

Fig. 13-48.

If source voltage is changed to 8 volts, C will discharge to 8 volts. The discharge current will produce a momentary pulse of voltage across R in the polarity shown.

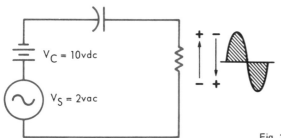

Fig. 13-49.

Here both a dc and an ac voltage are connected to the RC circuit. The ac signal causes the total voltage to vary between 8 and 12 volts. Therefore C will charge and discharge at the frequency of the ac generator.

The voltage appearing across R will rise and fall at the same frequency as the generator voltage. But, take particular note of Fig. 13-50.

Where the input signal varied around a dc level of 10 volts, the output signal varies around the zero level. The dc component has been removed. A capacitor blocks dc. In this respect, it is called a BLOCKING CAPACITOR.

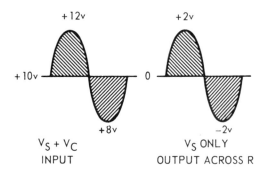

$V_S + V_C$
INPUT

V_S ONLY
OUTPUT ACROSS R

Fig. 13-50. The capacitor blocks the dc voltage, but permits the ac signal to pass.

Now let us examine this circuit from a mathematical point of view. C and R in series form an ac voltage divider. Refer to Fig. 13-51.

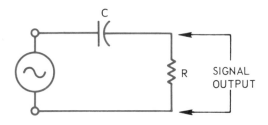

Fig. 13-51. R and C form a voltage divider for the ac signal.

The voltage division will depend upon the reactance of X_C at the frequency of the signal. The voltage output across R is the important consideration and we want as much of the ac voltage to appear across R as possible. Assuming a frequency of 1000 Hz and a value of C equals to .01 μ f, then:

$$X_C = \frac{1}{2 \pi fC} = \frac{1}{6.28 \times 10^3 \times 1 \times 10^{-8}}$$

$$\cong .16 \times 10^5 = 16,000 \ \Omega$$

If the value of R is ten or more times greater than X_C, then most of the voltage will appear across R. Assuming R = 160K Ω , then:

$$V_R = \frac{R}{R + X_C} \times V$$

$$= \frac{1.6 \times 10^5}{(1.6 \times 10^5) + (1.6 \times 10^4)} \times V$$

$$= \frac{1.6 \times 10^5}{17.6 \times 10^4} \times V = .91 \times 10V = 9.1V$$

It seems that almost all of the voltage does appear across R. Less than a volt has been lost as signal output. If R were made larger, then even more of the total output would be developed across R. In vacuum tube circuits, the tube has a high input impedance and is considered voltage controlled. This circuit then would be quite satisfactory. But look at the two stage amplifier using transistors in Fig. 13-52.

Also consider the signal voltage at the collector of Q_1. It truly has a choice of paths to go and will take the easiest path. It can go through R_C or through the coupling network which may be considered in parallel. If the impedance of the network is higher than R_C, then signal currents will go through R_C instead of to the next stage. So, therefore, the coupling capacitors used in transistor circuits must have values in the vicinity of 8 to 10 μf.

To prove the point, assume the input impedance of stage Q_2 is 500 Ω , then X_C would need to be 50 Ω or less. At 1000 Hz frequency:

$$C = \frac{1}{2 \pi fX_C} = \frac{1}{6.28 \times 10^3 \times 5 \times 10} \cong 3.2 \ \mu f$$

At lower frequencies which the amplifier would also amplify, X_C would be higher and a capacitor of 8 to 10 μf would be required to prevent loss of amplifier gain.

Note also in Fig. 13-52, that R_F and R_B are in parallel with the emitter-base circuit. It is

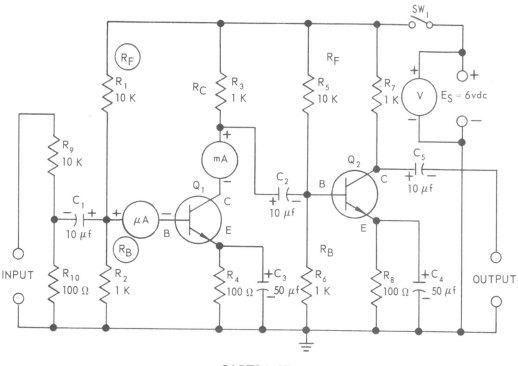

PARTS LIST

Voltmeter 0-25 Vdc.
Ammeter 0-10 mA dc.
Ammeter 0-0.1 mA dc.
Q_1, Q_2 — 2N649 Transistor.
R_1, R_5, R_9 — 10 K Ω, 1/2W.
R_2, R_3, R_6, R_7 — 1 K Ω, 1/2W.

R_4, R_8 — 100 Ohms, 1/2W.
R_{10} — 100 Ohms, 1 W.
C_1, C_2, C_5 — 10 μ f Electrolytic Capacitor.
C_3, C_4 — 50 μ f Electrolytic Capacitor.
C_6 — .01 μ f.
SW_1 — Switch SPST.

Fig. 13-52. Schematic and parts list for RC coupling.

desirable that the values of these resistors be sufficiently high so that the signal will not be bypassed around the EB junction.

The values of R_B and R_F were determined by the required bias and stability of the circuit. Also, the current drain from the source must be considered.

Circuit design with transistors can become quite complex and it frequently is a matter of give and take. Compare the output and input impedances of the transistors in the CE configuration. The input can be in the range of 500 to 1.5 K Ω and the output impedance in the range of 30 K Ω to 50 K Ω. This is a severe mismatch. With the RC coupling, the mismatch must be tolerated with its accompanying loss of power

gain. However, when cost is a factor, it may be cheaper to add another transistor stage to offset the loss due to mismatch rather than purchase a transformer for interstage matching.

DIRECT COUPLING

In many industrial circuits used today, it is necessary to amplify either very low frequency signals or it is required to retain the dc value as well as the ac value of a signal. Amplifier circuits using RC or transformer coupling will block out the dc component. The directly coupled amplified in Fig. 13-53 is an answer.

In this circuit, the collector of Q_1 is connected directly to the base of Q_2. The collector load resistor R_c also acts as a bias resistor for

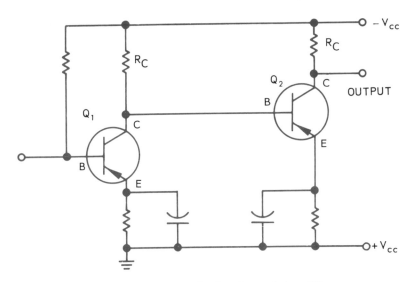

Fig. 13-53. The circuit of a directly coupled amplifier.

Q_2. Any change of bias current is amplified by the directly coupled circuit and it, therefore, is very sensitive to temperature changes. This disadvantage can be overcome with stabilizing circuits. Another disadvantage appears when one realizes that each stage requires a different bias voltage for proper operation.

Notice a complete transistor amplifier in Figs. 13-54 and 13-55.

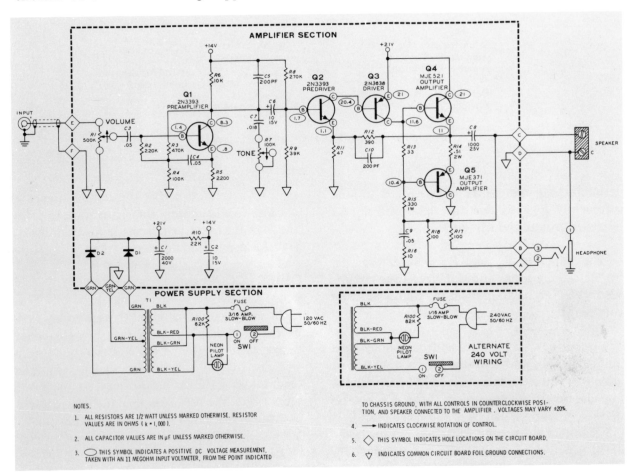

Fig. 13-54. Schematic of the Heathkit solid-state amplifier Model AA-18.

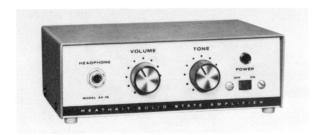

Fig. 13-55. Transistor amplifier. (Heath Co.)

THE DECIBEL

The output of audio amplifiers is converted into sound by a loudspeaker or a transducer. The unit that is used to determine relative sound level strengths by the human ear is the DECIBEL, db.

FREQUENCY DISTORTION

A properly designed amplifier should have sufficient gain for all frequencies for which it is intended. If some frequencies are amplified more, while others are attenuated, frequency distortion results. A common example of this type of distortion is the TONE CONTROL on your radio. Most people prefer the rich low tones in music, so provisions are made to remove high frequency tones for your listening pleasure.

HARMONIC DISTORTION

A harmonic frequency is the number of the harmonic times the fundamental frequency. For example, the second harmonic of 1000 Hz is 2000 Hz. The third harmonic of 5000 Hz is 15,000 Hz. Harmonic tones are desirable as a part of voice or music and identify the quality of the music. But harmonics generated within an amplifer (unrelated to music) are a form of distortion. If an amplifier is over-driven and its output wave is clipped, either by cutoff or saturation, the resulting wave is no longer a sine wave of a fundamental frequency. It is a composite wave representing the fundamental wave and several harmonics that were not present in the original sound. This is harmonic distortion.

PRECAUTIONS IN THE USE OF TRANSISTORS

1. Do not remove or replace transistors in circuits when the power is on. Surge currents may destroy the transistor.

2. Be very careful with close connections such as those found in miniature transistor circuits. A momentary short circuit can burn out a transistor.

3. When measuring resistances in a transistor circuit, bear in mind that your ohmmeter contains a battery and if improperly applied will burn out a transistor. It is also important to observe the correct polarity when measuring ohms for accurate readings.

4. Soldering the leads of transistors is a skill which you must develop. Heat will destroy the transistor. When soldering a transistor lead, grasp the lead between the transistor and the connection to be soldered with a pair of long nose pliers. These pliers will act as a heat sink and absorb the heat from soldering. Use a pencil type soldering iron of about 25 watts.

TRANSISTORIZED TELEPHONE AMPLIFIER

How many times have you wished that everyone in the room might hear the voice of friends or family over a long distance telephone call. The amplifier shown in Figs. 13-56 and 13-57, is a practical device and may be permanently attached to your telephone. A flick of the switch will activate the transistorized amplifier and you may then have a family telephone visit.

The telephone pickup T_4, Fig. 13-57, is connected to two transformer-coupled, common emitter stages which drive a push-pull power amplifier. Potentiometer R_7 is a gain control to adjust the volume of the amplifier. In the layout of the parts, keep transformers T_1 and T_2 as far from the input as possible to

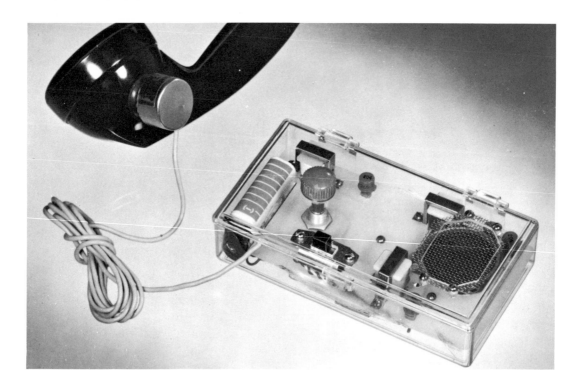

Fig. 13-56. Telephone amplifier.

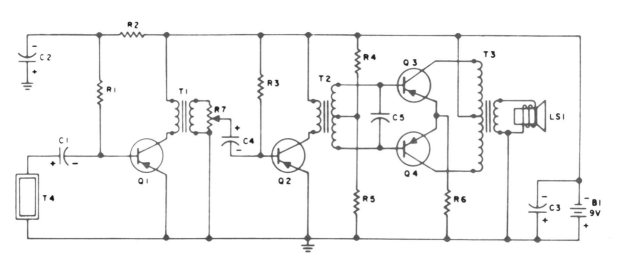

Fig. 13-57. Wiring diagram for telephone amplifier.   (Sylvania Electric Corp.)

PARTS LIST, TELEPHONE AMPLIFIER

B_1 – 9V battery
C_1, C_4 – 2 mfd, electrolytic, 15V
C_2, C_3 – 8 mfd, electrolytic, 15V
C_5 – 0.01 mfd, ceramic disc or mica, 100V
LS_1 – 3-6 ohm loudspeaker
Q_1, Q_2, Q_3, Q_4 – Sylvania 2N1265 Transistor (or RCA SK 3003)
R_1 – 470 K Ω, 1/2 watt
R_2 – 150 K Ω, 1/2 watt

R_3 – 270 K Ω, 1/2 watt
R_4 – 4.7 K Ω, 1/2 watt
R_5 – 68 Ω, 1/2 watt
R_6 – 12 Ω, 1/2 watt
R_7 – 100 K Ω, 1/2 watt potentiometer
T_1 – Argonne AR-104, Stancor TA-27 or equivalent
T_2 – Argonne AR-109, Stancor TA-35 or equivalent
T_3 – Argonne AR-119, Stancor TA-21 or equivalent
T_4 – Telephone pickup coil, Lafayette 99RG1970, Shield M-133, or equivalent

Fig. 13-58. Paging amplifier.

prevent possible feedback. Keep leads short. C_2, C_3 and C_5 are bypasses to prevent oscillation. If the circuit should oscillate, reverse the connections to the secondary of T_1.

In Fig. 13-56, the parts have been mounted on a small sheet of 1/8 inch plastic. The amplifier may be placed in a case of your own design.

A PAGING AMPLIFIER

Many times the electronics experimenter requires a small audio amplifier to connect to a microphone. We call such a device, Fig. 13-58, a paging or announcing amplifier, because it works well with a lapel microphone. It will boost your voice so that it may be heard in crowds for some distance. The circuit,

Fig. 13-59. Wiring diagram for paging amplifier. (Sylvania Electric Corp.)

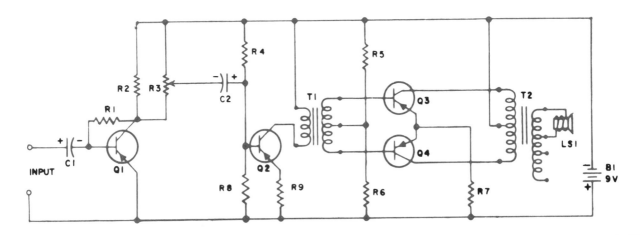

Fig. 13-59, has sufficient power to drive the small speaker. A gain control is provided by R_3.

In the paging amplifier three grounded emitter stages are connected in cascade. The last stage is a push-pull power amplifier. In the physical layout of parts, it is wise to keep R_3 some distance from transformer T_2 to prevent feedback.

These parts have been mounted on an 1/8 in. plastic sheet. Construct a new circuit in breadboard style and test it before final assembly and soldering. Then, try designing a modern case for it. This project represents several hours of recreation and education. Enjoy using it.

PARTS LIST FOR PAGING AMPLIFIER

B_1 – 9V battery
C_1, C_2 – 10 mfd electrolytic, 15V
LS_1 – any 4, 8, or 16 ohm loudspeaker
Q_1 – Sylvania 2N1265 transistor (or RCA SK 3003)
Q_2, Q_3, Q_4 – Sylvania 2N1266 transistor (or RCA SK 3005)
R_1 – 68 K Ω , 1/2 watt
R_4 – 120 K Ω , 1/2 watt
R_2, R_6, R_8 – 560 Ω , 1/2 watt
R_3 – 5 K Ω , 1/2 watt potentiometer
R_5 – 6.8 K Ω , 1/2 watt
R_7 – 100 Ω , 1/2 watt
R_9 – 10 K Ω , 1/2 watt
T_1 – Stancor TA-5 or equivalent
T_2 – Stancor TA-10 or equivalent

FOR DISCUSSION

1. Why does a tube using the grid leak depend upon an input signal for its bias?
2. What are some of the characteristics of a good high fidelity sound system?
3. What is the purpose of feedback in an amplifier?
4. Discuss the relative merits of RC coupling vs. transformer coupling.
5. Discuss the several kinds of distortion possible in an amplifier.
6. Why should transistors be temperature controlled by heat sinks and ventilation?
7. Discuss precautions in the use of and servicing of transistor circuits.

TEST YOUR KNOWLEDGE

1. State three methods used to bias a tube.
2. Name four types of coupling used to cascade amplifiers.
3. A tube has MU of 20 and r_p of 7000 Ω . What is the voltage gain when R_L = 5000 Ω ; when R_L = 10,000 Ω ; and when R_L = 50,000 Ω .
4. A tube requires a negative 4 volt bias. I_p =

5 mA. What value cathode resistor should be used?
5. What value bypass capacitor should be used in question 4 for 100 Hz?
6. Identify a Class A amplifier.
7. Name four characteristics of a Class A amplifier.
8. Identify a Class B amplifier.
9. Name four characteristics of a Class B amplifier.
10. Identify a Class C amplifier.
11. Name three characteristics of a Class C amplifier.
12. Draw circuit diagrams of the following.
    a. Two stages – RC coupled.
    b. Two stages – transformer coupled.
    c. Two stages – impedance coupled.
    d. A push-pull amplifier using transformers.
13. Draw a block diagram of an NPN transistor, with bias batteries connected with proper polarity.
14. Draw the common emitter amplifier circuit.
15. How is the gain in the common emitter stage computed?
16. Name four advantages of transistors over vacuum tubes.

Chapter 14

ELECTRONIC OSCILLATORS

The pendulum on a grandfather's clock which swings to and fro, marking the seconds of time as they become moments in history, should stimulate the inquiring mind of the young scientist. Why does the pendulum swing? The question is answered as the keeper of the clock restores its energy by winding the main spring with the familiar key. Has it kept a correct record of the passing time? The keeper makes adjustments on the length of the pendulum so the period of time required for one complete swing is synchronized with the fleeting seconds. The swinging pendulum may be described as OSCILLATING.

A brief review of Chapter 5 will rediscover how an alternating current or voltage is generated. In the study of electricity, we can consider an oscillating current as a current which is flowing back and forth, or varying first in one direction and then the other, from a common reference point or mean. A comparison of the clock pendulum and an oscillating voltage is graphically illustrated in Fig. 14-1. The amplitude of the voltage may be followed as it starts from its reference line and rises to maximum potential in one direction and falls to zero, then rises to maximum potential in the opposite direction and returns to zero. One CYCLE or chain of events has been completed. A continuation of the oscillation would be more of the same, a repetition of these cycles. The period of time which elapses during this cycle is called the PERIOD of cycle, and the number of these cycles occurring per second can be measured and described as the FREQUENCY in hertz. In our studies in Chapter 5

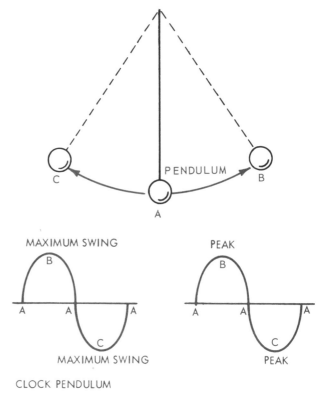

Fig. 14-1. A comparison between a swinging clock pendulum, and an oscillating current.

we discovered that the electricity in our homes and factories is an alternating or oscillating current at a frequency of sixty hertz. This alternating current is generated by dynamos driven by steam, water or atomic power.

In the study of electronics, voltages and currents of much higher frequencies are used and are generated by the VACUUM TUBE or the SEMICONDUCTOR (TRANSISTOR) USED AS AN OSCILLATOR. Actually these devices do not oscillate, but merely acts as a valve, which feeds energy to a tuned circuit to sustain oscillation. The oscillatory action in the tuned circuit was explained in Chapter 8. Now is a good time to turn back and review these pages. The basic block diagram of an oscillator is in Fig. 14-2.

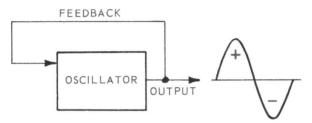

Fig. 14-2. A basic block diagram of a sine wave oscillator.

Two conditions must exist to sustain oscillations in the tuned circuit:
1. The energy fed back to the tuned circuit must be in phase with the first excitation voltage. The oscillator depends upon REGENERATIVE or POSITIVE feedback.
2. The amplitude of the feedback voltage must be sufficient so that the energy dissipated by the resistance of the circuit is replaced. Perpetual motion has not yet been discovered.

THE ARMSTRONG OSCILLATOR

The simple circuit of the vacuum tube Armstrong oscillator is diagramed in Fig. 14-3. From this circuit the basic theory of oscillators

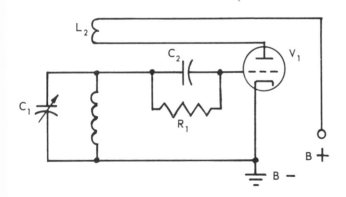

Fig. 14-3. Schematic of Armstrong oscillator.

may be understood. Notice the tuned-tank circuit $L_1 C_1$ which determines the frequency of the oscillator. Follow the sequence of events in the theory of this circuit.

Step 1. When the voltage is applied to circuit, current flows from B–, through tube, through tickler coil L_2 to B+. L_2 is closely coupled to L_1 and the expanding magnetic field of L_2 makes the grid end of L_1 positive and the grid of V_1 positive. C_1 charges to the polarity shown. The grid of V_1 also collects electrons and charges C_2 in the indicated polarity. See Fig. 14-4.

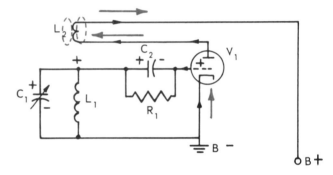

Fig. 14-4. Step 1. Armstrong oscillator.

Step 2. When V_1 reaches its saturation point, there is no longer a change of current in L_2 and the magnetic coupling to L_1 drops to zero. The negative charge on the grid side of C_2, no longer opposed by the induced voltage in L_1 drives the tube to cutoff. This rapid reduction of current through the tube and L_2 causes the grid end of L_1 to

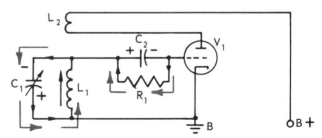

Fig. 14-5. Step 2. Armstrong oscillator.

become negative which INCREASES the negative bias on V_1. C_1 also starts its discharge through L_1 as the first half-cycle of oscillation. C_2 bleeds off its charge through R_1. See Fig. 14-5.

Step 3. The tube V_1 is held at cutoff until the charge on C_2 is bled off to above cutoff at which time the tube starts conduction and the cycle is repeated.

Points to remember in this theory of the ARMSTRONG oscillator include:

1. The voltage developed across L_1 first opposes and then adds to the bias developed by the RC_2 combination.

2. The energy added to the tuned-tank circuit L_1C_1 by the tickler coil L_2 is sufficient to offset the energy lost in the circuit due to resistance. The coupling between L_1 and L_2 is adjustable.

3. The combination RC_2 has a relatively long time constant and establishes the operating bias for the tube. V_1 is operated Class C.

An examination of the voltage wave form in Fig. 14-6 on the grid of V_1 will give further clarification. The shaded portion represents tube conduction. At point B the bias is very

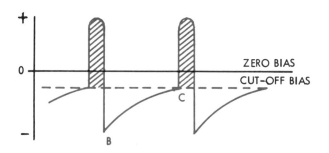

Fig. 14-6. Voltage wave form on grid of oscillator. Compare it to Steps 1, 2, and 3 previously discussed.

negative as the result of the charge on C_2 plus the induced voltage across L_1. Interval B to C represents the time that C_2 is discharging

through R until the cut-off point is reached and tube begins conduction for the next cycle.

THE HARTLEY OSCILLATOR

A very common oscillator used in radio receivers and transmitters is the HARTLEY OSCILLATOR. It has better stability than the Armstrong and the theory of operation is similar.

Since semiconductors are used more today in electronic circuits, the Hartley oscillator and the other oscillator circuits explain in this chapter will use solid state devices. One method to tell a Hartley oscillator from any other oscillator is by the tapped coil, L_1 and L_2 in Fig. 14-6.

The component parts in Fig. 14-7 are labeled to correspond to similar parts used in the description of the Armstrong oscillator in Figs. 14-4 and 14-5. Notice that the L_1 section of the coil is in series with the emitter-collector circuit and carries the total collector current.

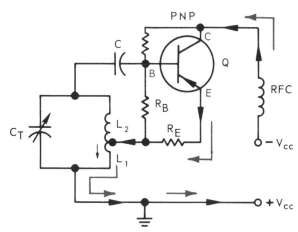

Fig. 14-7. A transistorized Hartley oscillator circuit.

The current I_E which includes I_C is indicated by the arrows. As the circuit is turned on, this current flowing through L_1 induces a voltage at the top of L_2 which makes the base of Q more negative which drives the transistor to saturation. At saturation, there is no longer a change of current and the coupling between L_1 and L_2 falls to zero. The less negative voltage at the base of Q causes the transistor to decrease in conduction. This decrease induces a positive voltage at the top end of L_2, which is reverse bias for the transistor and it is quickly driven to cutoff. The cycle is then repeated. The tank circuit is energized by pulses of current. The transistor alternates between saturation and cutoff at the same frequency.

The "at rest" bias condition of the transistor is established by R_B and R_E. The radio frequency choke, RFC, blocks the rf signal from the power source. In this circuit, note that the coil L_1 is in SERIES with the collector circuit of the transistor. It is named a SERIES-FED oscillator. In Fig. 14-8, a SHUNT-FED oscillator is shown. The operation is the same. Note that the dc path for the emitter-collector current is not through the coil L_1. The ac signal path, however, is through C and L_1. At point A, the two current components are separated and required to take parallel paths. Both oscillators receive their feedback energy by means of magnetic coupling.

THE COLPITTS OSCILLATOR

Feedback to sustain oscillation may also be accomplished by means of an electrostatic field as developed in a capacitor. If the tapped coil from the Hartley oscillator is replaced with a

split stator capacitor, a voltage of proper polarity may be fedback which causes the circuit to oscillate. This circuit has been named the COLPITTS OSCILLATOR and is illustrated in Fig. 14-9.

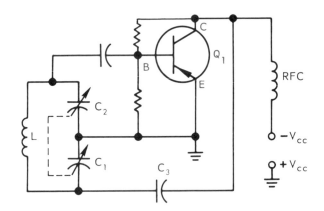

Fig. 14-9. Schematic diagram of a Colpitts oscillator.

The theory of operation is similar to the Hartley oscillator, except that the signal is coupled back to C_1 of the tank circuit through coupling capacitor C_3. A changing voltage at the collector appears as a voltage across the tank circuit LC_1C_2 in the proper phase to be a regenerative signal. The amount of feedback will depend upon the ratio of C_1 to C_2. This ratio is usually fixed and both capacitors C_1 and C_2 are controlled by a single shaft (ganged capacitor). The natural frequency of the oscillator is determined in the usual manner. The tuned tank consists of L and C_1 and C_2 in series. Notice that the circuit is shunt fed. Series feed is impossible due to the blocking of dc by the capacitors.

CRYSTAL OSCILLATORS

A circuit which exhibits a very high frequency stability is the crystal-controlled oscillator. Radio communications, broadcasting stations and other equipment demanding a fixed frequency with little drift or variation depend upon the crystal as a frequency stabilizer. In our earlier studies of the Sources of Electricity, the phenomena of producing an emf by mechanical pressure and/or distortion of

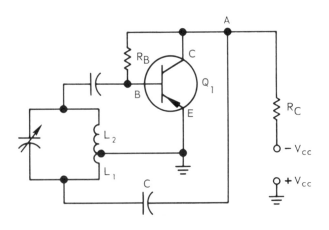

Fig. 14-8. A shunt-fed Hartley oscillator.

certain crystalline substances were discussed. The opposite is also true. A voltage applied to the surfaces of a crystal will produce distortion.

These effects are called PIEZOELECTRIC EFFECTS. When an electrical pressure is applied to a crystal, it will respond or oscillate at a frequency depending upon the size, thickness and kind of crystal used. Crystals are made from quartz, tourmaline, or Rochelle salts.

A crystal is used in Amateur Radio to control the frequency of the transmitter. These crystals may be purchased for fixed frequencies used in the amateur bands. At a commercial broadcasting station, crystals used to control the transmitter frequency are placed in <u>crystal ovens</u> which are very accurately maintained at an even temperature.

Before applying a crystal to stablize an oscillator, it is well to look at the ELECTRICAL EQUIVALENT of the crystal as in Fig. 14-10. A crystal is placed between two metallic holders. This forms a capacitor C_H

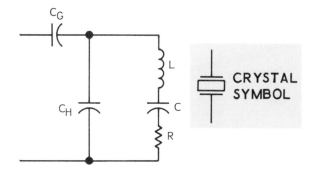

Fig. 14-10. The equivalent electrical circuit of a crystal and its schematic symbol.

with the crystal itself as the dielectric. C_G represents the series capacitance between the metal holding plates and the air gap between them as a dielectric. L, C and R represent the characteristics of the crystal. Of importance to our studies is the similarity of the equivalent crystal circuit to a tuned circuit. It will have a resonant frequency.

A transistorized crystal oscillator circuit is drawn in Fig. 14-11. Compare this circuit to

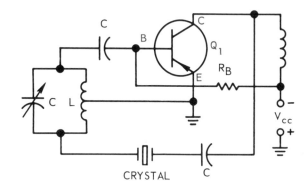

Fig. 14-11. A crystal controlled Hartley oscillator circuit.

Fig. 14-8. It is the same circuit with the crystal added to the FEEDBACK CIRCUIT. The crystal acts as a series resonance circuit and determines the frequency of the feedback currents. The tank circuit must be tuned to this frequency.

In the circuit of Fig. 14-12, the crystal is used in place of the tuned circuit in a Colpitts oscillator. This version has been named the PIERCE OSCILLATOR. Compare the circuit to Fig. 14-9.

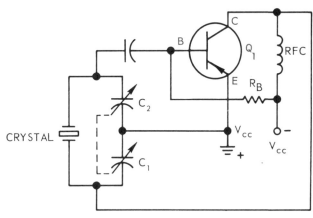

Fig. 14-12. Circuit of the crystal controlled Pierce oscillator.

The amount of feedback to energize the crystal again depends upon the ratio of C_1 to C_2. These capacitors form a voltage divider across the base-emitter of the transistor. The Pierce circuit is a very stable circuit under varying circuit conditions and changes.

POWER OSCILLATORS AND CONVERTERS

Sometimes it is desirable to convert dc to ac, and also, to convert ac to dc, transform it to a higher voltage and rectify it back to dc. A power oscillator may be used in this application. In Fig. 14-13 is a circuit of a converter.

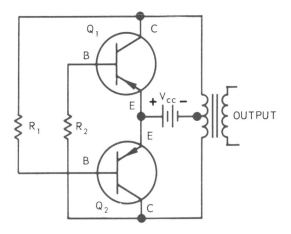

Fig. 14-13. The circuit of a power oscillator.

Actually, it is a push-pull oscillator. The collector load of each transistor is the primary of the transformer. The ac output will be found at the secondary. A suitable turns ratio may be used if higher or lower voltages are desired.

Only a slight imbalance in conductivity between Q_1 and Q_2 is required to start oscillation. This imbalance is always present due to components or temperature. The two transistors are either going toward saturation and cutoff respectively, or vice versa. Follow the action.

Assume Q_1 starts conducting. The voltage at C of Q_1 goes less negative, which makes the base of Q_2 less negative and drives Q_2 toward cutoff. A more negative voltage at C of Q_2 drives Q_1 base more negative and Q_1 reaches saturation. At saturation, when there is no change in current, the reactance of the transformer primary drops to zero with a corresponding decrease in collector voltage toward the value of V_{cc}. This more negative voltage

coupled to the base of Q_2 through R_1 starts Q_2 toward conduction and saturation. The transistors conduct alternately and the output is combined into a complete cycle at the transformer secondary output.

If the output from the transformer in Fig. 14-13 is connected to a rectifier and filter circuits, the output can again become dc.

CODE PRACTICE OSCILLATOR

For "fun and learning," the oscillator shown in Figs. 14-14 and 14-15, produces a variable tone which may be keyed for practicing the Morse Code. It will be useful to young people interested in radio communications, ham radio

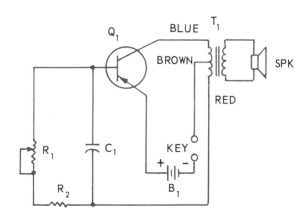

Fig. 14-14. Finished audio code oscillator in a shadow cabinet.

Fig. 14-15. Schematic diagram for code practice oscillator.

and Scouting. Simple musical tunes may be played with only a little practice by varying the tone control knob.

Once again the "shadow cabinet" is used in this project and the finished product looks like a commercial unit. The cabinet is made of 20 ga. galvanized iron and covered with imitation leather. Glue the leatherette to the metal. See Fig. 14-14.

PARTS LIST FOR CODE PRACTICE OSCILLATOR

C_1 – Capacitor .5 μ f
R_1 – Potentiometer 25K Ω
R_2 – 6800 Ω , 1/2W
T_1 – Transistor Output Transformer AR 119 or Thordarson TR 27
Q_1 – Transistor – CK-722 or RCA SK 3004
PM – Speaker 1 1/2" Philmore
S_1 – Switch, SPST
Binding Posts or phono jack, knob and small hardware

THE SQUAWKER

An individualized horn is something that has a lot of appeal to the youth with his or her own car. This transistorized horn was first designed

as a "fog horn" but has an equally startling effect on the customized car. See Fig. 14-16.

The circuit uses power transistors in a multivibrator configuration, Fig. 14-17. The amplifier stages are also used to generate the audio frequency oscillation by feeding the output of one transistor to the base of the other

Fig. 14-16. Completed squawker.

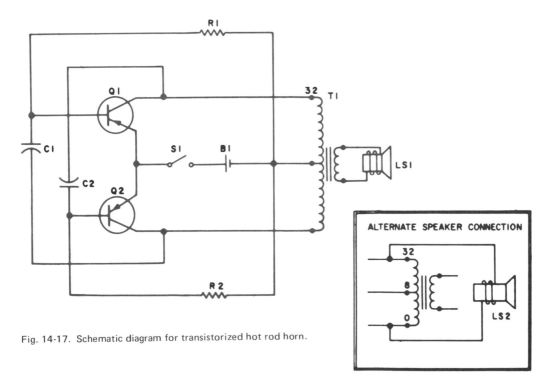

Fig. 14-17. Schematic diagram for transistorized hot rod horn.

transistor. Feedback is accomplished by C_1 and C_2. S_1 may be replaced by wires to the horn button. B_1 is the 12 volt battery in the car.

Any metal or wooden box may be used to house the horn. It is advisable to use a weatherproofed speaker if installed on a car or protect it against the weather by a plastic covering. No difficulty should be experienced in building the project, if the circuit diagram is carefully followed.

PARTS LIST FOR THE SQUAWKER

B_1 – 12 volt battery (1 ampere drain) conventional auto or boat battery, or 2 Burgess F4P1 batteries wired in series.

C_1, C_2 – 2.0 mfd, electrolytic, 25 volt

LS_1 – 8 ohm tweeter, Lafayette HK-3 or equivalent

Q_1, Q_2 – Sylvania 2N307 power transistor or RCA SK 3009

R_1, R_2 – 220 Ω , 1W

S_1 – Single-pole, single-throw pushbutton switch, normally open

T_1 – Lafayette TR-94 or equivalent

LS_2 – For alternate connection - 45 ohm paging trumpet; University CMIL-45, University MIL-45, or equivalent

ELECTROLUMINESCENT NIGHT LIGHT

A useful project, inexpensive and fun to own is shown in Figs. 14-18 and 14-19.

The electroluminescent light you will need in making this project may be purchased at your local hardware store. They are widely used as night lights and produce a warm and friendly greenish glow when plugged into the ordinary 115-volt ac outlet. This project has its own power supply, which is the conventional Hartley oscillator circuit you have studied. The output of the oscillator is transformer coupled

Fig. 14-18. The electroluminescent night light in a shadow cabinet.

to the light and is sufficient to operate the device without other sources of power.

The "shadow" cabinet used to house this light is a popular design. This one is made of 20 ga. galvanized iron on which an imitation leather is glued. This method of construction is easy and produces a "commercial look" to a project. The same methods may be used to build cabinets for radios, amplifiers and other projects. Try it! You will be rewarded by a project with a pleasing appearance.

PARTS LIST FOR NIGHT LIGHT

T_1 – Filament Transformer 6.3V CT, Triad F16X

S_1 – Switch SPST Toggle

C_1 – Capacitor 50 μ f-25V electrolytic

C_2 – Capacitor 8 μ f-15V electrolytic

R_1 – Resistor 2700 Ω , 1/2W

Q_1 – Transistor 2N241A Sylvania or RCA SK3004

B_1 – Battery 9V Eveready 264 or equivalent Sylvania Panelescent Light 120V–.02W Receptacle–Amphenol 61F1

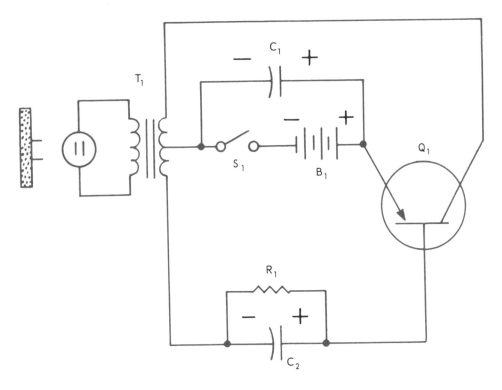

Fig. 14-19. The schematic diagram for electroluminescent night light.

FOR DISCUSSION

1. How does an electronic oscillator relate to a clock pendulum?
2. Can a piezoelectric crystal be used to produce electricity if pressure is applied?
3. Explain the operation of a Hartley oscillator.
4. What is the difference in a series-fed oscillator and a shunt-fed oscillator?
5. Why are crystal oscillators used today in many commercial transmitters?
6. How can a Colpitts oscillator be identified?

TEST YOUR KNOWLEDGE

1. Define cycle, period and frequency.
2. What two conditions must be satisfied to produce continuing oscillations?
3. What method of feedback is used in the Hartley oscillator?
4. What method of feedback is used in the Colpitts oscillator?
5. Draw a dc — ac converter with full-wave rectifier and π filter.
6. Draw the equivalent electrical circuit of a crystal and its schematic symbol.

Chapter 15

RADIO TRANSMITTERS

The human race has communicated for thousands of years. However, communications by electrical and electronic means has been possible only in the past century. Not only can one person communicate with other people by electronic means but he or she can also communicate with machines such as computers.

A basic communications model is shown in Fig. 15-1.

Anything that breaks or interrupts communications process is called NOISE. The most common place where noise enters the communications process is in the message channel. However, it can enter at any place. Good examples of noise in electronic communications is lightning and distortion.

A simpler communications model is shown in Fig. 15-2. The information is processed by

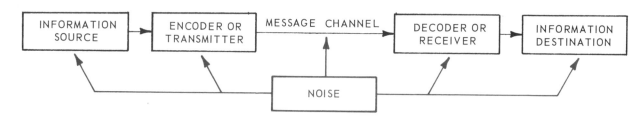

Fig. 15-1. A basic communications model.

Any communications process must have an INFORMATION SOURCE. This is the origin of the information to be communicated. Next, the information must be encoded or converted to an electrical signal. This second stage is often referred to as the TRANSMITTER. From the encoder or transmitter the communications message is fed through some CHANNEL. This can be electromagnetic waves, wires, air, or some other medium. After the message is fed through the particular channel, it is to be decoded. This stage picks up the transmitted message and converts it to some intelligent form in sound or picture (video). Many times the decoder stage is called the RECEIVER. Lastly the information is fed to the DESTINATION which could be a person or a machine.

the circuitry in the transmitter by the antenna. It is possible that the channel may be a wire. The receiver picks up, tunes, amplifiers, and reproduces the information.

Chapter 15 will discuss radio transmitters and in Chapter 16 you will learn about receivers.

FREQUENCY SPECTRUM

In our previous studies many references have been made to waves of certain frequencies. Mention has been made of power frequencies, audio and radio frequencies. It might be wise at this point to look over the frequency spectrum shown in Fig. 15-3 and satisfy our scientific

Radio Transmitters

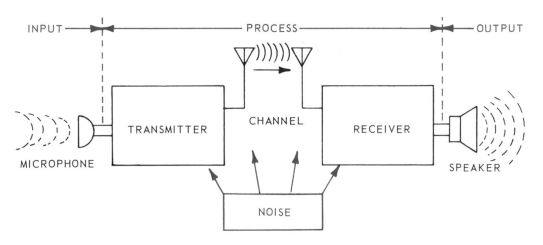

Fig. 15-2. Transmitter/receiver one-way communications model.

curiosity. This spectrum, although not in detail, will indicate the <u>bands</u> of frequencies used for communication by the many services including radio broadcasting, amateur radio, FM, TV, police and industry.

It must be remembered that the use of space in communications must be rigidly controlled, not only by our own Federal Communications Commission, but also by conferences involving other countries. There are only so many frequency allocations available. These must be assigned to satisfy the needs of all of the people for industrial, military and entertainment purposes and at the same time minimize any interference between the services.

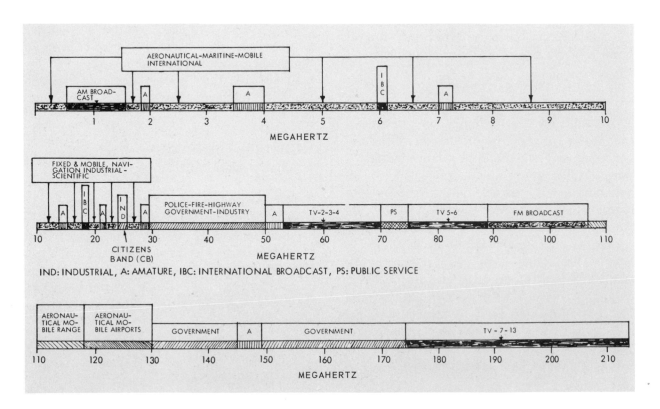

Fig. 15-3. An abbreviated frequency spectrum chart showing locations of several communications services.

Another general classification of waves is according to frequency. This will prove useful to you when discussing ranges of frequencies. See Table 15-1.

FREQUENCY	BAND
20-30,000 Hz	Audio Frequency
Below 30 KHz	VLF Very Low Frequency
30-300 KHz	LF Low Frequency
300-3000 KHz	MF Medium Frequency
3000-30,000 KHz	HF High Frequency
30,000 KHz–300 MHz	VHF Very High Frequency
300-3000 MHz	UHF Ultra High Frequency

Table 15-1

In this table of frequencies you will recognize the location of several familiar services. Notice that the AM broadcast stations use the HF range, whereas television may be found in the VHF and UHF ranges. A hi-fi operates in the audio frequency range. Electricity for the home is 60 Hz, or at the power frequency.

THE RADIO TRANSMITTER

We will now discuss the building of a radio transmitter. Any of the oscillators which have been discussed will produce a radio frequency wave, which if properly connected to an antenna system, would radiate energy into space. However, more powerful signals are desired, so some stages of amplification are needed to increase the amplitude of the oscillator wave so that it will drive a final power amplifier. The block diagram of a simple continuous wave (CW) transmitter is drawn in Fig. 15-4. The first block is the conventional crystal oscillator and then the final power amplifier. A power supply is provided for the oscillator and the final power amplifier.

The oscillator generates a sine wave alternating current at a desired frequency. This

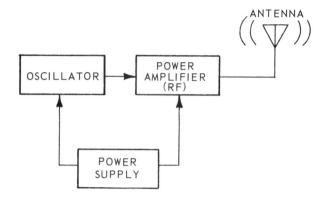

Fig. 15-4. Block diagram representing various stages of basic continuous wave (CW) radio transmitter.

signal is called the carrier wave. The carrier wave is then amplified by the rf power amplifier to the desired output wattage level. A power supply is required to provide the proper voltages and current needed to operate the oscillator and the rf power amplifier. The output is next fed to an antenna where the energy is radiated into space in the form of electromagnetic waves.

A continuious wave (CW) transmitter simply radiates energy into space without any intelligent message in the signal. The receiver picks up the signal and knows that there has been energy radiated from the CW transmitter antenna. So the basic intelligence with this type of transmitter is:

1. The transmitter is ON.
2. The transmitter is OFF.

One method to refine the uninterrupted CW transmitter is to switch it one and off according to some code (such as the Morse Code). This switching or keying process using a code can give the transmitted signal intelligence. Specific information can be sent from the transmitter to the receiver. A transmitter of this type is shown in Fig. 15-5.

The basic switched or keyed continuous wave transmitter can be improved by a buffer amplifier placed in between the oscillator and the rf amplifier. The buffer amplifier isolates

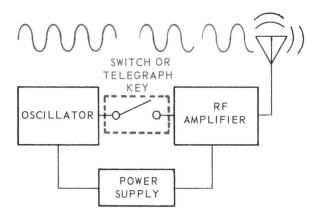

Fig. 15-5. Continuous wave transmitter with telegraph key.

the oscillator from the rf amplifier and keeps it from shifting off frequency. It also provides some amplification to the carrier wave.

Many CW transmitters use frequency multipliers to multiply the frequency produced by the basic oscillator. Frequency multipliers multiply the carrier wave by two (doubler) or three (tripler). These circuits operate from the principle of harmonics in the fundamental carrier frequency generated by the oscillator. The fundamental frequency is the basic one produced by the oscillator. Harmonic frequencies are multiples of the fundamental frequency.

MICROPHONES

The first problem in our study is the conversion of sound waves to electrical waves. The nature of a sound wave must be explored. Physicists describe a sound wave as condensations and rarefactions of the air. Your vocal cords set up vibrations in the air and these waves radiate out to all persons within hearing range. A device, called a MICROPHONE, is used to convert these sound waves to electrical audio waves of the same frequency and relative amplitude. In Fig. 15-6 three types of microphones are illustrated to familiarize the student with the interior of these devices.

A relatively simply and economical microphone takes advantage of the resistance of carbon granules. The theory may be followed by the sketch in Fig. 15-7. Granules of carbon are packed in a small container and electrical connections are made to each side. A transformer and a small battery are connected in series with the carbon. To one side of the container, called a "button," a diaphragm is attached. Sound waves striking the diaphragm cause varying degress of compression to the carbon granules which vary the resistance of the carbon. The varying resistance causes a varying

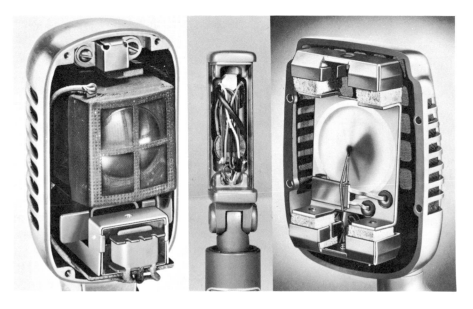

Fig. 15-6. Microphones used in broadcasting and general communications. (Shure Bros., Inc.)

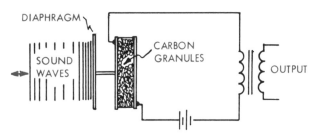

Fig. 15-7. In carbon microphone sound waves change resistance of circuit.

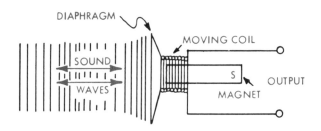

Fig. 15-9. Dynamic microphone. Electrical audio waves are produced by coil moving in magnetic field.

current to flow through the carbon button and the transformer primary. The output is a current which varies at the same frequency as the sound waves acting on the diaphragm. The CARBON MICROPHONE is a very sensitive device and has a frequency response up to about 4000 Hz. This is quite satisfactory for voice communication, but would reproduce music rather poorly. It does provide a good response for its intended frequencies and it is nondirectional. This means it will pick up sound from all directions.

A second type of microphone takes advantage of the piezoelectric effect of certain crystals. When sound waves strike a diaphragm, mechanical pressure is transferred to the crystal. The flexing or bending of the crystal produces a small voltage between its surfaces. This voltage is of the same frequency and relative amplitude as the sound wave. CRYSTAL MICROPHONES have a frequency response up to 10,000 Hz. They are rather sensitive to shock and vibration and should be handled with care. This "mike" is sketched in Fig. 15-8.

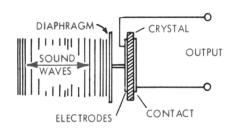

Fig. 15-8. Mechanical pressure produces electrical energy. The crystal microphone takes advantage of piezoelectric effect.

The DYNAMIC MICROPHONE or moving coil microphone is sketched in Fig. 15-9. As sound waves strike the diaphragm, they cause the voice coil to move in and out. The voice coil is surrounded by a fixed magnetic field. When the coil moves, a voltage is induced in the coil, (Faraday's discovery). This induced voltage will cause a current to flow at a frequency and amplitude similar to the sound wave causing the motion. It has a frequency response up to 9000 Hz. It is directional. It requires no external voltage for operation.

A very high quality microphone, called the VELOCITY MICROPHONE, is constructed with a corrugated ribbon of metal suspended in a magnetic field. Sound waves striking the ribbon directly cause the ribbon to vibrate. As the ribbon cuts the magnetic field a voltage is induced. Suitable connections at the ends of the ribbon bring the voltage out to terminals. This voltage varies according to the frequency and amplitude of the impinging sound waves. It is a relatively delicate "mike" and has a response above 12,000 Hz. When speaking into this microphone, one should either speak across its face or stand about 18 inches away or a "booming" effect is created.

MODULATION

When you turn on the radio or TV, you expect to hear music and the voices of your favorite program. The "dits and dahs" of the CW transmitter would have no meaning to the inexperienced operator. The process of combining or superimposing an audio wave on a radio frequency wave is called MODULATION. Sound waves must first be converted into electrical waves, amplified, and combined with the radio wave.

In one method, used in commercial broadcasting, the amplitude of the radio wave is made to vary at an audio frequency rate. This is amplitude modulation or AM. In a second method, the frequency of the radio wave is made to vary at an audio frequency rate called frequency modulation or FM. Both of these methods are illustrated in Fig. 15-10, showing first the rf and af waves separately and then the modulated waves.

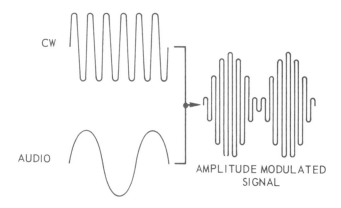

Fig. 15-11. A CW wave, an audio wave, and the resultant amplitude modulated wave.

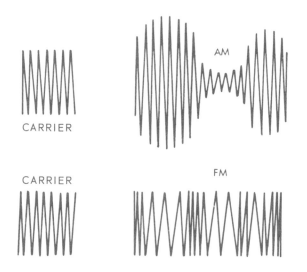

Fig. 15-10. Wave forms show amplitude modulation (AM) and frequency modulation (FM).

AMPLITUDE MODULATION (AM)

Most of us enjoy music and speech from our radios. We cannot understand the Morse Code. Intelligence superimposed upon a radio frequency wave is called MODULATION. Assume a radio transmitter is operating on a frequency of 1000 KHz and a musical tone of 1000 Hz is to be used for modulation. Referring to Fig. 15-11, the CW and the audio signal are illustrated. By a modulation circuit, the AMPLITUDE OF THE CARRIER WAVE IS MADE TO VARY AT THE AUDIO SIGNAL RATE.

Let us look at this modulation process another way. The mixing of a 1000 Hz wave with a 1000 KHz wave produces a sum and difference wave, which are ALSO IN THE RADIO FREQUENCY RANGE. These two waves will by 1001 KHz and 999 KHz. These

are known as SIDEBAND FREQUENCIES; the UPPER SIDEBAND and the LOWER SIDEBAND respectively. In Fig. 15-12, this modulation is shown as described.

The algebraic sum of the carrier wave and its sidebands results in the amplitude modulation wave. You should notice that the audio tone intelligence is present in both sidebands, as either sideband is the result of modulating a 1000 KHz signal with a 1000 Hz tone.

In Fig. 15-13, the waves are represented by their location on a frequency base. If a 2000 Hz tone was used for modulation, then sidebands would appear at 998 KHz and 1002 KHz. In order to transmit, using AM, a 5000 Hz tone of a piccolo or a violin, sidebands at 995 KHz and 1005 KHz would be required. This would represent a frequency band width of 10 KHz necessary to transmit a 5000 Hz musical tone. There just is not that much space in the spectrum to permit all broadcasters to transmit, and, if limited to musical tones of 5000 Hz, the output received and heard from your radio speaker would be VERY LOW FI. Indeed, it would be unpleasant to listen to. The BROADCAST BAND for AM radio extends between 535 KHz and 1605 KHz and is divided into 106 channels 10 KHz wide. A station is licensed to operate at a frequency in one of these channels. The channels are carefully allotted to stations at sufficient distances from each other to prevent interference. In order to improve the fidelity

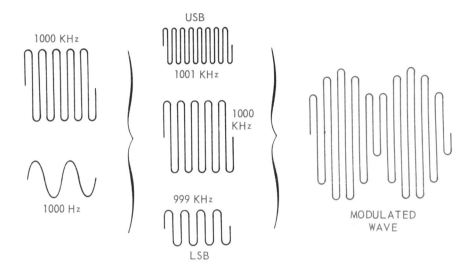

Fig. 15-12. Wave showing formation of sidebands and modulation envelope.

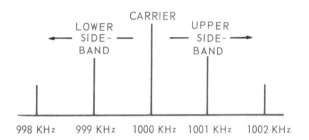

Fig. 15-13. Carrier and sideband locations for modulation tone of 1 KHz and 2 KHz.

and quality of music under these limitations of 10 KHz band occupancy, a filter called a VESTIGIAL SIDEBAND FILTER is used to remove a large portion of one sideband. You will remember that both sidebands contain the same information. By this means, frequencies higher than 5 KHz can be used for modulation. The fidelity is improved.

MODULATION PATTERNS

A radio transmitter is not permitted by law to exceed 100 percent modulation. This means that the modulation signal can not cause the carrier signal to vary over 100 percent of its unmodulated value. Examine the patterns in Fig. 15-14. Notice the amplitude of the modulated waves.

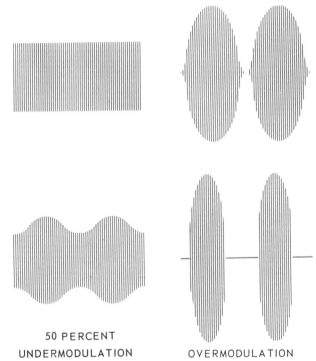

Fig. 15-14. Patterns of modulation for 0, 100 and 50 percent and overmodulation.

The 100 percent modulation wave variation is from zero to two times the peak value of the carrier wave. Overmodulation is caused when modulation increases the carrier wave to over two times its peak value and at negative peaks the waves cancel each other and leave a straight line of zero value. Overmodulation causes DISTORTION and INTERFERENCE called SPLATTER.

Percent of modulation may be computed by the formula:

$$\% \text{ MODULATION} = \frac{e_{max} - e_{min}}{2_{e_c}} \times 100$$

where,

e_{max} is the maximum amplitude of modulated wave.

e_{min} is the minimum amplitude of modulated wave.

e_c is the amplitude of unmodulated wave.

PROBLEM: A carrier wave has the peak value of 500 volts and a modulating signal causes amplitude variation from 250 volts to 750 volts. What is the percent of modulation?

$$\text{Solution } \% \text{ Mod.} = \frac{750 - 250}{2 \times 500} \times 100$$

$$= \frac{500}{1000} \times 100 = 50\%$$

SIDEBAND POWER

The dc input power to the final amplifier of a transmitter is the product of voltage and current. Let us see what power is required by a modulator. This formula may be used:

$$P_{audio} = \frac{m^2 P_{dc}}{2}$$

where,

P_{audio} is the power of the modulator.

m is the percentage of modulation expressed as decimal.

P_{dc} is the input power to the final amplifier.

PROBLEM: What modulation power is required to modulate a transmitter with a dc power input of 500 watts to 100 percent.

$$P_{audio} = \frac{(1)^2 \; 500 \text{ watts}}{2} = 250 \text{ watts}$$

This represents a total input power of 750 watts. Notice what happens under conditions of 50 percent modulation.

$$P_{audio} = \frac{(.5)^2 \; 500 \text{ watts}}{2} = 62.5 \text{ watts}$$

And the total input power is only 562.5 watts.

Where the modulation percentage is reduced to 50 percent, the power is reduced to 25 percent. This is a severe drop in power which decreases the broadcasting range of the transmitter. It is just good operation to maintain a transmitter as close to 100 percent modulation without exceeding 100 percent. You may wish to know why the term INPUT POWER has been used. That is because any final amplifier is far from 100 percent efficient.

$$\% \text{ EFF} = \frac{P_{out}}{P_{in}} \times 100$$

If a power amplifier had a 60 percent efficiency and a P_{dc} input of 500 watts, its output power would approach:

$$P_{out} = \% \text{ EFF} \times P_{in} = .6 \times 500 = 300 \text{ watts}$$

Ham radio stations are limited by law to 1000 watts input power. Their output power is considerably less. One important duty of transmitter engineers is to monitor the input power of their transmitters at the frequent intervals prescribed by law.

Considering a transmitter with 100 percent modulation with a power of 750 watts, our previous calculations show that 500 watts of this power is in the carrier wave and 250 watts added to produce the sidebands. Therefore, we can say that there are 125 watts of power in each sideband or one-sixth of the total power in each sideband. Now each sideband contains the same information and each is a radio frequency wave which will radiate as well as the carrier wave. So why waste all this power? In single sideband transmission, this power is saved. The carrier and one sideband is suppressed and only one sideband is radiated. At the receiver end the carrier is reinserted and the difference signal (the audio signal) is then detected and re-produced. This text will not cover the methods of sideband transmission and reception, but you may wish to make an independent study of this very popular communication system.

TRANSISTORIZED TRANSMITTERS

The vacuum tube no longer holds the prom-inent position as a power amplifier for trans-mitters. In recent years transistors have been developed with capabilities of handling large power requirements. Transistors can be found in audio circuits, oscillators and intermediate power amplifiers. Note the block diagram of a simple transistorized transmitter, Fig. 15-15.

Fig. 15-16. Walkie-talkie citizens band transceiver. (EICO)

table model citizens band radio is shown in Fig. 15-17. The citizens band frequencies assigned by the FCC are shown in Fig. 15-18.

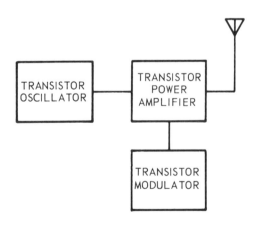

Fig. 15-15. Block diagram of a simple transistorized transmitter.

In Fig. 15-16, is a WALKIE-TALKIE trans-mitter and receiver which is now very popular for use in the citizens band. Transistors are used both for receiver and transmitter. It also contains a rechargeable battery and charger. A

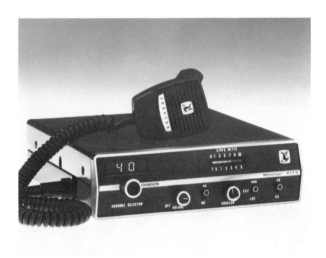

Fig. 15-17. Citizens band radio base station.

FREQUENCY MODULATION (FM)

FM radio has become a very popular method of radio communication. It employs an entirely different system of super-imposing intelligence

Channel	Frequency (MHz)	Channel	Frequency (MHz)
1	26.965	13	27.115
2	26.975	14	27.125
3	26.985	15	27.135
4	27.005	16	27.155
5	27.015	17	27.165
6	27.025	18	27.175
7	27.035	19	27.185
8	27.055	20	27.205
10*	27.075	21	27.215
11	27.085	22	27.225
12	27.105	23	27.255

* Channel 9 is the emergency channel.

Fig. 15-18. Citizens band channels and frequencies.

such as sound and music on the radio frequency wave. It enjoys its popularity because the capabilities of frequency modulation allow relatively high audio sound to be transmitted and still remain within the legal spectrum space assigned to the broadcast station. Also the general acceptance of stereophonic sound has been encouraged by FM transmission of dual channels of sound by multiplex systems. The commercial FM band is from 88 MHz to 108 MHz. A block diagram of a FM transmitter is shown in Fig. 15-19.

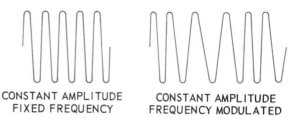

CONSTANT AMPLITUDE FIXED FREQUENCY   CONSTANT AMPLITUDE FREQUENCY MODULATED

Fig. 15-20. For FM the frequency of the wave is varied at an audio rate.

The amount of variation of frequency from each side of the center frequency is called the FREQUENCY DEVIATION and is determined by the amplitude or strength of the audio modulating wave. In Fig. 15-21 a small audio signal causes the frequency of the carrier wave to vary between 100.01 MHz and 99.99 MHz and the deviation is ± 10 KHz. In the second example a stronger audio signal causes a frequency swing between 100.05 MHz and 99.95 MHz or a deviation of ± 50 KHz. The stronger the modulation signal, the greater the FREQUENCY DEPARTURE and the greater the band occupancy.

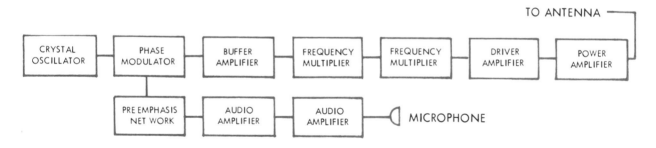

Fig. 15-19. The block diagram of a simplified FM transmitter.

Beginning with a constant amplitude continuous wave signal, the frequency is made to vary at an audio rate. This is shown graphically in Fig. 15-20.

Each broadcasting station is assigned a CENTER FREQUENCY in the FM band (92.1 to 107.9 MHz). This is the frequency to which you would tune your radio. Now study the wave forms in Fig. 15-21 to discover the exact meaning of FM.

The RATE at which the frequency varies from its highest to lowest frequency depends upon the FREQUENCY OF THE AUDIO MODULATING SIGNAL. Two graphic examples are found in Fig. 15-22.

If the audio signal is 1000 Hz, the carrier wave goes through its maximum deviation 1000 times per second. If audio signal is 100 Hz, then the frequency changes at a rate of

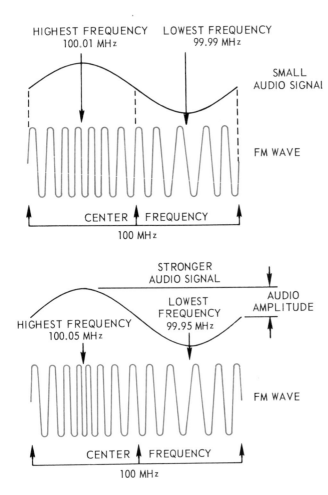

HIGHEST FREQUENCY
100.01 MHz

LOWEST FREQUENCY
99.99 MHz

SMALL AUDIO SIGNAl

FM WAVE

CENTER FREQUENCY

100 MHz

STRONGER AUDIO SIGNAL

LOWEST FREQUENCY
99.95 MHz

AUDIO AMPLITUDE

HIGHEST FREQUENCY
100.05 MHz

FM WAVE

CENTER FREQUENCY

100 MHz

Fig. 15-21. The amplitude of the modulating signal determines the frequency swing from center frequency.

100 times per second. Notice that the modulating frequency does not change the amplitude of the carrier wave.

When a signal is frequency modulated, there is also the formation of sidebands. However, the number of sidebands produced depends upon the frequency and amplitude of the modulating signal. Each sideband on either side of the center frequency is separated by the amount of the frequency of the modulating signal. This is illustrated in Fig. 15-23.

Notice that the power of the carrier frequency is considerably reduced by the formation of sidebands which take power from the carrier. The amount of power taken from the carrier depends upon the MAXIMUM DEVIATION and the modulating frequency.

Although a station may be assigned a center frequency and stays within its maximum permissible deviation, the formation of sidebands really is the determining factor on the bandwidth required for transmission. In FM, the bandwidth is specified by the frequency range between the upper and lower SIGNIFICANT sidebands. A significant sideband is one which has an amplitude of one percent or more of the unmodulated carrier.

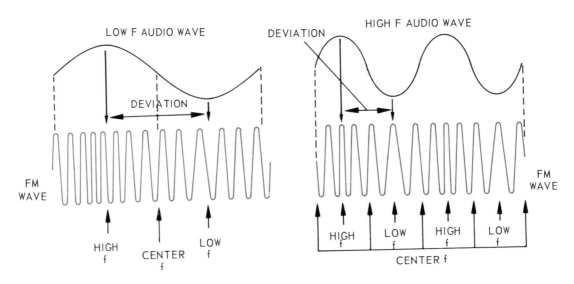

Fig. 15-22. The rate of frequency variation depends upon the frequency of the audio modulating signal.

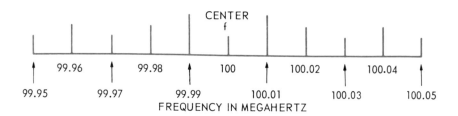

Fig. 15-23. Sidebands generated by a 10 KHz modulating signal on a 100 MHz carrier wave.

NARROW BAND FM

If maximum deviation of a carrier wave is limited so that FM wave occupies the same space as an AM wave carrying the same intelligence, it is called NARROW BAND FM. Some distortion is present in the received signal. It is quite satisfactory for voice communications, but not for high fidelity music.

MODULATION INDEX

This index is the relationship between the MAXIMUM CARRIER DEVIATION AND THE MAXIMUM MODULATING FREQUENCY and is expressed as:

$$\text{MODULATING INDEX} = \frac{\text{Maximum Carrier Deviation}}{\text{Maximum Modulating Frequency}}$$

By the use of this index, the number of significant sidebands and the bandwidth of the FM signal may be calculated. The complete index may be found in more advanced texts. Examples of the use of index are given in the following table:

Mod. Index	No. of Sidebands	Bandwidth
.5	2	4 x F
1	3	6 x F
5	8	16 x F
10	14	28 x F

Where F is the modulating frequency.

If the amplitude of a modulating signal causes a maximum deviation of 10 KHz and the frequency of the modulating signal was 1000 Hz, the index would be:

$$\text{M Index} = \frac{10,000}{1000} = 10$$

The FM signal would have 14 significant sidebands and occupy a bandwidth of 28 KHz.

PERCENT OF MODULATION

The percent of modulation has been arbitrarily stated as a maximum deviation of ± 75 KHz for commercial FM radio. For the FM sound transmission in television, it is limited to ± 25 KHz.

THE RADIO WAVE

A radiated wave into space may be visualized as waves rolling toward the ocean beach. A frequent comparison of the descriptions of radiating waves suggests the use of a calm lake. Throw a stone into the water. From the point where the stone entered the water, small waves radiate outward making concentric circular patterns. The stone created a disturbance, or set a wave in motion, which moved outward decreasing in amplitude as the distance from the center became greater. Consider this wave of water for a moment. Does the water move as a wave? No. The water moves UP to a crest and DOWN in a hollow or trough as the wave passes. The water has acted as the medium for transmitting the wave.

Scientists have discovered that radio waves travel through space at a speed of 300,000,000 metres per second or 186,000 miles per second. This is approximately the speed of light. Picture a wave at a frequency of one cycle per second originating from point A in Fig. 15-24 and traveling at the speed of 300,000,000 metres per second. At the beginning of the second cycle, the first wave has moved 300,000,000 metres away. If the frequency of the generated

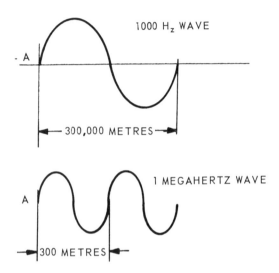

Fig. 15-24. A comparison between the distance, traveled during one cycle, of a 1000 Hz and a one megahertz wave.

wave was 1000 Hz, then the first wave would be 300,000 metres away by the time the second wave started. Continuing this explanation, the frequency is increased to 1,000,000 Hz or 1 mega Hz, then the first wave would have moved only 300 metres before the start of the second cycle. It seems, then, that a wave might be described not only by its FREQUENCY but also by its LENGTH. This is true, as shown in Fig. 15-25. The distance between the crests of the waves is termed the WAVE LENGTH. This conclusion may also be accepted: As the frequency increases, the wave length decreases. The Greek letter λ (lambda) signifies wave length in electronics and the mathematical relationship may be stated as:

$$\lambda \ \text{(in metres)} = \frac{300,000,000}{f \text{ (in Hz)}}$$

$$f \text{ (in Hz)} = \frac{300,000,000}{\lambda \text{ (in metres)}}$$

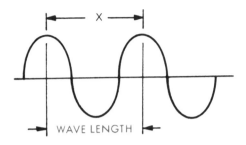

Fig. 15-25. Radio waves may be identified by their length and their frequency. X = one wave.

EXAMPLE 1. An amateur radio transmits on a frequency of 3.9 MHz. What is the wave length of the radiated waves?

$$\lambda \ = \frac{300,000,000}{3.9 \text{ MHz}}$$

3.9 MHz is converted to 3,900,000 Hz, then

$$\lambda \ = \frac{300,000,000}{3,900,000} \ = \frac{3000}{39} \ \text{or 77 metres}$$

Observe how the mathematics is simplified by using "powers of ten"

$$\lambda = \frac{3 \text{ x } 10^8}{3.9 \text{ x } 10^6} \ = \frac{3 \text{ x } 10^2}{3.9} \ = .77 \text{ x } 10^2 \ = \ 77 \text{ metres}$$

EXAMPLE 2. What is the frequency of a transmitter operating on 40 metres?

EXPLANATION:

$$f = \frac{3 \text{ x } 10^8}{4 \text{ x } 10} \ = .75 \text{ x } 10^7 = 7,500,000 \text{ Hz or} \\ 7.5 \text{ megahertz}$$

A more practical statement of the above formula may be made, since the foot is the common unit of measure and radio frequencies are usually in megahertz so,

$$\lambda \ \text{(in feet)} = \frac{984}{f \text{ (in MHz)}}$$

It would be wise to memorize this formula. It is used a great deal in the design of antennas for special frequencies.

RADIO WAVE PROPAGATION

In the study of electromagnetism, we learned that a conductor carrying an electric current was surrounded by a magnetic field. If the current were an alternating current, the magnetic field would periodically expand and collapse and change its polarity. Some of these magnetic lines of force will lose the influence of the conductor and radiate into space. These radiated electromagnetic waves are perpendicular to the direction in which they travel. Associated with any electromagnetic field is an electrostatic field which is perpendicular to the electromagnetic field, and is also perpendicular to the direction of radiation or travel. As a result, the radio wave consists of electromagnetic and electrostatic fields. Fig. 15-26 shows the relationship between these fields and direction

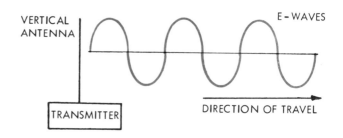

Fig. 15-27. A vertical antenna radiates a vertically polarized wave.

waves should be oriented the same as the transmitting antennas. At high frequencies the polarization changes during propagation to some extent.

The beginning student may ask a meaningful question, "Does the transmitting antenna radiate two waves?" The answer is found by the fact that without one there cannot be the other. A moving electrostatic field produces a moving electromagnetic field; and a moving electromagnetic field produces a moving electrostatic field. These conditions are true whether an actual conductor is present or not.

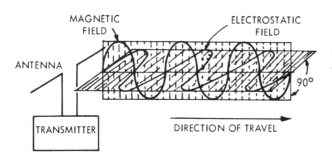

Fig. 15-26. The angular relationship between electrostatic and electromagnetic waves. They are perpendicular to each other and both are perpendicular to the direction of travel.

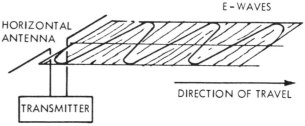

Fig. 15-28. A horizontal antenna radiates a horizontally polarized wave.

of travel. The position in which these waves radiate in respect to the earth is called POLARIZATION. In Fig. 15-27 the waves are radiated from a vertical antenna. Note that the electrostatic or E waves are in the same plane as the antenna yet perpendicular to direction of travel. The vertically polarized waves are perpendicular to the surface of the earth. In Fig. 15-28, the wave is radiated from a horizontal antenna. It is still perpendicular to the direction of travel, but is parallel to the surface of the earth. Generally speaking, the antenna which receives these

GROUND WAVES

The radiated waves from an antenna may be divided into two types. First, let us consider the GROUND WAVE, so called because it follows the surface of the earth to the radio receiver. The ground wave may be considered as having three parts:

1. The surface wave.

2. The direct wave which follows a direct path from the transmitter to the receiver.

3. The ground reflected wave which strikes the ground and then is reflected to the receiver.

These last two waves are combined and called the SPACE wave. The direct and reflected waves that constitute the space wave may or may not arrive at the receiver in proper phase. They may add together or cancel each other depending on distances traveled by each wave. Broadcast stations depend on the surface wave for reliable communications. As the surface wave travels along the surface of the earth, it will induce currents in the earth's surface which use up the energy contained in the wave. The wave becomes weaker as the distance over which it travels increases. An interesting side-light in the conductivity of salt water which is about 5000 times better than the earth. Overseas communication is very reliable when transmitters are near the coastline. These stations use high power and operate at lower frequencies than the familiar broadcast band.

SKY WAVES

The second radiated wave from an antenna is toward the sky. It is called the SKY WAVE. The sky waves strike a heavily ionized layer on the earth's atmosphere. This layer, called the IONOSPHERE, is located from 40 to 300 miles above the earth's surface, and is believed to consist of large numbers of positive and negative ions. This ionization is dependent upon the earth's position in relation to the sun and "sun spot activity." When radio waves strike an ionosphere, a part may be absorbed and a part may be refracted or bent back to the earth's surface. See Fig. 15-29. One of the fascinations of amateur radio results from the use of reflected waves as a method of communication. The "ham operator" calls "CQ" which is the general inquiry call to establish communication. The operator's reply may come from a distant station in another part of the world.

AM TRANSMITTER

It is very exciting to speak into a microphone connected to an AM transmitter and hear

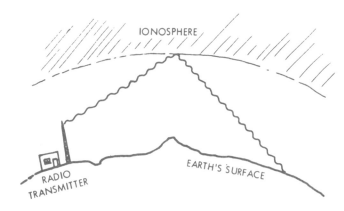

Fig. 15-29. Radio waves may bounce off the ionosphere back to the earth's suface.

yourself over your radio. This project will allow you to do this. The rf carrier range is 550-1500 KHz with an output power of less than 100 milliwatts. The transmitting distance depends on the environment. However, this distance should be less than 100 feet. Any AM receiver can be used to pick up the transmitted signal. A block diagram of the transmitter is shown in Fig. 15-30.

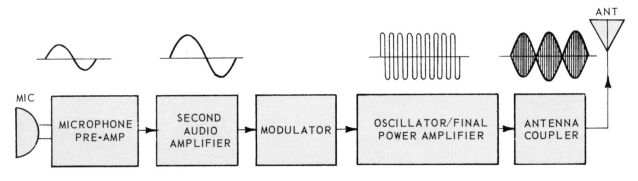

Fig. 15-30. Block diagram of AM transmitter.

The circuit used in this AM transmitter was provided by Graymark Enterprises, P. O. Box 54343, Los Angeles, California, 90054. It may be purchased as the Graymark Model 533 AM Transmitter. See Fig. 15-31 for photo and schematic.

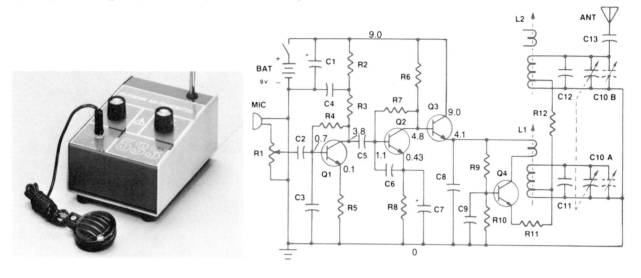

Fig. 15-31. Left. Completed AM transmitter. Right. Schematic for AM transmitter. (Graymark)

QTY.	SYMBOL	PART NO.	DESCRIPTION
1	R1	62389	Potentiometer, with switch, 50K
2	R2, R8	61412	Resistor, 470Ω, ¼-W, 10%
2	R3, R10	61415	Resistor, 4.7KΩ, ¼-W, 10%
1	R4	62390	Resistor, 470KΩ, ¼-W, 10%
1	R5	61420	Resistor, 100Ω, ¼-W, 10%
1	R6	61413	Resistor, 3.9KΩ, ¼-W, 10%
1	R7	62235	Resistor, 560KΩ, ¼-W, 10%
1	R9	61418	Resistor, 3.3KΩ, ¼-W, 10%
1	R11	62391	Resistor, 22Ω, ¼-W, 10%
1	R12	62392	Resistor, 10Ω, ¼-W, 10%
2	C1, C4	62393	Capacitor, electrolytic, 33μF, 16 wv
2	C2, C5	62394	Capacitor, disc, .1μF, 50 wv ("104")
1	C3	62365	Capacitor, disc, .0047μF, 50 wv ("472")
1	C6	62364	Capacitor, disc, .01μF, 50 wv ("103")
1	C7	61528	Capacitor, electrolytic, 4.7μF, 16 wv
2	C8, C9	62395	Capacitor, disc, .047μF, 50 wv ("473")
1	C10	62407	Capacitor, variable (tuning), 266pF
2	C11, C12	62397	Capacitor, disc, 22pF, 50wv ("22")
1	C13	62398	Capacitor, disc, 220pF, 50wv ("221")
2	L1, L2	62399	Coil, oscillator type
4	Q1-Q4	61533	Transistor, NPN silicon, 2SC372-Y
1		62400	Microphone, crystal lapel

QTY.	SYMBOL	PART NO.	DESCRIPTION
1		62401	Antenna, telescopic
1	J1	62402	Jack, microphone, 3.5mm type, NC
1		62403	Battery clip, for 9-volt battery
1		62405	Antenna bracket, steel
2		61548	Knob, black
4		61433	Rubber foot, black
1		62408	Rubber grommet, black
4	MS1	61267	Machine screw, large dia., slot head (3x5mm)
2	MS2	61061	Machine screw, small dia., slot head (2.6x4 mm)
1	MS3	62344	Machine screw, small dia., phillips head (2.6x6mm)
4	MS4	61264	Self-tapping screw
1	WS1	62410	Washer, shoulder, plastic
1	WS2	61258	Washer, flat, fiber
1		62412	Solder lug
35		61357	Soldering pin
1 set		62413	Hookup wire set, stranded, 1-red, 1-yellow
1		61174	Bare wire (buswire), single strand, #22
1		62414	Solder, rosin core, 60/40
1		62388	Printed circuit board
1		62416	Chassis, w/battery holder
1		62417	Cover, aluminum
1		62422	Breadboard panel, with schematic
1		—	Warranty card
1		61943	Instruction manual

Fig. 15-32. Parts list for AM transmitter. (Graymark)

See Fig. 15-32 for parts list. The test conditions for the initial operation of the transmitter are:

1. R_1 set at position "1".
2. C_{10} set at position "15" (full CW).
3. Antenna not extended.
4. Mike removed.
5. Power source: 9.0V dc.
6. All readings taken with 11M Ω TVM meter, with neg. lead connected to common ground (–).
7. Tolerance of all readings: ± 20%.

FOR DISCUSSION

1. Are the waves radiated to your TV antenna, horizontally or vertically polarized?
2. How is it possible for a ham radio operator to talk to distant countries?
3. Explain the basic block diagram of a transmitter.
4. Discuss the operation of the following microphones:
   a. Carbon microphone.
   b. Dynamic microphone.
   c. Crystal microphone.
5. What is the primary difference between AM and FM?
6. Discuss two basic methods of FM detection.

TEST YOUR KNOWLEDGE

1. What is the length of a wave which has a frequency of 1200 KHz?
2. A short wave broadcasting station has a wave length of 40 metres. What is the frequency of this wave?
3. Name and explain the two fields in a radiated radio signal.
4. What is the bandwidth for AM and for FM?
5. The frequencies used for commercial AM and FM broadcast bands are:
   AM _____ KHz to _____ KHz.
   FM _____ MHz to _____ MHz.
6. Draw the block diagram for a simple transistorized AM transmitter.
7. Define modulation.
8. The FM broadcast band lies between what two television channels in the radio frequency spectrum?

Chapter 16

RADIO RECEIVERS

One of the greatest inventions of all mankind is the radio. Like many other inventions, the development of the radio took place because of the work of many scientists. In 1864, James Maxwell theorized that electromagnetic waves existed. A little over twenty years later, in 1887, Heinrich Rudolph Hertz, confirmed this theory when he transmitted and received the first radio waves. The first continuous wave (CW) transmitter was developed in 1897 by Guglielmo Marconi.

Two major components that further helped in developing modern day radio was the diode, invented in 1904 by Alexander Fleming, and the "audion" triode amplifier tube, invented in 1904 by Lee DeForest. One of the most important scientists in the history of radio was E. H. Armstrong. He invented the regenerative radio circuit in 1913 and the superheterodyne

radio circuit in 1920. Major Armstrong is also credited with the development of the theory of much of our modern day FM radio.

In Chapter 15, a basic communications model showing the relationship between the transmitter and receiver was discussed. Let us look at this again in Fig. 16-1.

This chapter will discuss the AM and FM radio receiver.

THE AMPLITUDE MODULATED (AM) RADIO RECEIVER

The radio wave transmitted through space also carries information, such as voice and music, which has been combined with the carrier wave at the transmitter by the amplitude modulation process. These radiated electro-

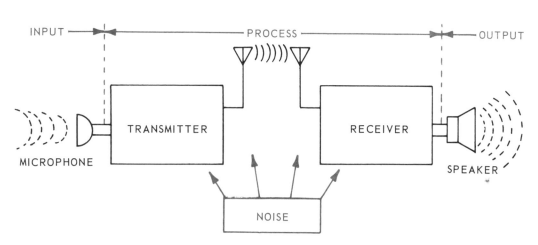

Fig. 16-1. Basic transmitter/receiver communications model.

magnetic waves cut across the receiving antenna and induce a small voltage in the antenna. The small radio frequency voltages are coupled to a tuned circuit in the receiver which selects the desired signal to be heard. After the selection, the voltages may be amplified, then demodulated. The term DEMODULATION means DETECTION or the separation of the audio portion of the signal from the carrier wave. The audio signal is then amplified to sufficient strength to drive a loudspeaker. A block diagram of this receiver is shown in Fig. 16-2. The modulated rf wave is shown as it passes the antenna; the audio wave is illustrated as the output of the detector. The increase in

schematic of the first stage of the TRF receiver. Following the signal of the antenna to the output of this circuit will materially aid the understanding of the operation. Radio signals of many frequencies from many radio transmitters are passing by the antenna. The induced voltage in the antenna causes small currents to alternate from antenna to ground and from ground to antenna through coil L_1. Since L_2 is closely coupled to L_1, the magnetic field created by the antenna current in L_1 transfers energy to L_2. The combination coils L_1 and L_2 are usually called the ANTENNA COIL and are wound on a common cylinder of cardboard. Fig. 16-4 illustrates two types used in transistor radios.

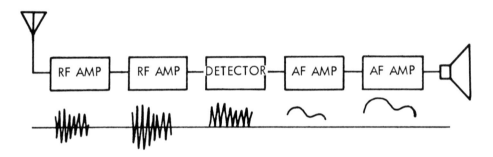

Fig. 16-2. A block diagram of a TRF (tuned radio frequency) receiver.

amplitude of the waves between the blocks is the result of stages of amplification. This type of receiver was once quite popular and is called the TRF or TUNE-RADIO-FREQUENCY receiver. For satisfactory operation of the circuit it was necessary to tune each stage to the correct incoming frequency of the signal. Some of the early radios had a series of tuning dials on the front panel which required skill and patience for proper adjustment. The development of the SUPERHETERODYNE receiver, which is used today, overcomes these early undersirable characteristics.

THE TUNING CIRCUIT

It has been stated that one of the functions of any radio receiver must be that of selection of the desired radio signal. Details of the "tank circuit" and flywheel action have been discussed in Chapter 8. Fig. 16-3 shows a

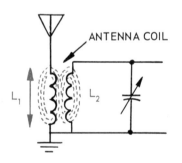

Fig. 16-3. The tuning or station selector section of the radio receiver.

Variable capacitor C_1 is connected across the terminals of L_2 and forms a tank circuit. C_1 may be adjusted to vary the RESONANT frequency of the tank. When the resonant frequency of the tank circuit is the same as the incoming signal, relatively high circulating currents develop in the tank. In other words, the tank circuit may be adjusted to give maximum

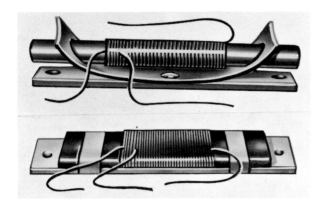

Fig. 16-4. Typical antenna coil used with transistorized receivers. (J. W. Miller Co.)

response for only one frequency. The L_2C_1 combination may only cover a certain range or band of frequencies. Most of our home radios cover only the broadcast band. For other groups of frequencies, such as short wave, another coil is switched in place of L_2, which changes the resonant frequency range of the circuit. This type of circuit is called a BAND SWITCHING circuit.

The ability of a radio receiver to select a single frequency and only one frequency is called SELECTIVITY. The ability of a receiver to respond to weak incoming signals is called SENSITIVITY. Both of these characteristics are highly desirable and special circuits and components have been devised to improve selectivity and sensitivity.

RF AMPLIFICATION

The radio frequency (rf) amplifier is usually the first stage that the received signal is fed to from the antenna. The rf amplifier is tuned to the frequency of the incoming signal and provides amplification or gain to this signal. Generally, rf amplifiers are narrow band amplifiers capable of amplifying only the band or frequencies to be picked up by that receiver. An example is the rf amplifier in an AM broadcast band receiver, capable of amplifying frequencies from 550-1500 KHz. A rf amplifier improves the selectivity of a receiver.

A typical rf amplifier is shown in Fig. 16-5, (RCA Transistor, Thyristor, and Diode Manual, SC-15, page 695). The tuning circuit is made up by C_1 and L_2 formed into a tank circuit. The tuned signal is fed into the secondary winding of L_2. The amplifier is the 2N1637 transistor. Some inexpensive radio receivers do not have rf amplifier stages. However, in very high quality radio receivers it is very likely to have more than two or more rf amplifier stages.

DETECTION

Assuming that the modulated radio frequency signal has been amplified by several stages of rf amplification, it now has sufficient amplitude for detection. Detection is the process of rectification and the removal of the rf wave, leaving the af wave only. See Fig. 16-6 which shows:

1. The rf carrier within the modulation envelope.

2. The rectified rf wave.

3. The audio wave when the rf wave has been removed.

Detection is necessary to recover the audio signal from the modulated rf carrier wave.

THE DIODE DETECTOR

One of the more common and quite satisfactory methods of detection employs the diode as a unilateral conductor (one direction). The characteristics of the diode were discussed in Chapter 11 and may be summarized again for purposes of review. When the anode of the diode is driven positive, electrons flow through the diode from cathode to anode. The diode conducts. When the anode is driven negative, the diode does not conduct. The schematic drawing in Fig. 16-7 represents the diode detector in its simplified form. The amplified modulated signal is supplied to the detector from the previous amplifier stage. When the

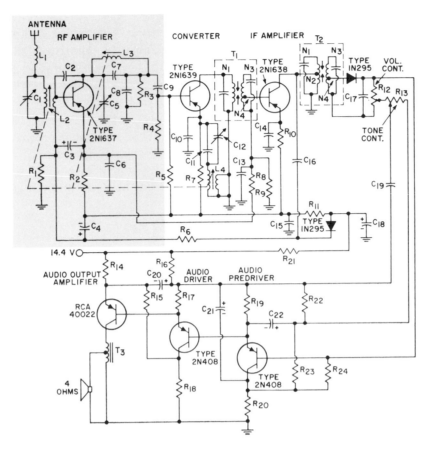

Fig. 16-5. 12-volt automobile radio receiver with RF amplifier.
(RCA Transistor, Thyristor, and Diode Manual, SC-15)

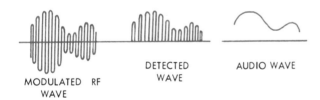

MODULATED RF WAVE

DETECTED WAVE

AUDIO WAVE

Fig. 16-6. The appearance of the wave forms as they are demodulated or detected.

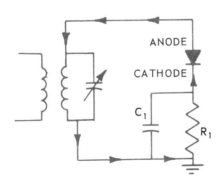

Fig. 16-7. The basic diode detector circuit.

input signal is positive, the diode conducts and a voltage is developed across the diode load resistor R_1. Conversely, when the incoming signal is negative, there is no current in the diode circuit. The diode has rectified or converted the modulated rf signal into pulses or waves of dc voltage which have a frequency and amplitude of the audio wave. This half-wave rectification may be understood by examination of the input and output wave forms in Fig. 16-8. The dotted curve in the output of the diode represents the AVERAGE dc value of the rectified voltage. In order to raise the average dc output voltage and cause it to conform to a more faithful reproduction of the input signal, a filter capacitor of the correct value C_1, is shunted across R_1. This capacitor charges to the peak value of the signal and prevents the voltage across R_1 from dropping to zero when the diode is not conducting. The improved output wave form as a result of

MODULATED
WAVE

DETECTED
SHOWING
AVERAGE VALVE

DETECTED
WITH
FILTER C_1

Fig. 16-8. The average value of the dc component is represented by the dotted line. The dc average is raised by adding the filter capacitor.

filtering is also shown in Fig. 16-8. The time constant of $R_1 C_1$ should be long when compared to the rf cycle, so that C_1 will not discharge to a low value. Conversely this time constant should be short when compared to the af cycle, so the voltage variations across the $R_1 C_1$ network will follow the audio frequency cycle.

THE SUPERHETERODYNE RECEIVER

To overcome many of the disadvantages of the tuned-radio frequency receiver, the superheterodyne circuit was developed. This heterodyning or mixing of signals provides a means of converting all incoming signals to a single INTERMEDIATE FREQUENCY which can be amplified with minimum loss and distortion. A block diagram of a superheterodyne receiver is illustrated in Fig. 16-9. The signal picked up by the antenna is first fed to a stage of radio frequency amplification (rf amp). The output of the rf amp is then fed to the MIXER or CONVERTER STAGE. Also the output of a LOCAL OSCILLATOR is fed to the converter.

When two signals are mixed together, four signals will appear in the output. These will include the original two signals, the sum of the two signals and the difference between the two signals. For example: A 1000 KHz signal is mixed with a 1455 KHz oscillator signal. Appearing in the output then will be:

1000 KHz original signal.
1455 KHz original oscillator frequency.
2455 KHz the sum of both frequencies.
455 KHz which is the difference between both frequencies.

The BEAT frequency of 455 KHz is of major importance in the study of the superheterodyne receiver. The desired station is selected by the tuning circuits of the receiver by turning the tuning knob, on the front panel, which varies the capacitance of the tuning circuit. Mechanically attached to the shaft of the tuning capacitor is another tuning capacitor which adjusts the frequency of the local oscillator, see Fig. 16-10. These capacitors, operating in step with each other, provide a change in oscillator frequency as the tuned frequency is changed, always maintaining a fixed difference or beat frequency of 455 KHz called the INTERMEDIATE FREQUENCY or if. The if output is then amplified by two stages of voltage amplification and fed to the DETECTOR. The output of the detector is an audio frequency voltage which is amplified sufficiently to operate the power amplifier and speaker. The wave forms at each stage are shown in Fig. 16-9.

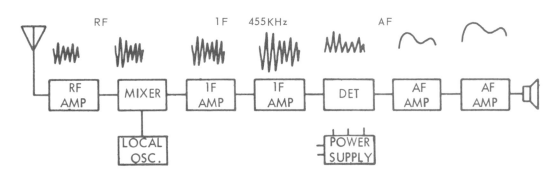

Fig. 16-9. The block diagram of the superheterodyne receiver.

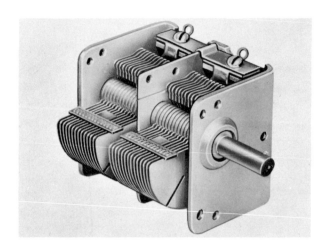

Fig. 16-10. A double section tuning capacitor.
(J. W. Miller Co.)

A six transistor transistorized AM superheterodyne receiver is shown in Fig. 16-11. This circuit is from the Graymark 806 AM six transistor radio. As the signal is picked up on the antenna, the rf amplifier is tuned to the incoming signal and amplified by the rf amplifier stage. The local oscillator is tuned to produce an unmodulated carrier signal of 455 KHz above the incoming signal. Both the local oscillator signal and the rf amplifier signal are fed into the converter (mixer) stage where the heterodyning principle takes place. The output of the converter/mixer is fed into the if transformer T_3 which is tuned to 455 KHz. Only this frequency, the if 455 KHz signal, passes on to the if amplifier. The if amplifier improves the selectivity and amplifies the 455 KHz intermediate frequency in T_4. CR_1 is the detector diode that is used to convert the modulated if signal to an audio wave. R_{11} is

BLOCK DIAGRAM

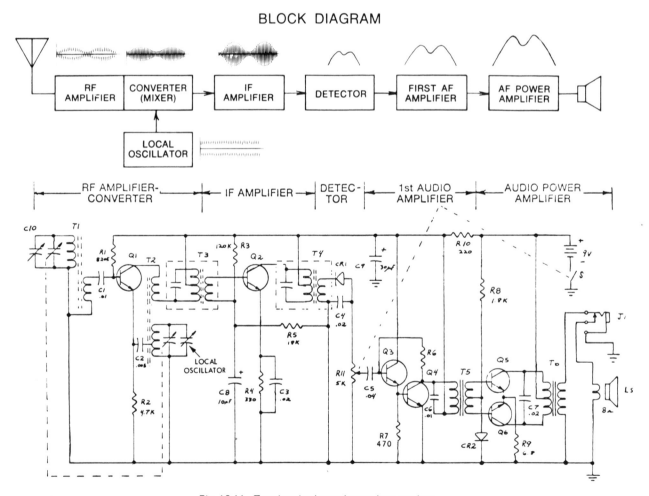

Fig. 16-11. Transistorized superheterodyne receiver.
(Graymark Enterprises, Inc.)

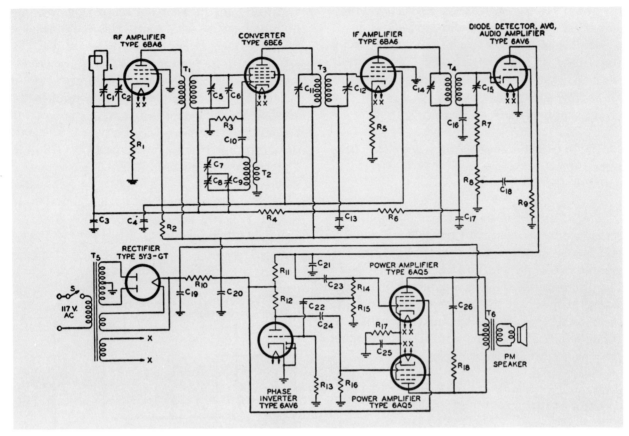

Fig. 16-12. The schematic diagram of a superheterodyne receiver. (RCA)

SUPERHETERODYNE RECEIVER
MATERIAL LIST

C_1, C_5, C_8 – Ganged tuning capacitors, 10-365 pF

C_2, C_6, C_9 – Trimmer capacitors, 4-30 pF

C_3, C_{13} – 0.05 μF, paper, 50V

C_4 – 0.05 μF, paper, 400V

C_7 – Oscillator padding capacitor, follow oscillator-coil manufacturer's recommendation

C_{10} – 56 pF mica

C_{11}, C_{12}, C_{14}, C_{15} – Trimmer capacitors for if transformers

C_{16}, C_{17} – 180 pF mica

C_{18}, C_{22} – 0.01 μF, paper, 400V

C_{19}, C_{20} – 20 μF, electrolytic, 450V

C_{21} – 120 pF mica

C_{23}, C_{24} – 0.02 μF, paper, 400V

C_{25} – 20 μF, electrolytic, 50V

C_{26} – 0.05 μF, paper, 600V

L – Loop antenna, 540-1600 KHz

R_1, R_5 – 180 ohms, 0.5 watt

R_2 – 12000 ohms, 2 watts

R_3 – 22000 ohms, 0.5 watt

R_4, R_6 – 2.2 megohms, 0.5 watt

R_7 – 100000 ohms, 0.5 watt

R_8 – Volume control, potentiometer, 1 megohm

R_9, R_{13} – 10 megohms, 0.5 watt

R_{10} – 1800 ohms, 2 watts

R_{11}, R_{12} – 220000 ohms, 0.5 watt

R_{14}, R_{16} – 470000 ohms, 0.5 watt

R_{15} – 8200 ohms, 0.5 watt

R_{17} – 270 ohms, 5 watts

R_{18} – 15000 ohms, 1 watt

S – Switch on volume control

T_1 – RF transformer, 540-1600 KHz

T_2 – Oscillator coil for use with 10-365-pF tuning capacitor and 455-KHz if transformer

T_3, T_4 – Intermediate-frequency transformers, 455 KHz

T_5 – Power transformer, 250-0-250 volts rms, 120 mA, dc

T_6 – Output transformer for matching impedance of voice coil to a 1000-ohm plate-to-plate tube load

the audio volume control. Q_3 is the first audio amplifiers. The final audio amplification is accomplished by the push-pull circuit using transistors Q_5 and Q_6. The final output electrical signal is converted from ac current to an audible sound wave by the speaker LS. The power supply is the 9 volt battery.

The schematic diagram of a complete tube type superheterodyne receiver is illustrated in detail in Fig. 16-12.

The rf amplifier stage before the converter serves a dual purpose. Not only does it amplify the incoming signal but the rf stage provides greater selectivity and IMAGE REJECTION. In the discussion of the heterodyning principle, it was explained that the incoming signal and the local oscillator signal were mixed together to produce a beat if of 455 KHz. Using the same frequencies, study this example:

Incoming signal	1910 KHz
Local oscillator	1455 KHz
if	455 KHz

It appears that the receiver could be tuned to 1000 KHz and yet respond to a frequency of 1910 KHz, if the tuning circuit were sufficiently broad to accept a part of this 1910 KHz signal. This is referred to as the IMAGE FREQUENCY and may be found by:

Desired station frequency ± 2 x if

frequency = Image frequency

Preselectivity of the rf amplifier reduces the response to the image frequency to a minimum. Images may be very troublesome problems to designers of receiver circuits.

The rf stage also acts as a buffer between the local oscillator and the antenna. The oscillator is, in fact, a small radio transmitter. It will radiate an rf signal into space that will interfere with other radio receivers nearby. The rf stage isolates the oscillator from the antenna, thereby reducing radiation.

Usually broadcast band receivers employ an if frequency of 455 KHz. This is not always true, however. Some communications receivers use other if frequencies. Frequently, a double conversion is made which requires two if frequencies. The technician should refer to the manuals to discover the if frequencies used in the equipment that is to be serviced. The next stage after the rf amplifier is the CONVERTER. This circuit uses a PENTA-GRID CONVERTER tube which means FIVE GRIDS. An analysis of the tube's operation shows its dual function. In Fig. 16-13, only the oscillator section is shown. Compare this diagram to Fig. 14-7 in Chapter

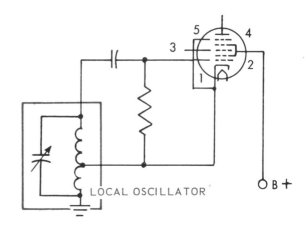

Fig. 16-13. Penta-grid converter showing the oscillator section.

14. Is is not the conventional HARTLEY oscillator circuit? Grids number 2 and 4 are connected together and serve as the PLATE for the oscillator. The incoming signal from the rf stage is fed to grid 3. Grid 5 acts as the suppressor grid. The tube serves as an oscillator and a mixer-amplifier. The amplified output of the heterodyned signals appears in the plate circuit. The if transformer, which couples the signal to the grid of the first if amplifier, is sharply tuned to 455 KHz and provides maximum response at that frequency. The other frequencies are rejected and bypassed to ground. An if transformer is illustrated in Fig. 16-14. After the if stage, the 455 KHz signal is fed into the detector stage. Refer to Fig. 16-12 and note that this stage uses a 6AV6 tube. A detailed detector circuit using this tube is shown in Fig. 16-15. This dual function tube

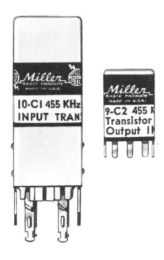

Fig. 16-14. Typical if transformers.

includes the diode detector and a triode voltage amplifier. Several new components have been added which require some explanation. The diode load resistor R_1 has been made a potentiometer, which serves as a volume control for the receiver. The desired amount of af voltage is selected and coupled through C_1 to the grid of the amplifier. C_2 is a small trimmer capacitor across the transformer secondary, which is adjusted during alignment for maximum response of the tuned circuit. Note that when the diode conducts, point A in Fig. 16-15 becomes negative in respect to ground. If the amplitude of the signal increases, point A will be correspondingly more negative. This voltage, which varies according to signal strength, may

be used to automatically control the volume of the receiver and is called the automatic volume control (AVC) voltage. The voltage is coupled, after proper filtering to remove the audio wave, to the grids of previous amplifier stages. If the incoming signal is strong, a strong negative AVC voltage decreases the gain of the amplifier stages. As a result, after one setting of the volume control on the receiver, the program reception remains fairly constant without further adjustment. AVC circuits are used in most radio receivers manufactured today.

Next the audio signal is fed into the audio amplifier stages. Usually there are two audio amplifier stages which are the FIRST AUDIO AMPLIFIER STAGE and the POWER AMPLIFIER STAGE. From the output of the power amplifier stage, the audio signal is fed to the transducer or reproducer to convert the electrical signal to audio air waves. Loudspeakers or earphones perform this amazingly well.

TRANSDUCERS

The purpose of the transducers or reproducer is to convert the electrical energy at audio frequencies in the final stage of the radio receiver into sound energy. Such a device is the LOUDSPEAKER or HEADPHONES.

One of the more common types of SPEAKERS is illustrated in Figs. 16-16 and 16-17. In this type a strong magnetic field is

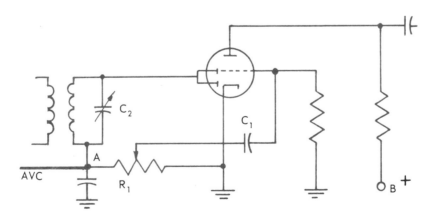

Fig. 16-15. The diode detector and first af audio stage contained in one tube. A volume control has been added.

structed with a permanent magnet is called a PM speaker. Such speakers are manufactured in a variety of sizes and types.

In the electrodynamic type speaker, the permanent magnet is replaced with an electromagnet, Fig. 16-18. The action is similar to the

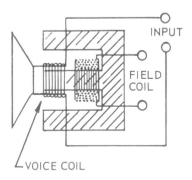

Fig. 16-18. In the electrodynamic speaker the permanent field is replaced by an electromagnetic field.

PM type. A strong source of direct current must be supplied to the electromagnet. This may be secured from the power supply. A common method is the use of the field coil of the speaker as the filter choke in the supply.

Due to increased interest in speakers by high-fidelity music enthusiasts, special sizes and novel types of speakers have been developed. These provide optimum response for specified bands of audio frequencies. Examples of these types are shown in Fig. 16-19. Low frequency speakers are described as WOOFERS; high frequency speakers are TWEETERS. The middle range is reproduced by the INTERMEDIATE speaker. Special filters and cross-over networks have been devised so that signals within specified ranges of frequencies are channeled to the speaker designed to reproduce the sound with maximum fidelity.

A cross-over network in its simplified form is diagramed in Fig. 16-20. Note that coil L is connected to the woofer. As the frequency of the sound increases, the reactance of L increases by the formula, $X_L = 2 \pi fL$. However, the tweeter is connected through C and as the

Fig. 16-16. A typical loudspeaker used in a radio. (Stromberg-Carlson)

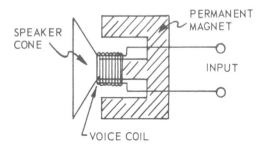

Fig. 16-17. This sketch shows the principle of operation of the permanent magnet (PM) speaker.

produced between the poles of a fixed permanent magnet. A small voice coil is suspended in the airgap and is attached to the speaker cone. When the audio alternating currents are connected to the voice coil, the interaction between the fixed field and the moving field causes the voice coil to move back and forth. This motion also causes the speaker cone to move back and forth, resulting in the alternating condensations and rarefactions of the air in the form of sound energy. A speaker con-

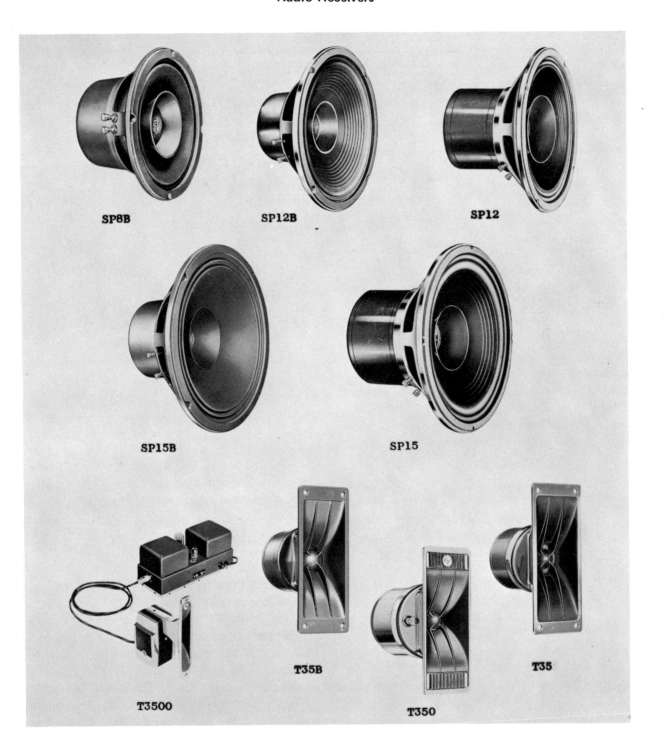

Fig. 16-19. An assortment of speakers designed for high-fidelity reproduction. (Electro-Voice)

frequency increases, the reactance of C decreases as, $X_c = \dfrac{1}{2\pi fC}$. By this circuit the frequencies are divided and reproduced by the speaker designed for best reproduction. Values of L and C may be selected for the desired cross-over point, usually between 400 and 1200 Hz. Note that at one frequency X_L equals X_c and the response is equal for both speakers at this cross-over frequency.

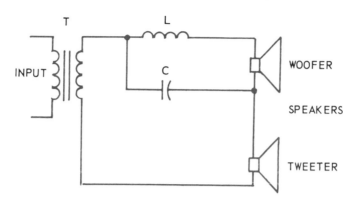

Fig. 16-20. A basic design of the cross-over network.

The HEADPHONES illustrated in Fig. 16-21 still remain as a popular and useful method of sound reproduction for private listening. The principle of operation is not too different from the speaker and may be understood by following the diagram in Fig. 16-22. On the poles of a small horseshoe magnet are wound coils of fine wire. The incoming audio signal is attached to these windings. Mechanically suspended before this magnet is a flexible soft-iron diaphragm. The magnetic force of the

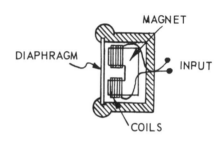

Fig. 16-22. Diagram shows construction of a typical headphone.

diaphragm creates sound energy of the same frequency and amplitude as the incoming audio signal within the limitations of the device.

TONE CONTROLS

People differ in respect to the amount of high or low frequencies preferred in their music. Most radios, TVs and audio amplifiers incorporate a front panel adjustment, so the desired tone may be selected for listening pleasure. The schematic diagram in Fig. 16-23 shows a common circuit used for this purpose. Capacitor C_1 has a low reactance for high audio frequencies and would bypass only the desired amount as selected by the listener. The value of C_1 is usually .05 μF and R_1 is 50,000 ohms. A tone switch may also be used providing three positions of tone control. Each position depends upon the reactance of its capacitor as a means of filtering or removing the high frequencies. Fig. 16-24 diagrams the tone switch control system, using three capacitors.

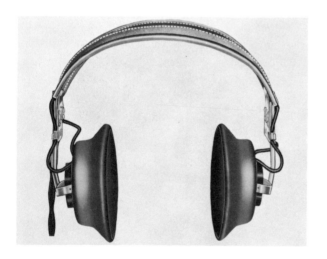

Fig. 16-21. The headphones provide private listening for communications receivers and a wide variety of other purposes.

permanent magnet holds the diaphragm under tension. The varying signal current through the coils either adds or subtracts from the magnetic field, and the diaphragm moves in and out as the magnetic attraction varies. The moving

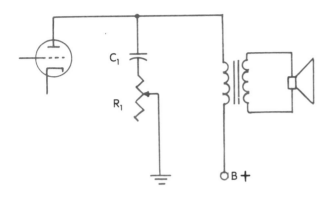

Fig. 16-23. A basic tone control circuit removes the higher frequencies from the speaker output.

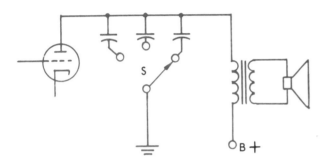

Fig. 16-24. This tone control has three positions only. The tone may be selected by the listener.

ALIGNMENT

The superheterodyne receiver contains several tuned circuits, including the primary and secondary windings of the if transformers, which must be tuned to resonance or maximum response for those signals which must be passed.

Variable trimmer capacitors connected in parallel with these coils provide the adjustment, which is called ALIGNMENT. Receivers generally need alignment after parts have been replaced in the service shop. Also the effect of gradual deterioration of parts due to age may be minimized by an occasional "peaking" or aligning the receiver. The basic instruments needed for the job include the SIGNAL GENERATOR and an output indicator such as an ac VOLTMETER. (Test instruments are discussed in Chapter 19.) Briefly, a signal generator produces either a modulated or unmodulated rf wave which may be selected by the controls on the panel. When an alignment job must be done, it is always best to consult the technical manual for that particular radio or TV. Consequently any discussion of alignment procedures must be general and refer only to the basic theory.

In Fig. 16-25, a diagram of only the signal circuit has been drawn of a superheterodyne receiver. The ac voltmeter is connected to the plate of the final power amplifier tube through a .1 μF-600V capacitor. This is the indicating device. The signal generator is set up to produce a 455 KHz modulated rf wave and is connected to the input grid of the converter tube, through a .001 μF capacitor. The local oscillator must be disabled by connecting a "jumper lead" between the rotor and stator plates of the oscillator tuning capacitor. Turn the attenuation control on the generator so that only a

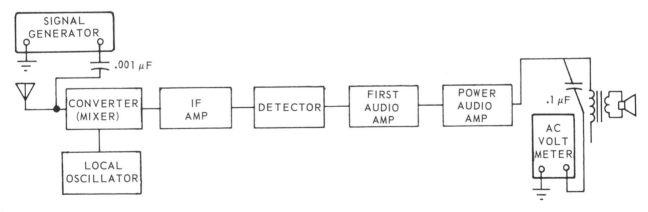

Fig. 16-25. This diagram shows only the signal path through a superheterodyne receiver. Connections for instruments are made for if alignment.

faint tone is heard in the speaker. Starting with the output of the last if transformer, adjust the trimmer capacitor, using a special alignment tool, for maximum reading on the voltmeter. Proceed toward the front end of the receiver by adjusting, in order, the secondary and primary of all if transformers for maximum output response. This completes the if alignment.

The oscillator section of the diagram in Fig. 16-25 contains both a trimmer capacitor and a series padder capacitor called the "600 PADDER." To adjust the oscillator, the generator is set at a modulated frequency of 600 KHz and attached to the antenna input terminals through a 250 pF capacitor. The receiver is tuned to 600 KHz and the 600 padder is adjusted for maximum response. Now the generator and receiver are tuned to the high frequency end of the band around 1400 KHz and the trimmer capacitor is carefully adjusted for maximum response. More complete instructions for the alignment of a particular radio will be found in the manual and will include antenna trimmer tuning and more precise oscillator adjustments.

THE FM RECEIVER

A block diagram of a complete FM receiver is illustrated in Fig. 16-26. Each block is labeled to designate its function in the system.

Basically the FM receiver is similar to the superheterodyne AM receiver except for three factors. The incoming signals to be tuned are from 88 to 108 megahertz. Secondly, the if frequency used in the FM radio is 10.7 megahertz. The same heterodyne principles are involved as with the AM receiver. The third difference is the method of detection a FM receiver uses.

FM DETECTION

In the AM radio, the detector had to be sensitive to amplitude variations of wave. A FM detector must be sensitive to frequency variations and remove this intelligence from the FM wave. In other words, the FM detector must produce a varying amplitude and frequency audio signal from a FM wave.

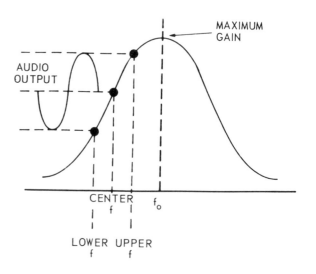

Fig. 16-27. These curves demonstrate slope detection.

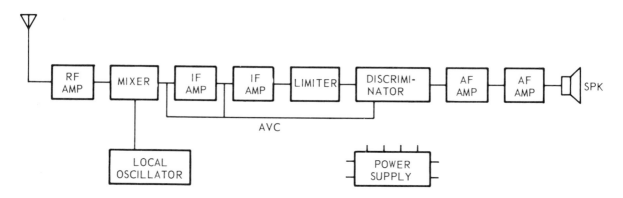

Fig. 16-26. The block diagram of a typical FM radio.

To approach this method of detection, consider the diagram in Fig. 16-27. Assume that a circuit has a maximum response at its resonant frequency. All frequencies other than resonance will have a lesser response. So if the center frequency of a FM wave is on the SLOPE of the resonant response curve, a higher frequency will produce a higher response in voltage and a lower frequency will produce a lower voltage response. The curves in Fig. 16-28 reveal that the amplitude of the output wave is the result of the maximum deviation of the FM signal and that the frequency of the audio output depends on the rate of change of frequency of the FM signal.

DISCRIMINATION

For ease in understanding, the discriminator in Fig. 16-28 uses three tuned circuits. In this circuit, L_1C_1 is tuned to the center frequency. L_2C_2 is tuned to above center frequency and L_3C_3 is tuned to below center frequency by an equal amount. At center frequency equal voltages are developed across the tuned circuits and D_1 and D_2 conduct equally. The voltages across R_1 and R_2 are equal and opposite in polarity and the circuit output is zero. If the input frequency increases above center, then L_2C_2 develops a higher voltage and D_1 conducts more than D_2 and unequal voltages develop across R_1 and R_2. The difference between these voltage drops will be the audio signal. The output therefore is a voltage wave varying at the rate of frequency change at the input and its amplitude depends upon the maximum deviation. The capacitors across the output of the discriminator are for filtering out any remaining radio frequencies.

The discriminator is redrawn in Fig. 16-29 as a typical circuit which you will encounter in FM receivers. L_1 and C_1 are tuned to center frequency. At frequency above resonance, the

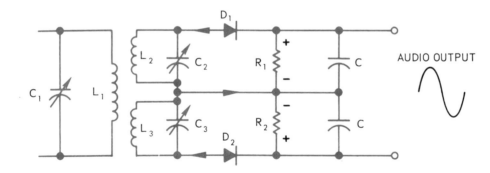

Fig. 16-28. A FM discriminator circuit using semiconductor diodes.

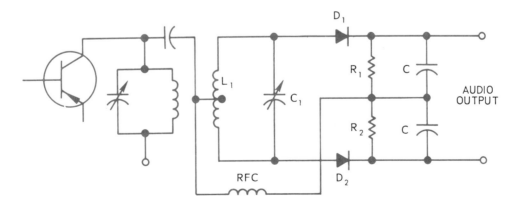

Fig. 16-29. A typical Foster-Seeley discriminator using special transformer designed for this purpose.

tuned circuit becomes more inductive; at frequencies below resonance, the circuit becomes more capacitive. The out of phase conditions produce resultant voltages which determine which diode will conduct. The output is an audio wave, the same as in the more elementary discussion.

This discussion is meant only to familiarize you with this type of detection. In advanced courses, you will study it in more detail.

It will interest you to know that each diode in the discriminator must have equal conduction capabilities which means that semiconductor diodes used must be in MATCHED PAIRS.

THE RATIO DETECTOR

Another type of FM detector is illustrated in Fig. 16-30. It is called the RATIO DETECTOR. In this circuit, the diodes are connected in series with the tuned circuit. At center frequencies, both diodes conduct during half-cycles. The voltage across R_1 and R_2 charges C_1 to output voltage. C_1 remains charged because the time constant of $C_1 R_1$ and R_2 is much longer than the period of the incoming waves. C_2 and C_3 also charge to the voltage of C_1. When both D_1 and D_2 are conducting equally, the charge of

C_2 equals C_3 and they form a voltage divider. At the center point between C_2 and C_3 the voltage is effectively zero.

A frequency shift either below or above center frequency causes one diode to conduct more than the other. As a result, the voltages of C_2 and C_3 will become unequal, but will still always total the voltage of C_1. This change of voltage at the junction of C_2 and C_3 is the result of the RATIO of the unequal division of charges between C_2 and C_3. This will vary at an audio rate the same as the rate of change of the FM signal.

Look at the charge on C_1 again. It is the result of the amplitude of the carrier wave or signal strength. It is charged by half-wave rectification of the FM signal. It is, therefore, a fine point to pick off an automatic volume control voltage to feedback to previous stages to regulate stage gain.

NOISE LIMITING

Another particular advantage of FM radio is the fact that the receivers are sensitive to and detect frequency variations and not amplitude variations. Most noise and interference to radio reception is in the form of NOISE SPIKES and are amplitude variations which have little effect

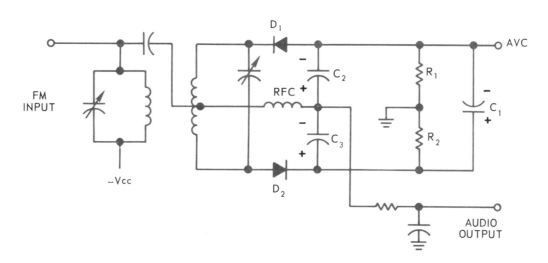

Fig. 16-30. A typical ratio detector circuit using semiconductor diodes.

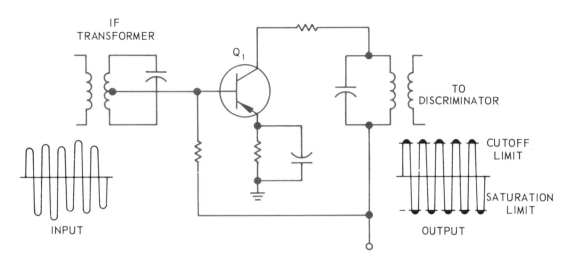

Fig. 16-31. A circuit from a limiter stage before a discriminator.

on the FM detector. Therefore, FM reception is relatively free of noise and disturbances. To hold the FM signal at a CONSTANT AMPLITUDE before detection in a discriminator circuit, a LIMITER is used as a previous stage. A schematic of a transistorized LIMITER STAGE is shown in Fig. 16-31.

A limiter is nothing more than an OVERDRIVEN amplifier stage. If the incoming signal reaches a certain amplitude in voltage, it drives the transistor to cutoff or to saturation when the voltage is opposite in polarity. At either of these points, gain can not increase. Consequently, the output is confined within these limits. Any noise spikes in the form of amplitude modulation would be clipped off.

CRYSTAL RADIO

Crystal radios have been built at home and in school for many years. The first experience of hooking a few parts together and hearing a local broadcast program may become a lifetime memory. To many young people, the home built crystal radio is the first step toward a life long interest in electronics. The little radio, Figs. 16-32 and 16-33, demonstrates some of the principles we have studied, such as tuning and detection.

Fig. 16-32. A practical crystal radio, using commercially manufactured parts.

PARTS LIST FOR CRYSTAL RADIO

T_1 — Antenna Coil: Miller 20A
C_1 — Variable Capacitor 365 pF, Philmore 1946 G
CR_1 — 1N34 Crystal Diode or RCA SK 3087
Headphones 2000 Ω
Binding Posts or Terminals

Commercial parts have been used in this project because its construction is an introduction to more sophisticated circuits. Early familiarization with components and con-

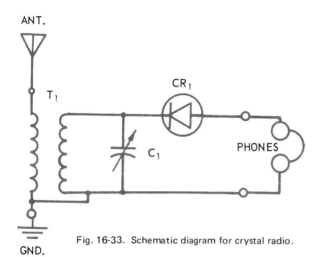

Fig. 16-33. Schematic diagram for crystal radio.

struction practice will pay dividends in more advanced projects. The radio may be built on a plastic sheet or a wooden base. It will require a good antenna for satisfactory operation.

FOUR TRANSISTOR RADIO

A four transistor radio that gives good reception on local stations and is relatively inexpensive to build is shown in Figs. 16-34, 16-35 and 16-36.

Station selection is accomplished by tuned circuit $L_1 C_1$. The signal is detected by crystal CR_1 and amplified by four transistor stages to drive the loudspeaker. This amplification also increases the sensitivity of the detector. The grounded-emitter configuration is used by Q_2, Q_3 and Q_4. C_4, C_5 and C_6 provide interstage coupling.

Referring to the photographs you will notice that tiny electrolytic capacitors are used, also the transistor tuning capacitor C_1 and midget speaker LS_1 permit a compact assembly. All parts have been mounted on a sheet of 1/16 in. plastic, but printed circuitry may be used.

FOUR TRANSISTOR RADIO PARTS LIST

B_1 — 6 volt battery
C_1 — 0.365 pF tuning capacitor
C_2, C_3 — 0.02 mfd ceramic disc or paper
C_4, C_5, C_6 — 1.0 mfd electrolytic, 15V
C_7 — 25.0 mfd electrolytic, 15V
CR_1 — Sylvania 1N64 or 1N34 diode or RCA SK 3087
L_1 — Ferri-loopstick antenna coil
LS_1 — 3-6 ohm loudspeaker
Q_1, Q_2, Q_3, Q_4 — Sylvania 2N 1265 transistor or RCA SK 3003
R_1, R_8 — 470 K, 1/2W
R_2, R_4, R_6 — 220 K, 1/2W
R_3, R_5, R_7 — 2.2 K, 1/2W
R_9 — 100 Ω , 1W
T_1 — Argonne AR-133 or equivalent

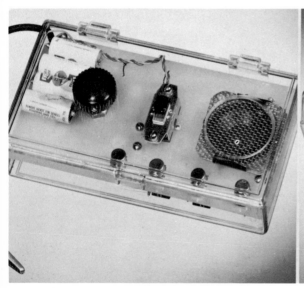

Fig. 16-34. Pocket size transistor radio with loudspeaker.

Fig. 16-35. Rear view of transistor radio showing parts placement.

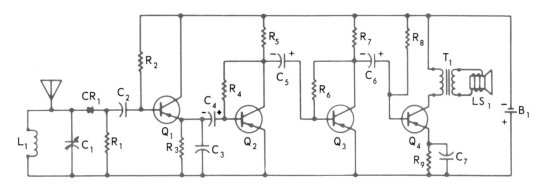

Fig. 16-36. Schematic diagram of transistor radio. (Sylvania)

It is advisable to build the radio first by using parts mounted on a breadboard. After satisfactory tests have been completed, a case may be designed to hold the radio.

SIX TRANSISTOR SUPERHETERODYNE RADIO

This is a superheterodyne receiver designed to receive AM (amplitude modulation) broadcast stations transmitting on frequencies between 550 and 1600 KHz. The rf amplifier-converter stage receives the modulated rf signal from the broadcast stations in your area. By tuning the radio, you select the modulated rf signal from one of the broadcast stations. In this stage, it is amplified and combined with the rf signal from the local oscillator in your radio to form an intermediate frequency (if) signal.

The if signal is then amplified in the if amplifier stage and passed on to the detector stage. The detector stage removes the carrier frequency, leaving only the af signal that was originally combined with the carrier frequency at the broadcast station; a process referred to as demodulation. This small af signal is amplified by the first af amplifier stage and then by the audio power amplifier stage, making it strong enough to operate the speaker.

The circuit used in this six transistor radio was provided by Graymark Enterprises. It may be purchased as their Model 806 six transistor radio.

The schematic for the six transistor radio is shown in Fig. 16-37. The parts list appears as Fig. 16-38. The radio is pictured in Fig. 16-39.

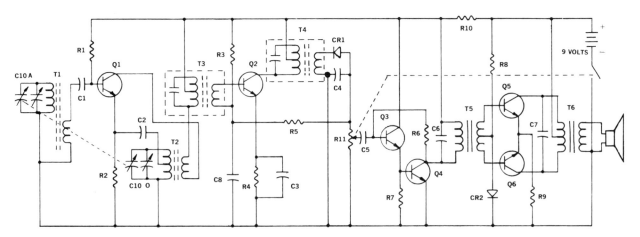

Fig. 16-37. Schematic for six transistor radio. Parts list is on following page. (Graymark)

QTY.	SYMBOL	PART NO.	DESCRIPTION
1	R1	62232	Resistor, 820K ohms, ¼W, 10%
1	R2	61415	Resistor, 4.7K ohms, ¼W, 10%
1	R3	62233	Resistor, 120K ohms, ¼W, 10%
1	R4	61421	Resistor, 330 ohms, ¼W, 10%
1	R5	62234	Resistor, 18K ohms, ¼W, 10%
1	R6	62235	Resistor, 560K ohms, ¼W, 10%
1	R7	61417	Resistor, 470 ohms, ¼W, 10%
1	R8	62236	Resistor, 1.8K ohms, ¼W, 10%
1	R9	62237	Resistor, 6.8 ohms, ¼W, 10%
1	R10	61417	Resistor, 220 ohms, ¼W, 10%
1	R11	62175	Potentiometer, 5K ohms*
2	C1, C6	62207	Capacitor, disc, .01μF
1		62207	Capacitor, disc, .01μF (for testing only)
1	C2	62153	Capacitor, disc, .005μF
3	C3, 4, 7	62212	Capacitor, disc, .02μF
1	C5	62208	Capacitor, disc, .04μF
1	C8	62238	Capacitor, electrolytic, 10μF
1	C9	62178	Capacitor, electrolytic, 30μF
1	C10	62239	Capacitor, tuning
1	Q1	62240	Transistor, silicon, 9600
1	Q2	62241	Transistor, silicon, 9600
2	Q3, Q4	62242	Transistor, silicon, 9630
2	Q5, Q6	62243	Transistor, silicon, 9616
1	CR1	62244	Diode, germanium (large)

QTY.	SYMBOL	PART NO.	DESCRIPTION
1	CR2	62245	Diode, silicon (small)
1	T1	62246	Antenna coil
1	T2	62247	Oscillator coil (red)
1	T3	62248	1st IF transformer (yellow)
1	T4	62249	2nd IF transformer (black)
1	T5	62250	Audio transformer, interstage (blue)
1	T6	62251	Audio transformer, output (red)
1	LS	62181	Speaker, 8-ohm*
1	EP1	62252	Earphone, magnetic
1	J1	62253	Earphone, jack
1		62162	Battery connector, 9V
1		62182	Antenna holder
1		62254	Knob, volume
1		62255	Knob, tuning
1		62256	Dial, for tuning knob
1		62165	Carrying strap, with ring
1		62189	Speaker washer, vinyl
3		62171	Machine screw, flathead
1		62170	Machine screw, small
2		62257	Self-tapping screw
1		62261	Solder, rosin core, 60/40
1 set		62258	Hookup wire, 2-orange, 1-green, 1-yellow
1		62259	Printed circuit board (PCB)
1 set		62260	Cabinet, front and back portions

*Preattached

Fig. 16-38. Parts list for six transistor radio.

Fig. 16-39. AM receiver. (Graymark Enterprise, Inc.)

Model 806 in Fig. 16-39 may be ordered from: Graymark Enterprises, Inc., P. O. Box 54343, Los Angeles, California 90054. For individual parts, quote part numbers from Fig. 16-38.

FOR DISCUSSION

1. How does the tuning circuit select a desired frequency?
2. Explain the process of heterodyning signals.
3. Draw the block diagram of a communication model and explain the function of each stage.
4. How does an AM receiver differ from a FM receiver?
5. Discuss the advantages and disadvantages of AM and FM.

TEST YOUR KNOWLEDGE

1. The process of removing the audio wave from the modulated rf wave is called _____ or _____ .
2. Name three types of detectors and draw basic circuits of each.

3. What two characteristics determine the quality of a tuning circuit?
4. Name three purposes of an rf amplifier.
5. What is the purpose of the capacitor across the load resistor in the detector?
6. What is the purpose of the AVC circuits?
7. What is the usual if frequency used in broadcast receivers?
8. What type of audio amplification is used in Fig. 16-11? Name the stages.
9. Draw a diagram and explain the operation of a PM speaker.
10. Explain the action of a tone control circuit.
11. What is the purpose of alignment?

Chapter 17

TELEVISION

In a brief space of a few years, television has grown from its infancy to a giant in the field of communications. Today, entertainment, education, on-the-spot news casts and commercial advertising are channeled into millions of homes as television has become a popular addition to family living.

Although it is beyond the scope of this text to study the detailed electronic circuits involved in the production, transmission and reception of television signals, it is important for students interested in the science of electronics to have an overall picture of television and related information.

THE BASIC TELEVISION CAMERA (BLACK AND WHITE)

What is a black and white picture? A picture may be considered as an infinite number of black and white dots, and dots of varying degrees between black and white. These dots are called picture elements. Examine a photo in

one of the local newspapers and these elements are clearly visible. As the television camera looks at a scene in the studio, it is looking at these many picture elements of varying intensities from black to white. The scene is focused on a photosensitive mosaic consisting of many photoelectric cells. Each cell responds by producing a voltage in proportion to the intensity of light. These voltages are amplified and used for modulation of the AM carrier wave, which is transmitted to the home receiver. A line drawing of an image orthicon is shown in Fig. 17-1. The scene before the camera is focused on the photo cathode by the conventional lens system used by cameras. The varying degrees of illumination cause electrons to be emitted on the target side of the cathode, actually forming an electron image of the scene before the camera. The target plate is operated at a high positive potential and the electrons from the cathode are attracted to the target. The target is made of low resistive glass and has a transparency effect. The electron image appears on both sides of the target plate.

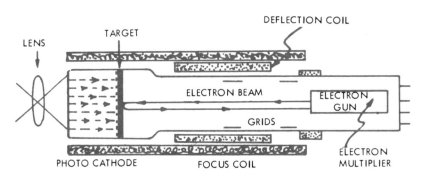

Fig. 17-1. A nontechnical sketch showing the interior arrangment of the image orthicon tube used in the television studio.

At the right in Fig. 17-1 is the electron gun which produces a stream of electrons. The electron stream is accelerated by the grids. The beam is made to scan left to right and from top to bottom by the magnetic deflection coils around the tube. The moving electron beam strikes the target plate and the electrons return to the electron multiplier section. The intensity of the electron stream returning to the multiplier is in proportion to the electron image and provides the desired signal current for electronic picture reproduction.

SCANNING

Scanning may be defined as the point-to-point examination of a picture. In the picture camera the electron beam must scan the electron image and respond to the point-to-point brilliance of the picture. The system of scanning used by television in the United States is the interlace system and consists of 525 scanning lines. The beam starts at the top left hand corner of the picture and scans, from left to right, the odd lines of the scanning pattern, such as lines 1-3-5-7 etc. At the completion of 262 1/2 lines, the electron beam is returned to the beginning by vertical deflection coils and the beam then scans the even numbered lines. One scan of 262 1/2 lines represent a FIELD and one set of the odd and even numbered fields represents a FRAME. The interlace system of scanning reduces picture flicker to a minimum. The scanning pattern is shown in an abbreviated form in Fig. 17-2.

The frame frequency has been set by the FCC as 30 hertz, which means that your TV receives 30 complete frames per second or 60 picture fields per second of alternate odd and even lines. A continuous, nonflickering picture is a result of your persistence of vision similar to your experience with a motion picture which is a successive showing of individual scenes at a rapid rate.

In order to produce 525 lines per frame at a frame frequency of 30 Hz, the horizontal deflection oscillator causing the electron beam

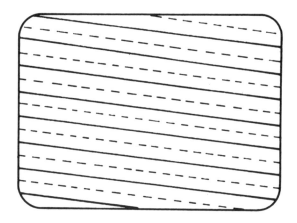

Fig. 17-2. Interlace system of scanning used in television in abbreviated form. There are actually 525 lines.

to move back and forth must operate at a frequency of 525 x 30 or 15,750 Hz. Likewise, the vertical deflection oscillator, which causes the beam to move from top to the bottom of the picture, must have a frequency of 60 cps. This is the field frequency.

Further examination of the scanning process reveals that the beam scans as it moves from left to right. After it has read one line, then it must quickly flyback to the left to start reading the next line, similar to the way you are reading this book. The FLYBACK time is very rapid, yet would show a visible retrace line in your picture. The picture must be BLACK during this retrace time. Also, when the beam reaches the bottom of the picture, the beam must be returned on the top in order to scan again. The picture must also be black during vertical flyback or trace lines would be visible.

The oscillators producing the scanning and flyback voltages for both the HORIZONTAL and VERTICAL SWEEP must produce a wave form as illustrated in Fig. 17-3. This is a saw-tooth wave form. Notice the relatively gradual increase in voltage during the sweep and then the rapid decrease during flyback. These voltages are applied to coils around the picture tube in the television called the DEFLECTION YOKE. The increase in the strength of the magnetic fields in the coils causes the electron beam to move. The deflection yoke and its schematic symbol are shown in Fig. 17-4.

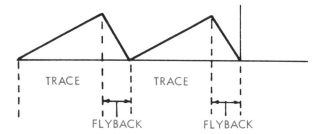

Fig. 17-3. The wave form of the voltages required for scanning and flyback. It is called a saw-tooth wave form.

It is obvious that the scanning of the picture at the studio and the scanning of the picture in your home television must be in step. In order to accomplish this, a very accurate pulse generator is used to TRIGGER the horizontal and vertical oscillators at the studio. This same pulse is also sent over the air and received by the television set. This pulse, known as the SYNCHRONIZATION PULSE or SYNC PULSE, triggers the oscillators in the receiver and keeps them at exactly the same frequency. The HORIZONTAL HOLD and VERTICAL HOLD controls on your TV are used to make slight adjustments so that the oscillators can lock-in with the SYNC pulses.

COMPOSITE VIDEO SIGNAL (BLACK AND WHITE)

The television signal received by your TV must therefore contain considerable information in addition to the picture (video) and sound (aural) information. Perhaps the easiest way to understand the complete video signal will be to follow the information of the signal as the camera scans a simple picture of a checkerboard in Fig. 17-5. Some standardization is necessary in the formation of the video signal so that your television may be used in any locality. These arbitrary standards are set up by the FCC and are used by all TV broadcasting stations. The amplitude of the modulation is divided into two parts. The first 75 percent is used to transmit video information and the remaining 25 percent for the sync pulses. Also a system of negative transmission has been adopted which means that the

Fig. 17-4. A deflection yoke which fits around the neck of the television picture tube. (Triad Transformer Corp.)

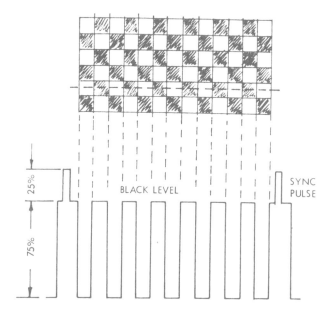

Fig. 17-5. The voltages developed as the camera scans one line across a checkerboard. The sync pulses are transmitted at the end of each line.

higher amplitudes of video information produce darker areas in the picture until at 75 percent the picture is completely BLACK. In Fig. 17-5 the percentage of video is indicated on the scale at the left. As the beam scans the next black bar, a similar action takes place. At the end of the line, the screen is driven back to the BLACK PEDESTAL or BLANKING LEVEL. During this blanking pulse the beam flyback occurs and also a sync pulse is transmitted in the blacker-than-black or infrablack region (upper 25 percent) for oscillator synchronization. A second line to be scanned would be a duplicate of the first unless the picture is changed. At the bottom of the picture a series of pulses trigger the vertical oscillator and keeps it synchronized. The illustration in Fig. 17-6 shows a composite video signal. The video information for one line between the blanking pulses consists of varying degrees of illumination between black and white.

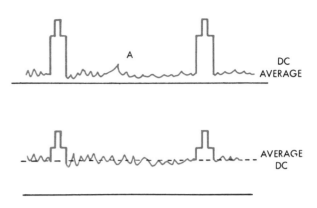

Fig. 17-7. A comparison of a dark and a light picture as it appears in the video signal.

Fig. 17-6. The composite video signal of one line scanned by the television camera.

A comparison between the two video signals in Fig. 17-7 would disclose that signal A is made up of mostly light objects or a daylight scene. The average overall brightness of the scene is another part of the information sent to your TV receiver from the transmitter and may be detected from the composite video signal.

THE BLACK AND WHITE PICTURE TUBE OR CRT

The average television viewer has a limited knowledge of the "insides" of the receiver. The viewers primary concern is with the picture appearing on the face of the picture tube. As students of electronics we will look behind the face of the cathode ray tube (CRT) and discover the mysteries of picture production. In Fig. 17-8 a sketch is produced which shows the functioning parts of the electron gun and deflection yoke. The theory of operation of the CRT is similar in most respects to the vacuum tube. Beginning at the left in Fig. 17-8 is the cathode and heater. Its purpose, of course, is the emission of a space cloud of electrons.

Surrounding the cathode is a metal cylinder, closed at one end except for a small hole or aperture, which controls the flow of electrons in the CRT. It is the control grid with which we are familiar. It also serves to concentrate or electronically focus the emitted electrons through the small hole in the shape of a beam. Grid No. 2 is operated at a higher positive potential and its strong attraction for the negative electrons causes the electrons to accelerate or speed up. Grids No. 3, 4, and 5 serve as further accelerating and focusing elements which results in a narrow stream or beam of electrons. The beam is moved from left to right, successively scanning the face of the picture tube from top to bottom, by the magnetic deflection coils around the neck of the tube. The inside of the face of the picture tube is treated with special fluorescent materials. The electron beam causes these chemicals to glow as the electrons strike the fluorescent screen. The brilliance is in proportion to the intensity of the electron beam. The picture is observed through the glass front of the tube. The inside of the bulb of the tube is coated with a conductive coating. External connections to this coating are made from a high voltage power supply. When the electron beam strikes the fluorescent screen, electrons are knocked off, called secondary emission. These electrons are collected by the high potential anode or conductive inner coating and thus complete the circuit.

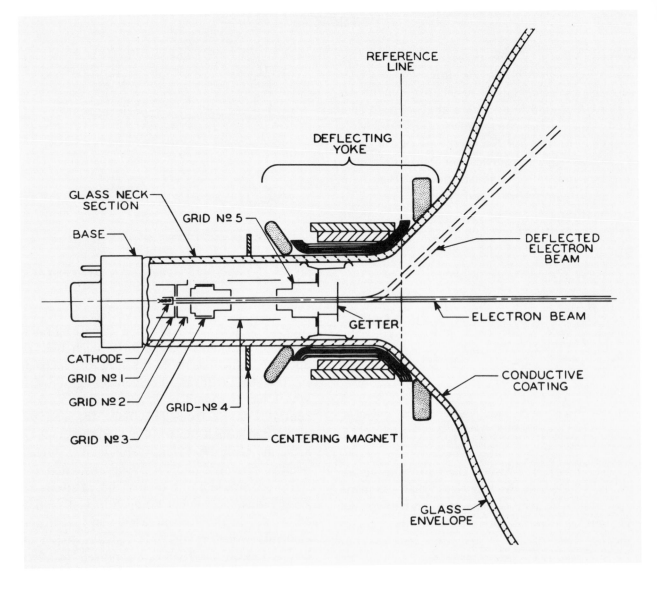

Fig. 17-8. Sketch of a typical CRT showing the electron gun.
(RCA)

A LESSON IN SAFETY: Extremely high and dangerous voltages are present at the external connections to this anode, usually from 8000 to over 20,000 volts. Extreme caution must be observed by the technician when working close to these connections. A wise person works with only one hand, so that an accidental shock will not be fatal. Keep one hand in your pocket!

Returning to the theory of vacuum tube operation, the control grid determines the flow of electrons through the tube. In the CRT this is also true. At zero bias the CRT is at maximum current and the screen, therefore, is bright or white. At cutoff bias the current is zero and the screen is black. The tube is operated at a selected bias on the control grid which may be controlled by a knob on the front of the TV called the BRIGHTNESS CONTROL. It adjusts the overall brightness level of the picture. When no picture is being received on the TV, the scanning electron beam may be observed by the lines on the TV screen called the RASTER. Turn your home TV to a vacant channel and observe this raster. Now adjust the brightness control from black to bright. The incoming detected video signal is applied to the grid of the CRT (sometimes to the cathode, depending on polarity of signal). The video signal adds to or subtracts from the bias on the tube resulting in a modulated or varying electron stream conforming to the picture information in the video signal. The picture is produced on the fluorescent screen.

The sharpness or focus of the electron beam may be adjusted by changing the voltages of the focusing grids. These controls are usually located on the rear of the TV and require occasional adjustments. Referring to Fig. 17-8 again, a centering magnet may be seen. Slight adjustments on this magnet may be made in the service shop when the picture is off center.

When the cathode in the CRT emits electrons, negative ions are also emitted. Since the ions have a greater mass than the electrons, they are not deflected as much, which could result in a brown spot about one inch in diameter forming in the center of the picture tube because of ion bombardment. Several methods are used to eliminate the ION SPOT. Some tubes employ a tilted electron gun arrangement

so that the ions hit the neck of tube. The electron beam is bent in its proper direction toward this screen by small magnets attached to the neck of the tube. These magnets are sometimes called an **ION TRAP** or **BEAM BENDER**. Aluminized picture tubes use an aluminum film on the electron gun side of the screen which protects the screen from ion bombardment as well as improving the sharpness of the picture.

THE BLACK AND WHITE TV RECEIVER

The block diagram in Fig. 17-9 represents the groups of components performing specific functions in the television receiver. This block representation is used to familiarize the student with the interrelationships between various sections of the television. The signal path through the various stages may be traced and the purpose of each group of components to the total reception of the TV program may be understood. The name of each block suggests its purpose in the circuit.

RF AMPLIFIER: As the name suggests, this tube serves a function similar to that in the superheterodyne radio. The incoming television

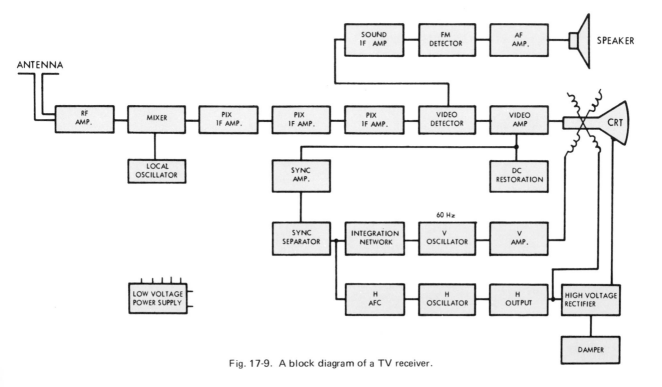

Fig. 17-9. A block diagram of a TV receiver.

signal is selected by switching fixed inductors into the tuning circuit. These carefully tuned circuits are necessary to provide uniform gain and selectivity for each television channel. In this stage the video signal with all its information is amplified and fed to the mixer.

MIXER: In the mixer stage the incoming video signal is mixed with the signal from a local oscillator to produce an intermediate frequency. The common if used in modern televisions is 45.75 megahertz. When the channel selector switch is turned on to a channel, not only is the tuning circuit changed, but also the frequency of the oscillator is changed, always producing the if of the correct frequency. A FINE TUNING control is provided which changes the frequency of the oscillator slightly in order to provide the best response. The above three blocks in the diagram, namely the rf amp, mixer and oscillator are combined in one unit called the TUNER or FRONT END of a television. These units are usually factory assembled and aligned. Adjustments should not be made on these units without correct instrumentation and thorough knowledge of procedures.

PIX-IF AMPLIFIERS: The output of the mixer stage, containing the 45.75 megahertz intermediate frequency with the video and aural information, is amplified by these stages. To provide a maximum frequency response of each stage up to 45.75 MHz, each stage must amplify a broad band of frequencies. Consequently the voltage gain of each stage is reduced and more stages of if amplification are required. In this system the aural information or sound is passed through the if amplifier with the video signal. It is called the "intercarrier system."

VIDEO DETECTOR: The output from the last if stage is fed to the detector or demodulator. The detection process is the same as in the radio. The video signal used to amplitude modulate the transmitted carrier wave is separated and fed to the next stage.

VIDEO AMPLIFIER: In these stages the demodulated video signal is amplified and fed to the grid (sometimes cathode) of the CRT. It is this signal which modulates or varies the intensity of the electron beam and thus pro-

duces the picture on the face of the CRT.

SOUND IF AMPLIFIER: Later in this chapter you will discover that the FM sound of the television program is separated from the video carrier wave by 4.5 megahertz. This produces a 4.5 MHz FM signal at the output of the video detector which is coupled to the sound if amplifiers. The FM sound signal is amplified in these stages.

FM AUDIO DETECTOR: This stage detects the frequency variations in the modulated signal and converts it to an audio signal.

AF AMPLIFIERS: These are the same as used in the conventional radio or audio system. The audio signal is amplified sufficiently to drive the power amplifier and the loudspeaker.

SYNC SEPARATOR: The output of the video amplifier is fed to this circuit which removes the horizontal and vertical sync pulses, which have been transmitted as part of the composite video signal. These sync pulses are used to trigger the horizontal and vertical oscillators and keep them "in step" with the television camera.

SYNC AMPLIFIER: This is a voltage amplifier stage to increase the sync pulses.

HORIZONTAL AFC: In this block the frequency of the horizontal oscillator is compared to the sync pulse frequency. If they are not the same, correction voltages are developed which change the horizontal oscillator to the same frequency.

HORIZONTAL OSCILLATOR: This oscillator operates on a frequency of 15,750 Hz and provides the saw-tooth wave form necessary for horizontal scanning of the picture.

HORIZONTAL OUTPUT: This stage shapes the saw-tooth wave form correctly for the horizontal deflection coils and drives the horizontal deflection coils. It also provides power for the high voltage rectifier.

HIGH VOLTAGE RECTIFIER: The output of the horizontal oscillator is used to shock excite the HORIZONTAL OUTPUT TRANSFORMER (HOT). High ac voltage developed by this autotransformer is rectified and filtered for the anode in the CRT. See Fig. 17-10.

DAMPER: This stage dampens out the oscillations in the HOT after retrace and also

Fig. 17-10. A typical horizontal output transformer (HOT). (Triad Transformer Corp.)

rectifies part of the ac for a boosted B+ VOLTAGE. Part of this damper voltage is used in the horizontal scanning under the name of "reaction scanning."

VERTICAL OSCILLATOR: The output of the sync amplifier is fed through a vertical intergration network to the vertical oscillator. The integrator, which is nothing more than a RC circuit, is designed with a time constant which builds up a voltage when a series or group of closely spaced sync pulses occur. The transmitted vertical pulse is a wide pulse and occurs at a frequency of 60 Hz. The vertical oscillator is triggered by this pulse and therefore keeps on frequency. The output of the vertical oscillator provides the saw-tooth voltage to the deflection coils which moves the beam from the top to the bottom of the screen and the flyback from bottom to top, at the end of each field.

VERTICAL OUTPUT: An amplifier for the output of the oscillator to provide the proper currents in the deflection yoke for vertical scanning.

SPECIAL PURPOSE CIRCUITS

Many refinements have been made on the basic television circuit, including automatic controls for many functions which would be tedious for the TV viewer to manage. It might be well to mention and define a few of these. Detailed explanations and circuitry may be found in many television text books.

DC RESTORATION: The average darkness or brightness of the TV picture (night or day scenes) is transmitted as the average value of the dc component of the video signal. If the video amplifiers are RC coupled, the dc value of the signal is lost. In this case, the average value of the video signal is taken from the detector and used to establish the bias on the CRT.

AUTOMATIC GAIN CONTROL: The AGC in the television serves the same function as the AVC in the radio receiver. Its purpose is to provide a relatively constant output from the detector by varying the gain of the amplifiers in previous stages. This is accomplished by rectifying the video signal to produce a negative voltage. This voltage is applied to the bias of the previous amplifiers to change their gain.

Fig. 17-11 shows the rear view of a color television set. Note the location of the various components

Fig. 17-11. Rear view of color television receiver.

COLOR TELEVISION

Color television was developed in the late 1940s. The current system used in the United States was pioneered by the RCA Laboratories.

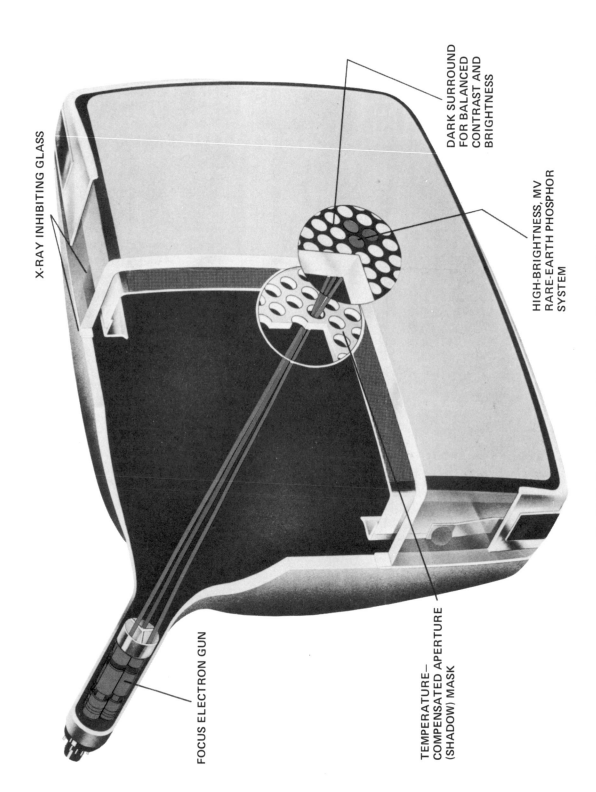

DARK SURROUND FOR BALANCED CONTRAST AND BRIGHTNESS

X-RAY INHIBITING GLASS

HIGH-BRIGHTNESS, MV RARE-EARTH PHOSPHOR SYSTEM

FOCUS ELECTRON GUN

TEMPERATURE— COMPENSATED APERTURE (SHADOW) MASK

Fig. 17-12. Basic type of color picture tube. (Sylvania-GTE)

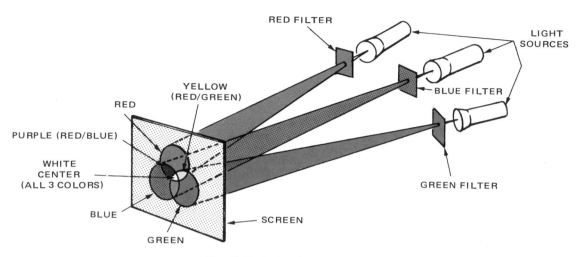

Fig. 17-13. Basic colors used in television.

In March, 1950, a color television demonstration was given in Washington, DC to the Federal Communications Commission personnel, news-personnel and other industrial personnel. As a result of this demonstration, an overwhelming reaction was given to the remarkable quality of the color television pictures.

One of the major inventions that made modern color television possible was the SHADOW-MASK PICTURE TUBE. Fig. 17-12 shows a picture of a shadow mask tube.

The three basic colors used in color television are red, blue, and green. See Fig. 17-13. By using combinations of these colors, any color can be produced on a television screen.

The first type of color picture tube to be produced commercially was the DELTA-TYPE tube. This was invented in 1950 and it is essentially the same tube that was used in the first color television demonstrations given by RCA Laboratories in Washington, DC.

The delta-type tube uses three separate electron guns placed in the neck assembly of the picture tube. See Figs. 17-14 and 17-15A.

As the electrons are emitted by the three guns they are directed toward the screen which

is filled with hundreds of thousands of triads of colors (red, blue, and green). In between the electron gun and the color producing screen of the picture tube, the three electron beams have to be focused through an aperture or shadow mask. This shadow mask assures that the electrons strike the dots where they are supposed to hit. As the scanning process takes place, a line is made by one complete scan (from left to right) of all three electron beams hitting all the series of dots across the screen. If all electron beams are adjusted properly, the result will be a white line. If only the red gun and green guns are on, a yellow line will be

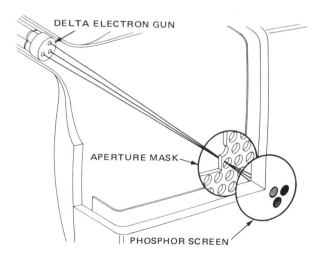

Fig. 17-14. Conventional delta gun, shadow mask, and tri-dot screen arrangement. (GTE, Sylvania)

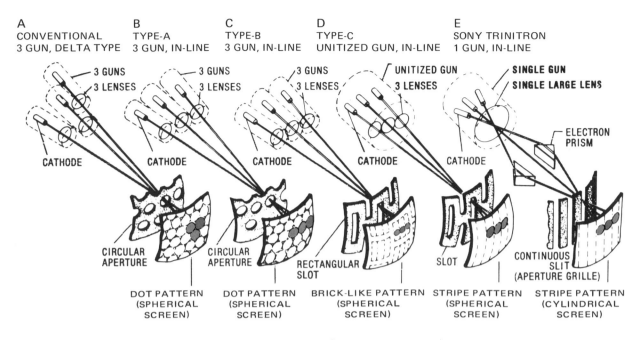

A	B	C	D	E
CONVENTIONAL 3 GUN, DELTA TYPE	TYPE-A 3 GUN, IN-LINE	TYPE-B 3 GUN, IN-LINE	TYPE-C UNITIZED GUN, IN-LINE	SONY TRINITRON 1 GUN, IN-LINE

Fig. 17-15. Color tube types. (Sony Corp. of America)

produced. Any colored line can be produced by the combination of intensity of electron beams from the delta assembly.

In-line gun assemblies were invented after the shadow mask tube. Fig. 17-15B, C, D, E, shows four in-line gun assemblies with different aperture grill and screen patterns. The type shown in Fig. 17-15C is used in many modern television receivers. Also the type shown in Fig. 17-15E is used by Sony Corp. of America in its line of color television receivers. It is called a **TRINITRON** tube and it provides a very sharp picture.

For some time in the future, the conventional shadow mask delta-type tube will be used as well as in-line tubes. In the not too distant future, the in-line tubes may overtake the delta-type tube in popularity.

A color television receiver is a very complicated electronic instrument. It has to be capable of producing both color picture and also black and white. The incoming signal has to be COMPATIBLE with black and white sets also. Fig. 17-16 shows the circuitry of a modern color television receiver. Fig. 17-17 shows a block diagram of a popular color television receiver.

Fig. 17-16. Component view of color television receiver. (Quasar Corp.)

266

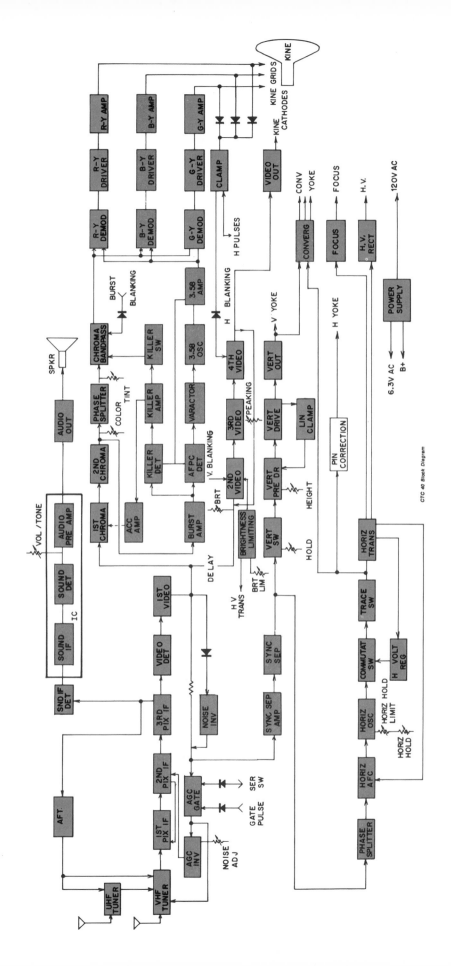

Fig. 17-17. RCA CTC-40 Color receiver diagram. (RCA)

VERY HIGH FREQUENCY (VHF)
TELEVISION CHANNELS (2-13)

CHANNEL NUMBER	FREQUENCY BAND (MHz)	VIDEO CARRIER FREQUENCY	AURAL CARRIER FREQUENCY
2	54-60	55.25	59.75
3	60-66	61.25	65.75
4	66-72	67.25	71.75
5	76-82	77.25	81.75
6	82-88	83.25	87.75
7	174-180	175.25	179.75
8	180-186	181.25	185.75
9	186-192	187.25	191.75
10	192-198	193.25	197.75
11	198-204	199.25	203.75
12	204-210	205.25	209.75
13	210-216	211.25	215.75

Table 17-1. Frequency allocations.

THE TELEVISION CHANNEL

The Federal Communications Commission has assigned a channel or section of the radio frequency spectrum for each television channel. The two types of television channels are very high frequency (VHF) and ultra high frequency (UHF). Each channel is 6 MHz wide. The VHF channels, 2 to 13, are shown in Table 17-1. Let us examine Channel 4 in detail, Fig. 17-18. The basic video carrier frequency is 67.25 MHz. Reviewing Chapter 15 you will recall that when an rf carrier wave is amplitude modulated,

sideband frequencies appear which represent the sum and difference between the carrier frequency and the modulating frequencies. To transmit a real sharp video picture, frequencies up to 4 MHz and more are necessary for modulation. These would combine in Channel 4 to give a band occupancy necessary for the modulated signal from 67.25 – 4MHz or 63.25 MHz to 67.25 + 4 MHz or 71.25 MHz, or a total channel width of 8 MHz. But the FCC only allows 6 MHz and a compromise must be made. In commercial TV broadcasting the upper sideband is transmitted without attenuation. The lower sideband is partly removed by a vestigial-sideband filter at the transmitter. The curve in Fig. 17-18 represents the typical response characteristics of the TV transmitter. The sound is transmitted as a frequency-modulated signal at a center frequency 4.5 MHz above the video carrier.

In Channel 4 the sound is at 71.75 MHz.

Ultra high frequency (UHF) television band covers from channel 14 to 83. As in very high frequency (VHF), the bandwidth of each channel is 6 megahertz. The same frequencies bandwidths are used for picture carrier as VHF.

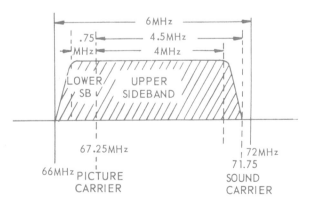

Fig. 17-18. Location of picture carrier and sound in Channel 4. Channel is 6 MHz wide.

The UHF channels used for commercial television, are set by the Federal Communications Commision (FCC) in the United States. Table 17-2 shows examples of where the ultra high frequency television channels falls along the radio spectrum.

ULTRA HIGH FREQUENCY (UHF)
TELEVISION CHANNELS (14 - 83)

CHANNEL NUMBER	FREQUENCY BAND (MHz)
14	470-476
15	476-482
16	482-488
17	488-494
18	494-500
19	500-506
(And so on to UHF channel 83-6 MHz per channel)	
83	884-890

Table 17-2. UHF channels.

TELEVISION OF THE FUTURE

There are many predictions for improving television. Large screen projection television is already in existence. However, it is anticipated that in the future your television picture tube will hang on the wall as a picture or portrait does.

Three dimensional television using holographic images may be possible in the next twenty five years. These devices will produce life size, three dimensional images that will literally fill the living room or family room. Laboratory research is currently being done on this process.

FOR DISCUSSION

1. Name and discuss the function of each of the following controls in a TV receiver. State the circuit which is regulated by each control:

   a. Horizontal hold.
   b. Brightness.
   c. Vertical hold.
   d. Fine tuning.
   e. Channel selector.
   f. Channel selector.
   g. Vertical linearity.
   h. Horizontal linearity.
   i. Height control.
   j. Contrast.
   k. Width control.

2. Explain the meaning of negative transmission used in television in the U.S.
3. How does the vertical integration network separate the vertical sync pulse?
4. If dc amplifiers were used in the video section, would dc restoration be necessary?
5. Why are both UHF and VHF channels needed for television?
6. Why have most television manufacturers changed from the use of vacuum tube to solid state circuits?
7. Discuss the various types of color picture tubes.
8. Discuss how you think television will be like in the year 2000 A.D.

TEST YOUR KNOWLEDGE

1. What is the frequency of the horizontal oscillator in a TV?
2. What is the frequency of the vertical oscillator?
3. What is meant by the interlace scanning system?
4. How many horizontal lines are in one field?
5. Which circuit and major components produce the high voltage for the picture tube?
6. In television, which methods of modulation are used for the video and sound transmission?
7. What information is contained in the composite video signal?
8. What is the purpose of the ion trap?
9. At center frequency or carrier frequency, what is the value of the audio voltage in a FM discriminator circuit?

10. Draw a block diagram of the television receiver.
11. What frequency is the PIX-IF in television?
12. What frequency is the sound if?
13. What are the three basic colors used in color television?
14. What is the purpose of the shadow mask in the color picture tube?
15. What frequency is Channel 10 in the VHF bands?

Chapter 18

ELECTRONICS IN INDUSTRY

The visible evidence of progress in electronics technology is apparent in radio and television and in space age developments.

Almost any manufactured product we use today has in some way been made better or more economically through the use of electronics. Automation in industry means electronic control of machine operation, grading and counting of products and inspection. Only a few years ago many people performed long and tedious operations which now may be accomplished with only a few highly skilled technicians. This means shorter working hours and more leisure time; higher wages and a bountiful supply of useful products for everyone to enjoy. Now, highly complex computers do the work of thousands of people involved in mathematical computations and accounting. The computer can complete work in a few minutes which previously required weeks and months of manpower. Yes, we do live in an electronic age. Average citizens would be amazed, if they did know, how involved their lives have become in electronics and how dependent they are upon this science.

The purpose of this chapter is to acquaint the student with a few of the more common electronic applications in industry. Advanced studies in this specialized field will be rewarding. Career opportunities for skilled technicians seem almost unlimited in this technical age.

PHOTOELECTRIC CONTROLS

In our study of the principles of the vacuum tube, we discovered that certain materials, when heated, would "boil out" or emit electrons. This type of material was used as the cathode in the vacuum tube. This type of emission was called THERMIONIC EMISSION.

Other materials exhibit a similar characteristic when exposed to light. Materials which are sensitive to light include zinc, potassium, and other alkali metals. This type of electron emission has been named PHOTOELECTRIC EMISSION. When a light strikes a surface of these materials, the light energy, if sufficiently strong, will cause free electrons to be emitted from the surface. This is a very useful discovery. Circuits may be designed which will take advantage of the intensity of light to control certain industrial operations, counting, sorting and inspection.

There are three classes of light sensitive materials which are used in light control circuits:

1. PHOTOEMISSIVE CELLS OR PHOTOTUBES. These are gas filled or high vacuum type tubes with two elements, i.e. cathode and anode. This tube is illustrated in Fig. 18-1 with its schematic symbol. Light

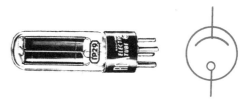

Fig. 18-1. The phototube and its schematic symbol.

shining on the cathode will cause electrons to be emitted. When a positive voltage is applied to the anode, it will attract the electrons and current will flow through the tube. This current may be used to control another vacuum tube and relay circuit.

2. PHOTODIODES. These materials show a decrease in resistance and consequently an increase in current when light shines on them. In Fig. 18-2 a photodiode is diagramed. This is a PN junction photodiode connected with reverse bias. Only a small amount of reverse current will flow, in the order of a few microamperes. This is termed

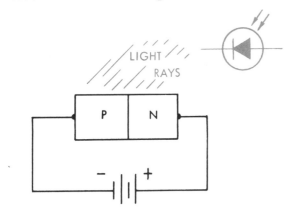

Fig. 18-2. The block diagram and symbol of a photodiode.

the DARK CURRENT. When a light strikes the surface of the diode, the current will increase due to the diffusion of minority carriers across the junction. A minority carrier in a P crystal is an electron; in an N crystal, a hole.

3. PHOTOVOLTAIC CELLS. This cell has many uses. It will develop a voltage across its surfaces when exposed to light. This discovery came from the research on transistors by the Bell Telephone Laboratories. Walter H. Brattain, co-inventor of the transistor, remarked when he first observed this experiment:

"There was this man . . . . . . and he had a little chunk of black stuff with a couple of

contacts on it and when he shone a flashlight on it, he got a voltage . . . . . . I didn't believe it."

Fig. 18-3 shows an assortment of solar cells used to convert sunlight into electricity. A very familiar application is the light meter. Such an instrument is indispensable to the photographer, who uses the meter to determine the intensity of the light.

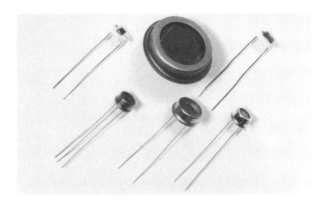

Fig. 18-3. Solar cells — P on N silicon.

In Fig. 18-4 a bank of solar cells is used to power a telephone communications system for the Bell System. The symbol for the photovoltaic cell is shown in Fig. 18-5.

Fig. 18-4. Sun cells used to power telephone communications system. (Bell Telephone Laboratories)

Fig. 18-5. Symbol for photovoltaic cell.

BASIC PHOTO CONTROL CIRCUITS

In the diagram, Fig. 18-6 the schematic of the basic photo control circuit is shown. This circuit can be used as an automatic door bell or counter. The door bell can be connected to the relay contacts, or an impulse type counter may be connected across the relay.

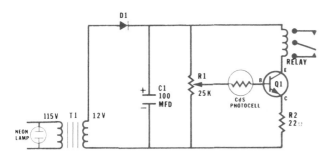

Fig. 18-6. Basic photo control circuit.   (Radio Shack)

TIME CONTROL CIRCUITS

In industry there are many needs for precise timing operations. In many cases the timing is done by mechanical devices and switches. The electronic timer is a very useful device, whether used to control a photo enlarger in a photographic laboratory, or a huge welding machine in industry. Basic time SCR-UJT control circuit is shown in Fig. 18-7.

RADIO CONTROL

The use of radio frequency waves to control machinery and devices in distant locations is becoming increasingly more popular. A radio-

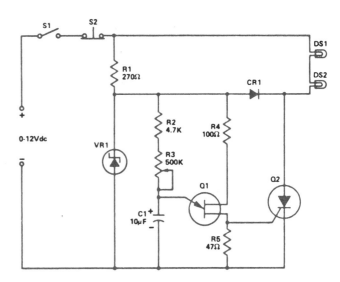

Fig. 18-7. A simple SCR-UJT timer circuit.
(Lab-Volt, Buck Engineering, Co.)

controlled garage door opener is illustrated in Fig. 18-8. The car is equipped with a small radio transmitter. When keyed by the driver, the transmitter sends out a modulated rf wave to the garage receiver, which activates a relay and starts the door opening mechanism.

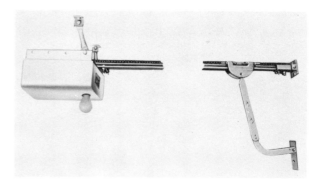

Fig. 18-8. This garage door opener is radio controlled.
(Perma-Power)

The aerospace industry uses radio frequency waves to control the flight pattern of new aircraft ideas in model form, Fig. 18-8A. The model is tested before going to the tremendous expense of building and testing full size aircraft. In this way, the aircraft can be tested to the complete limits of their capability without endangering the life of a test pilot.

Fig. 18-8A. Radio controlled miniature aircraft used to determine the shape of the space shuttle re-entry vehicle.
(NASA)

Military applications of remotely piloted vehicles (RPVs) is developing into a new technology. Radio control now provides adequate command and control of unpiloted aircraft over hostile environments at altitudes up to 70,000 ft. for reconnaissance purposes. RPVs, Fig.

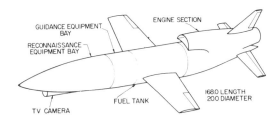

Fig. 18-9. One of the newest RPV (remotely piloted vehicle). This craft permits the reconnaissance of enemy territory with the pilot safely located many hundreds of miles away.

18-9, equipped with miniature TV cameras, can accurately deliver weapons on ground targets with the control pilot safely located miles away.

The hobby of controlling boats, planes and cars, Fig. 18-10, is followed enthusiastically by millions of persons. These hobbyists are able to put their craft through intricate maneuvers, on command, by using radio equipment that can be purchased fully assembled or in kit form, Fig. 18-11.

The Federal Communications Commission (FCC) has set aside specific frequencies on the 27 MHz, 53 MHz and 72 MHz bands for exclusive use of radio controlled models.

NOTE: It is necessary to have a CLASS C operator's license from the FCC before radio control equipment can be operated on the 27 and 72 MHz bands. A technician class or higher amateur radio operator's license is required to operate on the 53 MHz band.

THEORY OF OPERATION

The radio control unit shown in Fig. 18-11, was constructed from a kit, and can be constructed as either a 3 or 4 channel unit. That is,

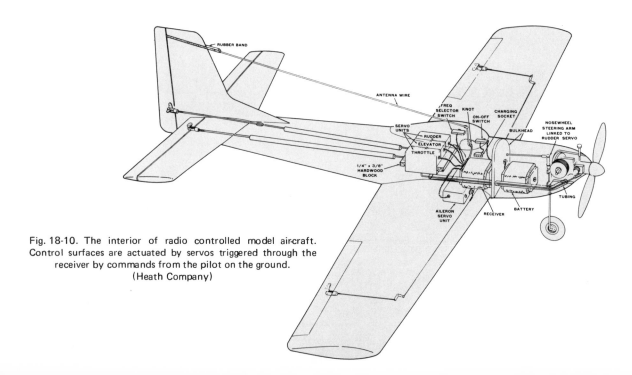

Fig. 18-10. The interior of radio controlled model aircraft. Control surfaces are actuated by servos triggered through the receiver by commands from the pilot on the ground.
(Heath Company)

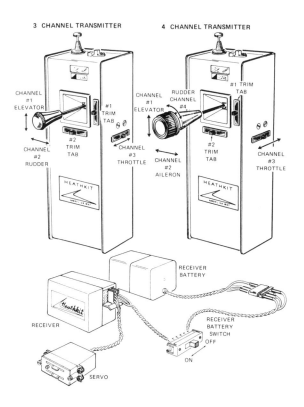

Fig. 18-11. Radio control units can be purchased already assembled or constructed from well-designed kits. Shown are 3 and 4 channel digital proportional systems. Units with up to 8 channels are available. (Heath Company)

it can be used for the remote control of three or four devices (usually servos) depending upon the type of circuit constructed. It is a digital guidance system that operates on a pulse-width comparison principle. Pulse modulation of the rf carrier is crystal controlled.

When all controls are in neutral, the transmitter sends out a frame of five pulses that are repeated continuously every 16,000 microseconds. See Fig. 18-12, WAVEFORM A. Each

pulse in the frame is 350 microseconds wide, and all pulses in the frame, except the first one, normally start 1500 microseconds after the start of the previous pulse. This is called a "fixed frame rate" because the time interval between the first pulse in one frame and the first pulse in the next frame is always 16,000 microseconds. This cannot be changed.

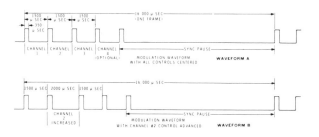

Fig. 18-12. Typical waveform of a 3 or 4 channel RC unit. (Heath Company)

The time frame between pulses is 1500 microseconds and can be increased or decreased as much as 500 microseconds. The infinitely (immeasureable) variable width between the individual pulses controls the amount of servo travel (how far the control surface will move). A tiny change in pulse width, wider or narrower, produces a small amount of servo travel in the appropriate direction. A large movement of the control stick will produce a large change in pulse width and gives a correspondingly greater amount of servo travel. Other channels are not affected, Fig. 18-12, WAVEFORM B.

The modulated rf carrier can be seen in Fig. 18-13. It shows that the rf carrier is turned off during the 350 microsecond pauses, but is on all

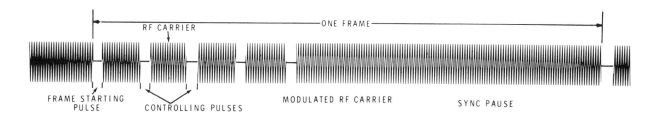

Fig. 18-13. Modulated rf carrier of a 4 channel RC circuit. (Heath Company)

other times. Off time is minimized to reduce the possibility of interference that will cause erratic and unwanted servo movement (called "glitches" by the RC hobbyist). The schematic for the 3 channel digital proportional transmitter is shown in Fig. 18-14.

The pulse train, upon being detected by the receiver, is amplified. It triggers the decoder circuits which "read-out" the individual pulses to the appropriate servos. For example: The pilot wants the model plane to make a climbing right turn. The elevator control (channel No. 1) is pulled down slightly for a gradual climb and the rudder control (channel No. 2) is moved to the right. The control stick movements change the pulse lengths of channels No. 1 and 2. Pulse length of the other channel is not affected. The decoder circuit in the receiver notes the change in the pulse length of channel No. 1 and instructs the elevator servo how far up it must move the plane's elevator. Almost instantaneously, the decoder is also telling the rudder

servo how far it is to move the rudder to the right. Control surface movement is proportional to the stick movement.

Control movements are held until the control stick is returned to neutral or moved to a new position. The decoder circuits are reset in synchronism with the transmitter signal by the long sync pause.

ELECTRONIC HEATING

In recent years industry has discovered many methods of heating both conductive and non-conductive materials by using high frequency alternating currents. These methods have provided a means to quickly heat, to a controlled temperature and depth, many materials for specific industrial applications.

INDUCTION HEATING

Induction heating is used for materials which conduct electricity. In its more simple form it

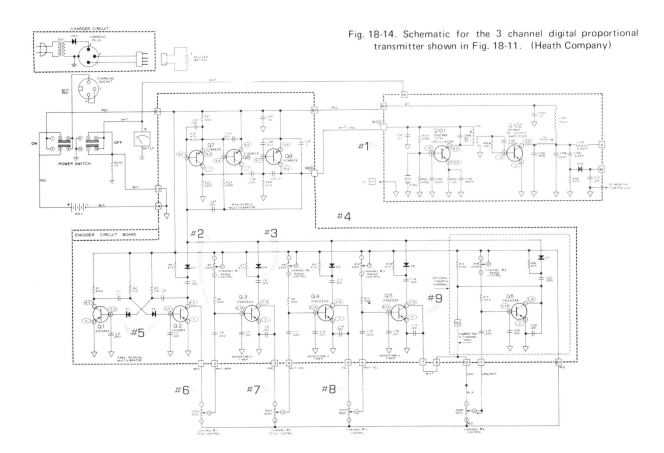

Fig. 18-14. Schematic for the 3 channel digital proportional transmitter shown in Fig. 18-11. (Heath Company)

consists of only a transformer with a single turn or short circuited secondary winding as drawn in Fig. 18-15. The alternating current in the primary winding induces a current in the secondary. In earlier studies we learned about the I^2R power loss in a transformer, generator or motor. In the induction heater the power

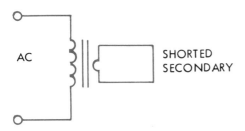

Fig. 18-15. The short circuit secondary winding demonstrates the principle of induction heating.

loss, as a result of the high currents, serves a useful function of heating the material. More specifically, the secondary of the transformer may be the material to be heated, and the induced currents are the eddy currents within the material. See Fig. 18-16. As the frequency of the alternating current is increased, the eddy currents have a tendency to flow on the surface of the material. This phenomena is known as SKIN EFFECT. It is useful in industrial processes. In many cases it is necessary to heat the surface of a material for surface hardening, yet leaving the interior of the material softer and stronger. In induction heaters, vacuum tube oscillators are used for the generation of high frequency currents. These oscillators do not differ, circuitwise, from oscillators used in radio transmitters, except in their ability to produce

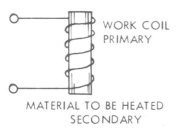

Fig. 18-16. Heat is produced in the material when it is inserted in the work coil.

high power. In some industrial applications thousands of watts of power are required. A conventional oscillator circuit used in an induction heating circuit is drawn in Fig. 18-17.

The tuned circuit L_1C_1 determines the frequency of the oscillator. Energy is fed back to the grid of tube V_1 from the tickler coil L_2. Even though this circuit appears like a conventional oscillator used in radio communications, it must handle high voltages and frequently many thousands of watts in power. All parts must be rugged and well insulated.

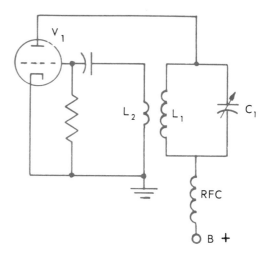

Fig. 18-17. A basic oscillator circuit used in induction heating.

Examples of induction heating may be found in heat treating processes, brazing, tin plating and other applications requiring a high degree of temperature control and heat penetration, as well as cleanliness and economy.

An induction furnace for the heat treating of steel is illustrated in Fig. 18-18. The photos show front and rear view of a 10 KW induction heating unit. Work to be heated is placed in the work coil, as seen in the front view, which acts as a primary winding of a transformer. The resulting eddy currents in the work produce heat as they overcome the resistance of the work. The depth of the heating may be controlled by changing the frequency of the induction heating. Lower frequencies produce

Fig. 18-18. An induction furnace used in heat-treating steel in industry.

greater depth of heat; surface heating is accomplished by higher frequencies. This is one way by which electronics serves us. Improved methods in manufacturing bring to the consumer better products at lower prices. Improved methods in industry provide more healthful and satisfying working conditions, plus higher wages and shorter working hours.

DIELECTRIC HEATING

When nonconductive materials must be heated, they may be placed between plates as a dielectric material used in the familiar capacitor. Fig. 18-19 illustrates this type of heating. When a high ac voltage is applied to the plates, the heating within the insulating material or dielectric is the result of molecular distortion and friction. The high frequency voltage applied to the plates causes the molecules of the dielectric to vibrate and rub together and heat is produced by intermolecular friction. This may be compared to hysteresis loss in the iron core of a transformer, and is termed dielectric

hysteresis. This type of heating requires higher frequencies than the induction heating, usually from 1 MHz to 50 MHz.

Interesting applications of dielectric heating are in the electronic oven for cooking, electronic gluing of plywoods and furniture, heating and shaping of plastics, and medical treatments requiring deep heat and the producing of artificial fever in patients.

MAGNETIC AMPLIFIERS AND REACTORS

The terms magnetic amplifier and saturable reactor are sometimes used interchangeably, but actually the magnetic amplifier consists of the saturable reactor plus other associated components such as rectifiers, resistors and transformers. Before actually looking at the saturable reactor, a brief review of magnetism and inductive reactance should be made. Repeat the experiments as illustrated in Fig. 7-3. The student must realize that the reactance of a coil varies according to the permeability of the core.

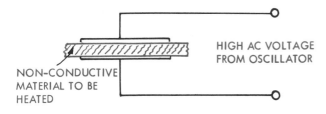

Fig. 18-19. In dielectric heating the material serves as the dielectric of a capacitor.

In the case of the dimmer circuit in this experiment the core was changed from air to iron by inserting the iron core in the coil. It is, therefore, a correct assumption that the impedance of a coil may be varied by varying the permeability of the core. Also the permeability as stated in Chapter 4, is

$$\mu = \frac{B}{H}$$

where,

μ = Permeability

B = Flux density

H = Magnetizing force

These are the principles on which the saturable reactor is based.

The basic circuit of the saturable reactor shown in Fig. 18-20, requires some explanation. To one side of the transformer is connected a variable dc voltage. The load, ac power source and the other side of the reactor are connected in series. When no dc voltage is applied to the control side, the impedance of the load

windings is maximum and the current through the load is minimum. As the dc control voltage is increased, the magnetic flux density of the core increases and its permeability decreases. This reduces the impedance of the load windings and the current through the load increases. This is a very useful device and has many industrial applications. Its advantage lies in the fact that a small direct current may be used to control a relatively large ac power circuit. Theaters use a device of this sort for dimming the lights. Otherwise, large rheostats would be necessary which would be cumbersome and wasteful of power.

This basic circuit is not practical for an obvious reason. The ac in the load circuit would induce a voltage in the dc control side through transformer action. One common solution to this problem is found in the three-legged saturable reactor illustrated in Fig. 18-21. The dc control produces a magnetic field as indicated by the solid arrows. The ac load circuit produces magnetic fields as indicated by the dashed lines. Note that in the center core the ac fields are opposite and equal and therefore cancel each other. The fields in the

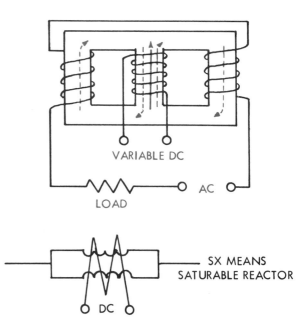

Fig. 18-21. The three-legged or tertiary wound reactor diagram.

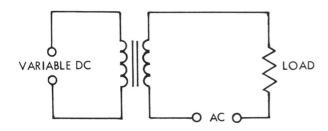

Fig. 18-20. The basic circuit of the saturable reactor.

two outside cores are the resultant between the ac and dc fields. (On one side the fields are opposing, on the other side the fields are additive.) The resulting impedance of the two series load windings is the sum of the impedance. Since impedance is a function of "rate of change" in flux density, a low dc control current permits maximum change and therefore maximum impedance. A high dc control current permits a minimum change and minimum impedance.

MAGNETIC AMPLIFIER

A detailed study of magnetic amplifiers is beyond the scope of this text. Only a simple explanation will be included, such as the circuit illustrated in Fig. 18-22. The amplifier includes other parts, such as the saturable reactor, rectifier and load resistor. This circuit produces

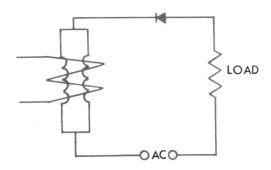

Fig. 18-22. The magnetic amplifier circuit.

a controllable dc voltage across the load R_L. It is a simple half-wave rectifier circuit. The magnitude of the voltage appearing across the load is a function of the impedance of the load windings, which in turn is varied by the dc control windings. This circuit may be converted to a controlled full-wave rectifier output by using the bridge type rectifier in Fig. 18-23.

The amplifying characteristic shown by the magnetic amplifier has already been mentioned. Further amplification may be achieved by a feedback system whereby part of the load current change may be inductively coupled to the control circuit in the proper phase. This is

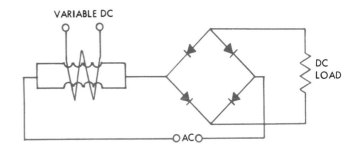

Fig. 18-23. The magnetic amplifier using a full-wave bridge rectifier for controlled dc voltage.

commonly called regenerative feedback. In this case windings are included on the saturable reactor for feedback as shown in the symbol in Fig. 18-24.

Interested students should refer to the bibliography in the Appendices of this text for a list of books on Magnetic Amplifiers and Applications.

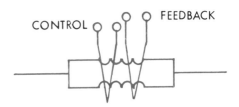

Fig. 18-24. The schematic diagram of the reactor with control and feedback windings.

ULTRASONIC CLEANING

An ultrasonic washer is shown in Fig. 18-25. Sound waves, generated by an oscillator, are

Fig. 18-25. An ultrasonic washer for cleaning small parts and mechanisms.

transmitted to the water by a transducer that invisibly agitates the water and cleans without the use of soap, detergents or hot water.

The principle involved has already found many applications in industry for cleaning small and intricate mechanisms and parts.

THERMOELECTRIC COOLING

Your home of the future will probably contain a refrigerator which freezes or cools your food in absolute and mysterious silence. No motors or refrigerants will be required. Cooling will be by "thermoelectric refrigeration." This phenomenon is produced when an electric current is passed through the proper kinds of semiconductor materials. Thermoelectric refrigeration is characterized by its simplicity, ruggedness and freedom from

maintenance. Your "Home of the Future" refrigerator may consist of a series of cabinets and drawers as illustrated in Fig. 18-26. These amazing futuristic appliances will be another electronic contribution to better living.

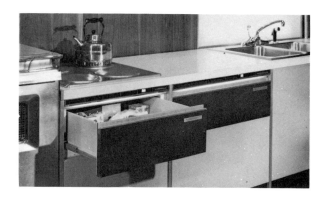

Fig. 18-26. Thermo electric refrigeration featured in the House of the Future.

FOR DISCUSSION

1. List all the ways that you can think of by which electronic circuits have improved our way of living.
2. Suggest some applications for the photo electric control system.
3. Suggest some applications for the electronic timing circuits.

TEST YOUR KNOWLEDGE

1. Name three types of photo sensitive materials used in light control circuits.
2. Name two types of electronic heating.
3. Draw the basic magnetic amplifier circuit using half-wave rectification.
4. Draw the basic oscillator circuit used in induction heating. How does it differ from the oscillator in a radio transmitter?

Chapter 19

TEST INSTRUMENTS

The basic instrument meter movement used in the measurement of voltage, current and resistance was discussed in detail in Chapter 10. Actual measurement problems were used to show some of the disadvantages of these meters. Among these disadvantages is that by nature the meter will load a circuit. Increasing the sensitivity of the meter in ohms/volt reduces the error resulting from loading to a tolerable amount. An improvement upon the basic meter is the vacuum tube voltmeter (VTVM) which has an input resistance of 10 megohms and higher and the current drawn from the circuit under test is negligible. The D'Arsonval rectifier type meter, when used for measuring high frequency alternating currents also has a shunting effect due to the capacitance of the rectifiers. In the VTVM the shunting effect is eliminated and voltages at frequencies up to 100 MHz may be measured.

THE VTVM

The basic circuitry of the vacuum tube voltmeter is drawn in Fig. 19-1. The tube V_1 is biased to cutoff and the meter in the plate circuit reads "zero." When a positive voltage is applied to the grid, the tube starts conduction and the meter in the plate circuit is calibrated to read in volts. This plate circuit meter is really an ammeter, but its deflection is the result of voltage applied to the grid.

The same type of circuit may be used for measuring ac. Since the tube is biased to cutoff, only the positive half of the ac cycle causes current to flow in the plate circuit. The negative half-cycle drives the grid more negative beyond cutoff. The meter actually measures the average value of the ac wave, but may be calibrated to indicate average, rms or peak voltages applied to the grid.

A more sophisticated VTVM employs multiplier resistors to increase its range and a balanced dc amplifier. This circuit is shown in Fig. 19-2. Consider the circuit first with no voltage applied to the grid of V_1. The bridge may be balanced by adjusting R_1 until the meter reads "zero." A negative voltage applied through the voltage divider network to the grid of V_1 upsets the balanced condition of the bridge since V_1 would be conducting less than V_2. The unbalanced state creates a difference in

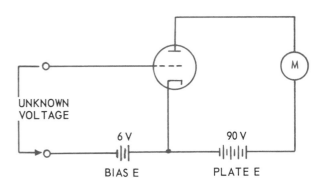

Fig. 19-1. Theory of the VTVM may be compared to a simple vacuum tube circuit.

voltage drop between R_2 and R_3 which causes current to flow through the meter. The meter is calibrated to read in volts. The switching arrangement permits the measurement of either negative or positive voltages.

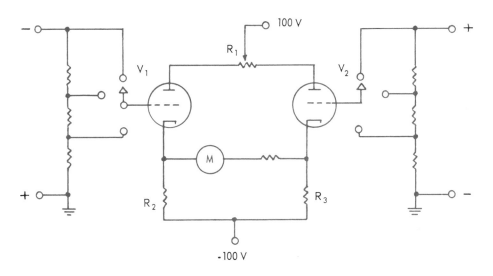

Fig. 19-2. Practical VTVM circuit used in many commercial meters.

In order to measure ac voltages with this VTVM a diode rectifier probe must be used, Fig. 19-3. Another type of probe uses a series resistor with a value between 1 and 5 megohms. This resistor prevents detuning or loading of circuits to be measured. Another probe, used in the measurement of ac voltages, employs a small series capacitor which reduces the input capacitance of the meter. It also reduces detuning effects when making ac measurements. Still another probe uses a crystal diode which extends the high frequency response of the meter.

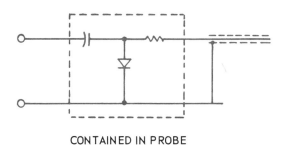

CONTAINED IN PROBE

Fig. 19-3. Circuit diagram of an ac probe used with a VTVM when measuring ac voltages.

The VTVM is a useful service and test instrument. Its advantages include its high sensitivity and high impedance input. It may be used to measure voltages in high impedance circuits, amplifier gains and signal tracing. A typical instrument is illustrated in Fig. 19-4. This instrument measures ac voltages, both negative and positive dc voltages and resistances. The right hand switch sets the meter on its proper function and the left hand switch selects the range. Potentiometers are provided for "ohm's adjust" and "zero adjust" for bridge balancing.

Fig. 19-4. Typical VTVM used by technicians. (Dynascan Corp.)

THE VOLT-OHM-METER (VOM)

A very simple meter to use in electronic circuits is the volt-ohm-meter (VOM). It basically consists of a circuit that has a meter, a power source, a variable resistor (for zeroing the meter) and a load. There are two basic types of

VOMs . . . the series and the shunt. Fig. 19-5 shows a series VOM, Fig. 19-6 shows a shunt type VOM.

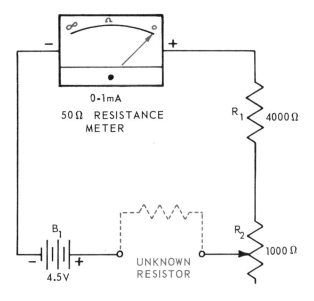

Fig. 19-5. A series VOM.

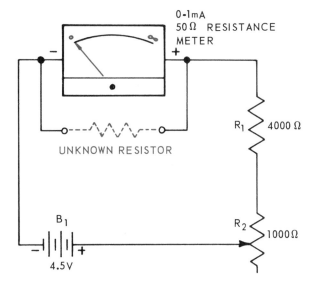

Fig. 19-6. Shunt VOM.

A commercial volt-ohm-meter is shown in Fig. 19-7. Volt-ohm-meters have the advantage of being portable and low in cost. They do have a low input resistance (in ohms-per-volt) on the lowest voltage range which can cause serious accuracy problems.

Fig. 19-7. Volt-ohm-meter. (Simpson Electric Co.)

With the development of the field-effect transistor, the VOM was redesigned to overcome the low input impedance problem. The FET-VOM can be used to measure ac and dc voltage, ac and dc current, resistance and decibel ratings. Fig. 19-8 shows a modern commercial FET-VOM.

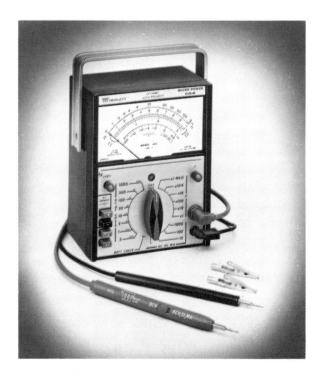

Fig. 19-8. Field effect transistor, volt-ohm-meter. (Triplett Corp.)

DIGITAL METERS

Digital multimeters are becoming more and more popular in the electronics field. These meters measure ac and dc current, ac and dc voltage, resistance with excellent accuracy. Many are portable, however, some may be operated from ac line voltage.

The digital multimeter has the ability to provide instantaneous visual display of the unknown circuit value. Costs of these multimeters have reduced drastically since their first appearance in the test instrument market. Figs. 19-9 and 19-10 show some typical digital multimeters.

Fig. 19-9. Digital multimeter. (Simpson Electric Co.)

Fig. 19-10. Digital multimeter. (Ballantine Laboratories Inc.)

An interesting digital multimeter for hand-held testing is shown in Fig. 19-11. This unit is completely self-contained and can be used anywhere.

THE OSCILLOSCOPE

The oscilloscope permits the technician to observe the wave forms in an electronic circuit.

It is a valuable service instrument. Although a detailed description of the oscilloscope is beyond the scope of this text, the student should be familiar with some of the basic principles and the operating controls.

Fig. 19-11. Hewlett-Packard Model 970A digital multimeter. (Hewlett-Packard Corp.)

The display of wave forms is produced upon the face of the cathode ray tube (CRT) in a fashion similar to your television. An electron beam from the gun in the CRT is made to sweep from left to right by the saw-tooth voltages produced by the sweep oscillator. Review Chapter 16. This produces a single line (usually green) on the face of the tube. If an ac voltage is applied to the vertical deflection circuit, then a wave is produced on the CRT. This wave represents the instantaneous voltage during the cycles of the ac input.

The horizontal sweep oscillator is adjustable through a wide range of frequencies. Assume that it is adjusted at 60 Hz. It is then producing 60 straight lines per second across the face of the CRT. The time for one sweep would be 1/60 of a second.

Now a 60 Hz ac voltage is applied to the vertical input terminals. This wave starts at zero, rises to maximum positive and returns to zero, then decreases to maximum negative and returns to zero. This all occurs during one cycle and the period of this cycle is 1/60 second. Therefore, the horizontal sweep and the input signal voltage are synchronized. <u>One wave</u> appears on the screen. If the horizontal sweep were set at 30 Hz, which is only one half as fast as the 60 Hz input signal, then <u>two complete</u> waves would appear on the screen. The horizontal frequency may be adjusted so that the wave forms appear <u>stationary</u>.

OSCILLOSCOPE FAMILIARIZATION

The function of the basic oscilloscope controls will be described, using the illustration in Figs. 19-12 and 19-13.

Fig. 19-12. A dual trace oscilloscope. (Heathkit)

Fig. 19-13. A triggered single trace oscilloscope.
(Hickok Teaching Systems)

INTENSITY: This controls the brightness of the trace on the screen. It is combined with the on-off switch.

FOCUS: Used to adjust the electronic focusing of the electron beam in order to make a sharp trace on the screen.

V-CENTERING: To move the beam up or down, so that it can be located in the center of the screen.

H-CENTERING: To move the beam right or left so that the wave pattern may be centered on the screen.

BAND RESPONSE SWITCH: This switch provides either a wide or narrow band response.

V-POLARITY: The pattern may be inverted by this switch.

SWEEP VERNIER: Fine frequency adjustment for horizontal sweep oscillator.

H-GAIN: Increases the length of pattern by expanding it horizontally.

V-RANGE: Selects the voltage range for measurement.

V-CAL: To make calibration adjustments so that peak voltages may be read directly on the scope. Set wave form peaks to the required height.

SWEEP: Selects the range of frequencies for the sweep oscillator.

SYNC: Switch for either internal or external synchronization pulses. These pulses are used to lock input signal and horizontal sweep for a stationary pattern.

V-INPUT: Terminal for input to vertical response circuits.

GND: Ground.

Z-AXIS: A signal may be applied to this terminal to modulate the intensity of the trace.

PHASE: Used in conjunction with "line sync"

so that sweep oscillator will be synchronized correctly.

SYNC TERMINAL: To this terminal an external sync pulse may be applied.

H-INPUT: The horizontal sweep may be switched from the selective internal oscillator, to 60 Hz line or to an external oscillator by this switch.

CALIBRATION SCALE: This is a plastic scale which fits over the face of the CRT. The instrument is calibrated so that these scales may be used for direct measurements of the peak value of wave forms.

Complete the activity with the "scope" outlined at the end of this chapter. Experience is the best teacher.

THE SIGNAL GENERATOR

Signal generators may be described as instruments which produce an electrical wave at a given amplitude and frequency. The generator employs electronic circuitry and vacuum tubes or transistors in the conventional oscillator circuits which you have studied. They may be classified according to their range of frequencies, such as audio and radio frequency generators. In either case the generator produces a signal of known frequency and amplitude which may be used by the technician in the service and adjustment of circuits. An rf generator set at the intermediate frequency 455 KHz is used to align the superheterodyne radio. The generated signal is fed through the if stages and the transformers are tuned to give maximum response. See Fig. 19-14. The audio generator may be used to check the audio sections of radios and TVs as well as the response of the now popular hi-fi sound systems. See Fig. 19-15.

These two generators are combined in one instrument. See Fig. 19-16. To use a generator of this type, and all types are similar, the output cable is connected to the desired output

Fig. 19-14. An rf signal generator. (Dynascan Corp.)

Fig. 19-15. Audio frequency (sine and square wave) generator. (Dynascan Corp.)

Fig. 19-16. An rf and af signal generator. (Hickok Teaching Systems)

jack, rf or af. The bandswitch is set to the band which includes the desired frequency. The tuning dial is adjusted to that frequency. The multiplier switch must be considered on this instrument. The amplitude of the output will depend upon the multiplier factor selected. The ATTENUATOR varies the amplitude of the signal. The signal selector switch is set to the desired af or rf output. When in the modulation (MOD) position, the rf wave is modulated with a 400 Hz audio sine wave. This instrument will be used in "Activities with the oscilloscope" at the end of this chapter.

TUBE CHECKERS

An instrument to test the condition of vacuum tubes and transistors is a "must" in every electronics laboratory. With the possible exception of the multimeter, it will be the most used piece of equipment. Tube failures in radios and televisions account for the major share of service problems. Considering the vacuum tube, there are three general faults which may develop and a good checker should provide a means of checking them.

1. The emission of the cathode. As tubes age in service, the ability of the cathode to emit electrons deteriorates. Most tube checkers employ a meter which reads, "Good," "Doubtful," "Bad" and the condition may be quickly determined.

2. Interelectrode short circuits between the elements in the tube must be discovered.

3. The transconductance of the tube which is a measurement of the electron flow through the grid type tube.

Provisions for checking rectifiers and diodes must also be included in the instrument.

Tube checkers vary over a wide range of quality and price. The determining factor is the accuracy and dependability required for specific applications. Checks on tubes may be static tests or dynamic tests. In static tests constant voltages are applied to the elements of the tube and the dc plate current is measured. The information secured from this test is not particularly valuable since the tube is not performing under simulated operating conditions. A more meaningful test is the dynamic test, during which an actual signal is applied to the grid of the tube. In the dynamic test a comparison is made between the change in plate current as a result of a change in grid voltage. This quotient reads directly on the meter in "micromhos" of transconductance.

A popular type of electron tube tester is shown in Fig. 19-17. This particular tester is automatic in many respects. A card is selected from the file representing the tube to be checked. The card is inserted in the instrument and all proper connections and voltages are made to the tube when the lever on the left is placed in testing position. Other testers require

Fig. 19-17. A modern tube tester. (Dynascan Corp.)

a manual setting of several levers and dials for each tube to be checked. There are many variations to the basic tube checker and the operator should become thoroughly acquainted with the particular instrument in use. Detailed instructions will always be found in the manual supplied with the tester.

TRANSISTOR TESTERS

Commercial transistor testers are available to test every kind of semiconductor device produced. Transistor testers may be of the type that classify transistors into "good" or "bad" types. Also, they may be of the more sophisticated variety that measures the inherent characteristics of a particular transistor. The later type of transistor testers usually display the output data on a meter or an oscilloscope screen.

Many transistor testers will test the transistor in the circuit. Laboratory testers usually require out-of-the-circuit testing, however, more exact standards are met with these. Many commercial transistor testers also can be used to perform diode tests. See Fig. 19-18.

In the sweep generator a part of the modulating voltage is brought out to a terminal, which may be connected to the EXTERNAL SYNC terminal of the oscilloscope. This is done to synchronize the sweep circuits of the scope with the generator. See Fig. 19-19.

Fig. 19-19. Sweep generator with markers. (Dynascan Corp.)

Fig. 19-18. Transistor tester. (Hickok Teaching Systems)

THE SWEEP GENERATOR

The primary function of any generator is to simulate an actual signal, which may be applied to a circuit for measurements of performance and the location of malfunctions. The sweep generator is another valuable instrument for servicing FM radios and TVs. This generator consists of two oscillators:

1. A fixed oscillator, the frequency of which is manually controlled.

2. A variable frequency oscillator which frequency modulates the fixed rf signal. As a result the signal "sweeps" back and forth between the limits of the modulation. This sweep width is set by the amplitude of the control voltage.

When this instrument is connected to the if stages of an FM receiver, the output may be observed on an oscilloscope which will show the response pattern of the stages over the band of frequencies used in modulated signal.

FREQUENCY METERS

The frequency meter is discussed at this point to reinforce some of the lessons learned in the earlier studies of basic electronics. In discussions of tuning your radio, you will recall that it was necessary to tune the tank circuit of the radio to the frequency of the incoming signal. The tuning circuit provided a means of selecting the desired broadcasting station to be received. The ABSORPTION FREQUENCY METER operates on the same principle and is represented in its basic form in Fig. 19-20. A commercial type is illustrated in Fig. 19-21. During operation coil L is loosely coupled to the oscillating circuit under measurement. Capacitor C is adjusted for maximum response or absorption of energy. This point is indicated by the brilliance of the lamp in series with the

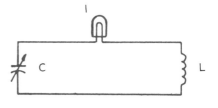

Fig. 19-20. Circuit for an absorption frequency meter.

Fig. 19-21. A commercial absorption frequency meter.
(James Millen Co.)

circuit. The dial for capacitor C is calibrated in frequency and may be read directly. An instrument of this type is made to cover a limited range of frequencies. The range is fixed by the values of L and C. Usually the technician has a set of these absorption meters to cover the radio frequency spectrum.

ACTIVITIES

The following is presented so that you may become familiar with the oscilloscope and signal generators.

1. Connect the output terminals of an af signal generator to the V-input terminals of an oscilloscope.

2. Turn on scope. Adjust intensity and focus to produce a sharp visible line.

3. Adjust H and V positioning control until line is in center of screen.

4. Adjust H gain control to produce a line about two inches long in center of screen.

5. Set V gain to lowest position and sync switch to "Internal Sync."

6. Set sweep frequency range to lowest range.

7. Set audio generator to 100 Hz. Turn generator on and observe pattern on screen.

8. Turn fine frequency control on scope until one sine wave pattern appears stationary.

9. Adjust V gain and notice the change in amplitude of the wave form.

10. Adjust output of generator and notice change in amplitude.

11. Adjust fine frequency control so that two complete sine waves appear on the screen.

AUDIO GENERATOR

1. Return to the single sine wave pattern and then increase audio generator to 200 Hz. Two sine waves should appear on screen. Slight frequency adjustment may be necessary.

2. Increase generator to 400 Hz. Change frequency range on scope and adjust frequency control to a single sine wave pattern.

RF GENERATOR

1. Connect output of rf generator to the V terminals of the scope. Set frequency range of scope to include 400 Hz. Set rf generator to 1000 KHz.

2. Turn instruments on and adjust gains and positioning until a rectangular green pattern about 2 inches wide and 1 1/2 inches high appears on the screen. (This is the pattern of the rf wave. The frequency is so high that individual waves cannot be distinguished.)

3. Turn rf generator to MOD position and notice that the amplitude of the rf wave now follows a 400 Hz wave pattern.

NOTICE: If your scope has a sync control, increase in a clockwise direction sufficiently to obtain a stable stationary pattern. This will probably be between 25 to 35 percent of the maximum sync.

FOR DISCUSSION

1. What are the advantages of the VTVM over the VOM?
2. Why are digital meters predicted to revolutionize the future in test instruments?
3. Discuss the major uses of the oscilloscope.
4. Explain how an audio signal generator can be used to test an amplifier circuit.
5. Which test instruments are likely to be found in an electronic service person's shop?
6. Discuss vacuum tube failures and what tests tube checkers perform.

TEST YOUR KNOWLEDGE

1. What is the major differences between a VTVM and a VOM?
2. Name the ways a transistor tester shows whether a solid state device is good or bad.
3. The two basic types of signal generators are _____ and _____.
4. The intensity control on the oscilloscope is comparable to what control on the television?
5. Name two types of VOMs.
6. Tube testers may be classified as _____ and _____.

Chapter 20

INTEGRATED CIRCUITS, DIGITAL AND LINEAR

An integrated circuit (IC) is a complete electronic circuit in one package. Usually this package includes transistors, diodes, resistors and capacitors with the connecting wiring and diodes. The integrated circuit was invented in 1958 at Texas Instruments.

One major advantage of the integrated circuit is its size. It would be possible to wire all the individual or discrete components used in an IC, however, they would take up much more space. With the size advantage of ICs, it is possible to have high capacity computers that can literally fit in a suitcase.

Integrated circuits are very reliable as compared to individual components. There is another advantage of lower cost.

An example of a typical circuit found in an integrated circuit is found in Fig. 20-1. This circuit is an audio amplifier that could be used in automobile radios.

Integrated circuits can be classified into two basic types:

1. Digital.

2. Linear.

Digital integrated circuits are used as switches and operate in either on or off conditions.

Linear integrated circuits are used as amplifiers and have variable outputs. The remainder of this chapter will discuss these two types of integrated circuits.

DIGITAL INTEGRATED CIRCUITS

Digital integrated circuits handle digital information by using switching circuits. They work as a result of combining gating circuits and flip-flop circuits. Let us look at gates and how they operate by using binary logic.

THE BINARY NUMBERING SYSTEM

Electronic circuits can be made to act in only two discrete states: ON and OFF. This two state system called BINARY can be compared to a single-pole, single-throw (SPST) switch as shown in Fig. 20-2.

When the switch, S_1, is off, the condition may respresent a 0 in the binary numbering system. Likewise, when switch S_1, in Fig. 20-2 is on, this can represent the 1 in the binary numbering system. There are only two conditions to the circuit: ON or OFF.

In actual circuits, it is not very practical to build large electronic logic circuits using manual switches. They do, however, provide an easy basis for understanding other switchable electronic components. The most common elec-

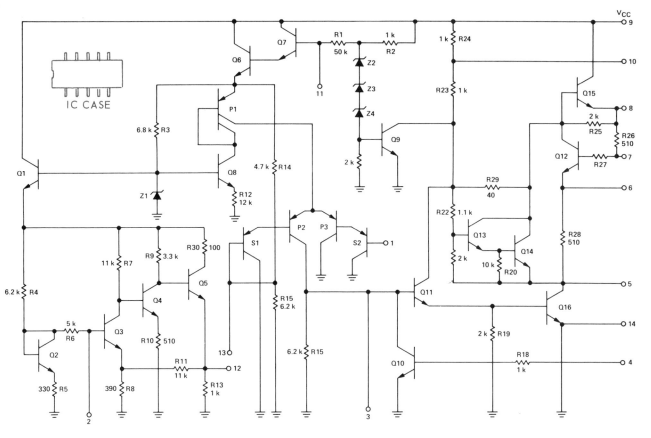

Fig. 20-1. Schematic diagram of MC1385. (Motorola)

tronic component that can be used as a switch is a transistor. These can allow current to flow through them which can be an ON or 1 state or they can be in cutoff condition representing the 0 in the binary numbering system. Simple circuits made up of diodes, transistors, and resistors can be used to perform the basic logic functions. Since the development of the inte-

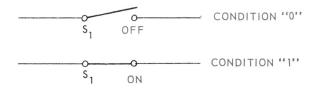

Fig. 20-2. A single-pole single-throw switch as a binary device.

grated circuit (IC), tens or hundreds of circuits can be designed and fabricated into a single semiconductor chip .

With only one SPST switch, you can only count up to a maximum of 1. Suppose you wanted to count higher, how can the binary system be used for a number such as 45 or 79?

A basic concept governs the process of counting in any system. This concept is: YOU MUST RECORD DIGITS, ONE AFTER EACH OTHER, FOR EACH COUNT UNTIL THE COUNT EXCEEDS THE TOTAL NUMBER OF INPUTS AVAILABLE: THEN YOU SHOULD START A SECOND COLUMN AND CONTINUE COUNTING. In the binary system, the numbers are represented as shown in Fig. 20-3.

The digit 0 in the decimal system is 0 in the binary system. Likewise the digit 1 in the decimal system is 1 in the binary system. However, the number 3 in the decimal system is represented by the number 11 in the binary system. The binary number for 6 is shown in Fig. 20-4. Thus the binary number for 6 is 110.

2^5	2^4	2^3	2^2	2^1	2^0
32NDS	16THS	8THS	4THS	2'S	1'S
6TH NUMBER	5TH NUMBER	4TH NUMBER	3RD NUMBER	2ND NUMBER	1ST NUMBER

Fig. 20-3. The binary numbering system.

2^2	2^1	2^0
4'S	2'S	1'S
1	1	0

←— BINARY NUMBERING FOR DECIMAL NUMBER 6

Fig. 20-4. Binary numbering for decimal 6.

Fig. 20-5 shows the decimal conversion to binary numbers for the digits 0-25.

The more complex decimal number of 79 is shown in binary form in Fig. 20-6. Note that the binary number for 79 is 1001111.

LOGIC GATES

Electronic switching circuits that can be used with digital inputs are called LOGIC GATES. By using digital and linear circuits, decisions can be made. The circuits we will now discuss are the basic building blocks used in a large computer. They are important in helping the computer make the calculations and decisions it can do so well in extremely short periods of time. The basic types of logic gates are:

1. AND GATES.

2. OR GATES.

3. NOT GATES.

4. NAND GATES.

5. NOR GATES.

DECIMAL			BINARY NUMBER				
HUNDREDS	TENS	UNITS	2^4	2^3	2^2	2^1	2^0
			16'S	8'S	4'S	2'S	1'S
		0					0
		1					1
		2				1	0
		3				1	1
		4			1	0	0
		5			1	0	1
		6			1	1	0
		7			1	1	1
		8		1	0	0	0
		9		1	0	0	1
	1	0		1	0	1	0
	1	1		1	0	1	1
	1	2		1	1	0	0
	1	3		1	1	0	1
	1	4		1	1	1	0
	1	5		1	1	1	1
	1	6	1	0	0	0	0
	1	7	1	0	0	0	1
	1	8	1	0	0	1	0
	1	9	1	0	0	1	1
	2	0	1	0	1	0	0
	2	1	1	0	1	0	1
	2	2	1	0	1	1	0
	2	3	1	0	1	1	1
	2	4	1	1	0	0	0
	2	5	1	1	0	0	1

Fig. 20-5. Decimal to binary conversion table.

2^6	2^5	2^4	2^3	2^2	2^1	2^0
64THS	32NDS	16THS	8THS	4THS	2'S	1'S
1	0	0	1	1	1	1

→ 79

Fig. 20-6. Binary number for 79.

AND GATES

Let us start with the AND gate to see how each of these logic gates operate. The AND gate is an electronic circuit that takes YES and NO inputs. It makes certain basic decisions with the output also expressed in YES or NO.

As previously discussed, YES or NO data is binary in nature. This can indicate that there are two levels of current (or no current) in a circuit. Also, YES or NO can be described as 1 or 0. The symbol for an AND gate is shown in Fig. 20-7.

The operation of an AND gate is best described as an electronic circuit that will produce an output of 1 (YES) if <u>all</u> of the inputs are 1 (YES). A very simple circuit with

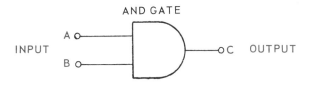

Fig. 20-7. AND gate symbol.

switches that shows the equivalent of an AND gate is shown in Fig. 20-8. Note that both switches have to be ON or S_1 <u>and</u> S_2 both have to be ON to give a complete circuit to cause the lamp to burn. An electronic AND gate used in a computer or counting circuit is composed of transistors or diodes in an integrated circuit or wired individually. This circuit using SPST switches was only used for explanation purposes.

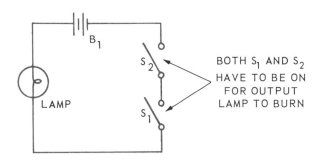

Fig. 20-8. Simple electrical circuit that simulates AND gate.

A binary table that explains the operation of the AND gate is shown in Fig. 20-9. This type of table is called a TRUTH TABLE.

The AND gate is used to determine the presence of YES signals or 1's on both the inputs, A and B. When this occurs, the output signal will be 1. See Fig. 20-9.

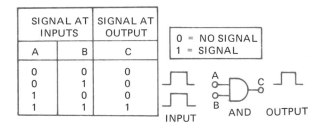

SIGNAL AT INPUTS		SIGNAL AT OUTPUT
A	B	C
0	0	0
0	1	0
1	0	0
1	1	1

0 = NO SIGNAL
1 = SIGNAL

Fig. 20-9. AND gate truth table.

OR GATES

The OR gate will provide an output signal of 1 (YES) when either one <u>or</u> the other of its inputs is 1 (YES). Likewise, if any of the inputs are 0 in an OR gate, the output is a 0 (NO). The OR circuit is normally used in electronic circuits to see if any input is present. The symbol for an OR gate is shown in Fig. 20-10.

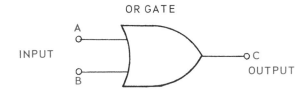

Fig. 20-10. OR gate symbol.

SIGNAL AT INPUTS		SIGNAL AT OUTPUT
A	B	C
0	0	0
0	1	1
1	0	1
1	1	1

0 = NO SIGNAL
1 = SIGNAL

Fig. 20-11. OR gate truth table.

The truth table for an OR gate is shown in Fig. 20-11.

NOT GATES

The NOT gate is often referred to as an "inverter" because this is what it does in a circuit. The NOT gate is put into a circuit to invert the polarity of the input signal. If the input signal is 1 or YES, the output signal will be 0 or NO. Likewise if the input signal is 0, the output will be 1. The symbol for the NOT gate is shown in Fig. 20-12. Note in Fig. 20-12 that

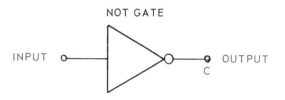

Fig. 20-12. NOT gate symbol.

the NOT gate symbol has only one input lead. Also there is a small circle at the end of the triangle in the symbol. Again the inverter simply inverts the input signal to another opposite value in the output. A truth table for a NOT gate is shown in Fig. 20-13.

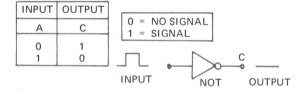

INPUT	OUTPUT
A	C
0	1
1	0

Fig. 20-13. NOT truth table.

NAND GATE

All logic gates are combinations of the basic gates: AND, OR, and NOT. The NAND is a negative AND gate. It is made up of an AND gate and a NOT gate that inverts the output. Sometimes it is called a NOT-AND gate. The NAND gate symbol is basically an AND gate symbol with a circle at the end. See Fig. 20-14.

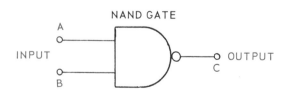

Fig. 20-14. NAND gate symbol.

The NAND truth table shown in Fig. 20-15 is also the reverse of the AND truth table.

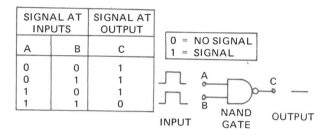

SIGNAL AT INPUTS		SIGNAL AT OUTPUT
A	B	C
0	0	1
0	1	1
1	0	1
1	1	0

Fig. 20-15. NAND truth table.

NOR GATE

The NOR gate is to the OR gate what the NAND gate was to the AND gate. It gives the opposite (or negative) results of the OR gate. It is made up of an OR gate and a NOT gate (inverter). The symbol for a NOR gate is shown in Fig. 20-16.

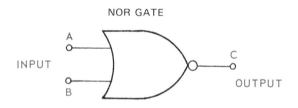

Fig. 20-16. NOR gate symbol.

The NOR circuit can be used if you want to see if there is any kind of input. If there is no input, the output will be 1. Conversely, if there is an input, there will be no (0) output. This effect is shown in the truth table in Fig. 20-17.

SIGNAL AT INPUTS		SIGNAL AT OUTPUT
A	B	C
0	0	1
0	1	0
1	0	0
1	1	0

Fig. 20-17. NOR truth table.

Compare the information in tables shown in Figs. 20-15, and 20-17.

Since the development of the integrated circuit, it is possible to have a number of logic circuits on one chip the size of a postage stamp or smaller. Most calculators and digital watches use the basic technology described in this chapter. See Fig. 20-18 for a modern calculator

Fig. 20-19. A handheld calculator that is programmable. (Hewlett-Packard)

Fig. 20-18. A desk top calculator with printer, plotter, and card reader. (Hewlett-Packard)

with a printer, plotter, and card reader. Fig. 20-19 shows a handheld calculator. Users of the Hewlett-Packard Model 65 pocket-sized calculator can write and edit their own programs, which are recorded on tiny magnetic cards. Cards can be inserted into machines for performing complex programs, then saved for use as often as needed. The HP-65 is the first pocket-sized calculator with full programming capability.

LINEAR INTEGRATED CIRCUITS

Linear integrated circuits are the type that amplify signals or regulate them. They may be compared with digital ICs which act as

switches. The output of a linear integrated circuit is not abrupt as "on" or "off," however, it is smooth. The primary function of linear ICs is to amplify – to increase the gain of the circuit which means to increase the voltage, current or power. In any linear integrated circuit, output always remains proportional to the input level. This circuit is abbreviated LIC.

OPERATIONAL AMPLIFIER

One important type of linear integrated circuit is the OPERATIONAL AMPLIFIER, or "OP-AMP." An operational amplifier is a high

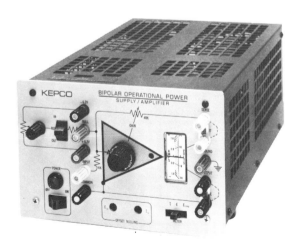

Fig. 20-20. Typical bipolar power supply using an operational amplifier frequently used as power amplifier. (Kepco, Inc.)

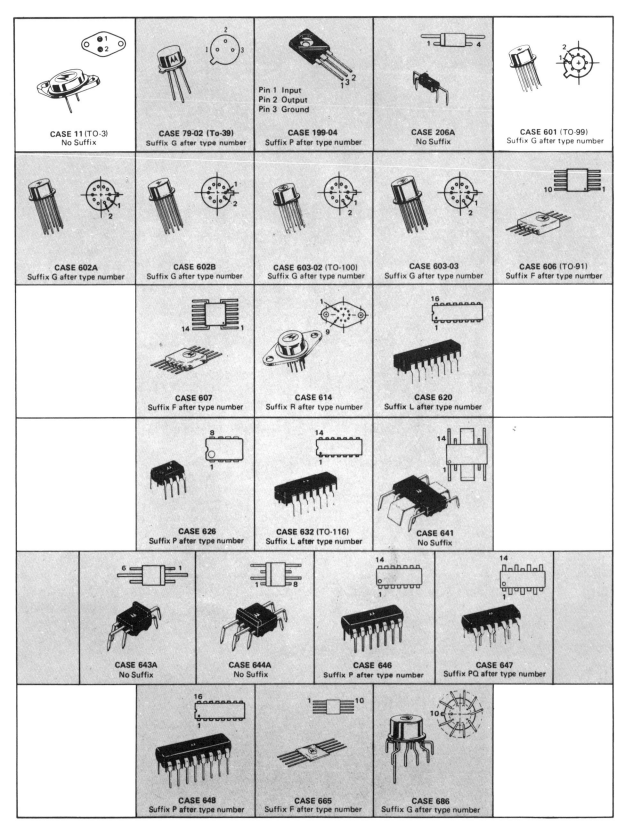

Fig. 20-21. Linear IC packages. (Motorola, Inc.)

gain linear integrated circuit that uses direct-coupling. You will find that operational amplifiers may be used in circuits such as audio and video amplifiers, power supplies, or as oscillators. See Fig. 20-20. Operational amplifiers also may be used in computers to perform an amplifier function.

Linear integrated circuits come in many shapes and sizes. Fig. 20-21 shows some examples of different package designs for integrated circuits.

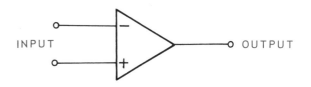

Fig. 20-22. Operational amplifier (OP-AMP) symbol.

Symbol for an operational amplifier is shown in Fig. 20-22. Note the resemblance between this symbol and digital IC gate symbols.

FOR DISCUSSION

1. What are the advantages of integrated circuits over individual transistors?
2. Name the basic types of integrated circuits.
3. What is the difference between a NOT and a NOR logic gate?
4. What are the basic differences between the operational amplifier circuit symbol and the NOT gate?
5. What advances do you think are ahead in the computer field in the next 25 years?

TEST YOUR KNOWLEDGE

1. The digital number that represents binary number, 10101, is _____.
2. What is the truth table for an AND gate?
3. What is the primary differences between the digital numbering system and the binary number system?
4. What company was responsible for inventing the integrated circuit?
5. A digital integrated circuit performs what function in a circuit?

Chapter 21

CAREER OPPORTUNITIES

IN ELECTRONICS

Electronics offers some excellent career opportunities. The electronics industry is looking for bright young men and women to take advantage of rewarding satisfying careers in this field.

What do you want to do when your education is completed? What vocation will you choose to assure an interesting and rewarding future for you and your family? What type of education should you acquire in preparation for your selected future? Are you qualified by interest, desire and capacity to take advantage of the educational opportunities within your reach today? These questions can only be answered by you. To help you make an intelligent decision your parents, your school guidance counselors, your teachers and friends in business and industry stand ready to give you information gained from years of experience. Do not hesitate to ask questions.

Newspaper headlines and editorials indicate the explosive discoveries and developments in science and technology. Without question, ELECTRONICS, together with other fields of science, will continue to play a major role in our modern civilization. In the field of communications the familiar telephone, radio and TV are supplemented with the advanced techniques of satellites. Ships and aircraft arrive safely at their destinations when such navigational aids as LORAN, SONAR, and RADAR constantly supply needed information to their captains.

The defense of our nation depends upon the great research and development of our space and rocket program. Today's industry is electronically controlled. Automation is displacing thousands of production workers each year. Electronic computers now do work in business and accounting, and in mathematical and scientific calculations in only a few minutes, which previously required months of work. Electronics has entered the field of medicine and has become an important instrument in the hands of your physician and hospital. As each new electronic miracle is born and given to our people, thousands of highly skilled technicians must be trained to maintain and service it.

CAREER CLUSTERS

The United States Office of Education (USOE) has identified fifteen career clusters. These clusters are merely categories for classifying jobs. The fifteen USOE clusters represent over 20,000 different jobs such as television service person, electronics engineer, appliance installer, instrument repair person, etc. Careers in electronics cuts across all fifteen career clusters. See Fig. 21-1.

Another method of classifying careers in electronics is by using the categorizing of GOODS AND SERVICES. See Fig. 21-2. The goods producing industries are those that produce electronic components, devices and systems. These could be resistors, capacitors,

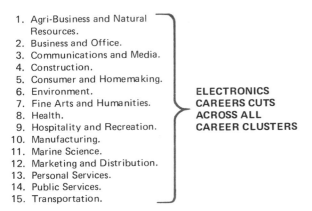

1. Agri-Business and Natural Resources.
2. Business and Office.
3. Communications and Media.
4. Construction.
5. Consumer and Homemaking.
6. Environment.
7. Fine Arts and Humanities.
8. Health.
9. Hospitality and Recreation.
10. Manufacturing.
11. Marine Science.
12. Marketing and Distribution.
13. Personal Services.
14. Public Services.
15. Transportation.

ELECTRONICS CAREERS CUTS ACROSS ALL CAREER CLUSTERS

Fig. 21-1. Career clusters.   (U.S. Office of Education)

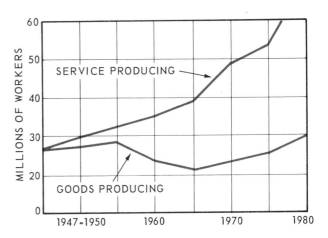

Fig. 21-3. Goods vs. service producing occupations. (U.S. Bureau of Labor Statistics)

radios, televisions, digital watches, calculators, security systems for the home, etc. Services are those careers that involve the installation of electronic devices or systems; the maintaining and repairing of these devices or systems; and the changing of these devices or systems to update or improve them.

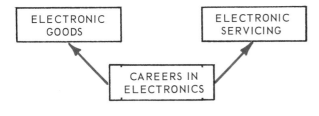

Fig. 21-2. Goods and servicing in electronics.

The United States Bureau of Labor Statistics estimates that employment has been shifting from the goods-producing industries in the past few years. Another way of saying this is to state that we need more people to service electronic devices and systems than we need to produce them. This is shown for all service workers in Fig. 21-3 including those who work in all service fields including transportation, trade, finance, real estate, government and others.

CAREERS IN ELECTRONIC SERVICING*

The field of electronic servicing offers many advantages and opportunities to the person who is at the point of choosing a career. For the

young man or woman interested in this field, training is often available as an integral part of regular high school curriculum. The knowledge and skill required to be a successful service technician is well within the ability of the average high school student. The person who is no longer in high school will find training offered at a variety of public and private technical schools. Privately operated training programs are frequently of fairly short duration, one year or less. This is because the training is intense to provide the greatest amount of training in the shortest possible time. See Fig. 21-4.

Salaries vary depending on location and experience. And, of course, there is the possibility of going into one's own business, in which case, income is limited only by imagination and ability.

Electronic service technicians enjoy the freedom of being able to find employment in almost every community in the nation. They may often choose the size and type of service organization that they like best. Service tech-

* This information prepared by the Electronics Industries Association (EIA) and it is available in a Career Guidance Brochure titled "Futures Unlimited," free. Write to EIA, 2001 Eye Street, NW, Washington, D.C. 20006

product only, such as color television or perhaps high fidelity equipment. On the other hand, they may prefer the variety associated with servicing the whole range of products. See Fig. 21-5.

Similarly, the technician may like working with the products of any one manufacturer, or this person might wish to service the equipment produced by many different companies.

YOUR EDUCATION

A wise philosopher once stated, "There are many roads to the top of the mountain." This is true also at technical levels of employment in electronics. A high school education is required as a foundation on which to build further skills and knowledge. Your specialized training may come from a two year Junior or Community College offering an associate degree in technology. You may wish to enroll in one of the many fine residence schools in electronic technology located throughout our country. See Fig. 21-6. Some high school graduates have entered industry directly and received their

Fig. 21-4. Students in an Electronics Industries Association servicing seminar.

nicians are employed by small shops, large service companies, huge retail stores, and by electronic manufacturers and distributors. Or, they may choose to be self-employed. The technician may choose to specialize in one

Fig. 21-5. A service technician using modern test equipment.
(Quasar Corp.)

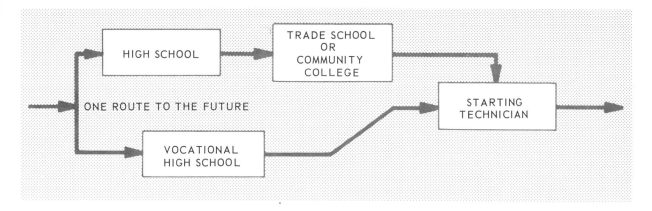

Fig. 21-6. A career map for a service technician.
(Electronics Industries Association)

specialized education in the training facilities maintained by large companies. Employment is not an end in itself. One must constantly read and study to keep abreast with this rapidly changing industry. Advancement and security will come through self-improvement.

In the fields of science and engineering, a college degree is almost a necessity, with more advanced degrees becoming increasingly important. There are many state and private universities offering engineering degrees. Consult your school counselor for locations and en-trance requirements. A new job classification is presently emerging, frequently called an engineering assistant or technologist. This classification requires the skills of the technician and a B.S. degree from an approved institution. Undergraduate studies sometimes include some business administration.

The chart shown in Fig. 21-7, shows in graphical form some of the major job categories and the usual educational requirements. Careful study of the chart will help you plan your high school and college courses.

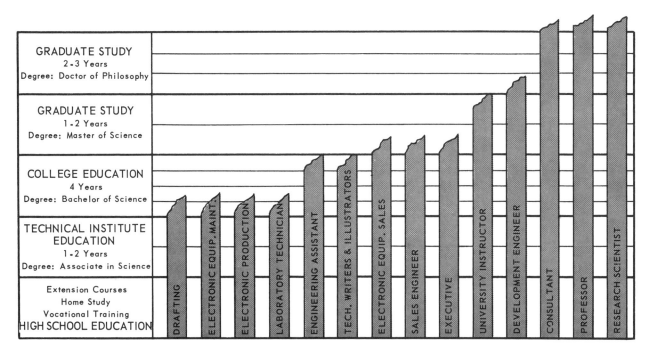

Fig. 21-7. Educational requirements for careers in electronics. (Reprinted from "Electronics Your Chance to Shape the Future," published by Electronics Industry Association and the IEEE.)

YOUR DECISION

The electronics industry offers a great many employment opportunities. You have reviewed the education requirements and technical skills demanded. Will you be successful in reaching your goal? No one can accurately predict your personal success and fulfillment of your personal satisfactions. There are, however, some guide lines which may influence your decision:

1. Since electronics is a science, a healthy interest in scientific subjects during your high school career will give you direction.

2. Mathematics is the tool of the scientist and engineer. If you like these subjects and do well with them, a career in electronics may be a wise choice.

3. Many able scientists and engineers started with an early interest in electronics as a hobby. Experimentation with electronics and amateur radio are leisure time educational activities which will point toward success in this chosen vocation.

4. Discuss your abilities with your counselor. There are a number of valid interest inventory tests which you may take to discover your interests and capacity for learning this scientific field.

5. The fact that you are now reading and studying this text is a fairly accurate indication of your interest in electronics.

6. Have you elected some courses in electronics in high school? Did you like them? Did you do well.

The choice is your choice in the final analysis. Study hard and prepare yourself well. Do not study only electronics, for your contacts with industry and fellow workers demand an understanding of languages of communication, social studies, how to get along well with your co-workers, and your ability to contribute to the success of your company. Remember that many of the greatest discoveries in recent years have resulted from team effort.

AUTHOR'S ACKNOWLEDGEMENTS

We would like to acknowledge and express our appreciation and thanks to these associates who so willingly supplied their assistance and knowledge in the preparation of these texts:

1. 1964, 1968, and 1975 Editions:
   Robert Mathison, Gary Van, James Collins, Leon Pearson, and John Lienhart, Electronic technicians at Chico State College. Bobbie Fikes, and Moore Smalley, Graduate students in electronics at San Jose State College (Illustrations by Charles Tyler, San Jose City College). Virtue B. Gerrish (wife of Mr. Gerrish) for typing the original manuscript.
2. 1977 Edition:
   Mr. Neal S. Hertzog, Marketing Office Manager, Buck Engineering Co. and Mr. Gerald J. Kane, Director of Engineering, Buck Engineering Co. Also Mrs. Shirley Miles and Ms. Carol Keeton for typing this manuscript. In addition, we appreciate the help and support of El Mueller, Quasar Electronics Corp. Special thanks is due to Carl Giegold for photography and to Carrie Dugger and Ed, Cammie and Toy Dugger for their support and patience.

Also to the many industries for their generous supply of illustration and product information, without which this text could not have been written. Special credit is due:

Allen-Bradley Co.; Ballantine Laboratories, Inc.; Barley and Dexter Laboratories; Bell Telephone Laboratories; Bud Radio, Inc.; Centralab, Globe Union, Inc.; Cincinnati, Milacron; Daystrom, Weston Instruments Div.; Delco Products, General Motors; Delco Remy, General Motors; DeVilbiss Co.; Dynascan Corp.; Electronic Industries Association; Electronic Instrument Co. (EICO); Fosdick Machine Tool Co.; General Electric Co.; Graymark Enterprises, Inc.; Gulton Industries, Inc.; Hammarlund Mfg. Co., Inc.; Heath Company; Hewlett-Packard, Inc.; Hickok Teaching Systems, Inc.; International Rectifier Corp.; James Millen Co.; J. W. Miller Co.; Lab-Volt, Buck Engineering Co.; Lafayette Radio Electronics; Lindberg Engineering Co.; Liquid Xtal Displays Inc.; L. S. Starrett, Co.; Minneapolis-Honeywell Regulator Co.; Montgomery Ward; Motorola Semi-Conductor Products, Inc.; National Automatic Tool Co.; Ohmite Mfg. Co.; Optima Enclosures; Pacific Telephone Co.; Perma-Power; Philco Corp.; Plastoid Corp.; Potter & Brumfield, AMF Inc.; Quasar Electronics Corp.; Radio Corporation of America (RCA); Radio Shack Corporation; Raytheon Company; Science-Electronics, Inc.; Shure Bros., Inc.; Simpson Electric Co.; Sony Corp. of America; Sprague Products Co.; Stromberg-Carlson; Sylvania Electric Products, Inc. (GTE); Texas Instruments Inc.; The American Radio Relay League; The Electric Storage Battery Co.; The Johnson Co.; The Murdock Corp.; The Superior Electric Co.; Triad Transformer Corp.; Triplett Corp.; Ungar Electric Tools; Union Carbide Consumer Products Co.; Union Switch & Signal Co.; United States Bureau of Labor Statistics; United States Office of Education (USOE); United Transformer Corp; Vactec, Inc.; Westinghouse Electric Corp.; Welch Scientific Co.

REFERENCE SECTION

Appendix 1
SCIENTIFIC NOTATION

In the study of Electronics it is necessary to work with extremely small quantities. As a student of mathematics you will realize that the multiplication and division of numbers involving many zeros and decimal points will cause errors unless caution is exercised at all times. The use of "power of ten" or Scientific Notation is a rapid and easy method. Let's see how it works:

$$1 \times 10^0 = 1$$
$$1 \times 10^1 = 10$$
$$1 \times 10^2 = 100$$
$$1 \times 10^3 = 1000$$
$$1 \times 10^4 = 10,000$$
$$1 \times 10^5 = 100,000$$
$$1 \times 10^6 = 1,000,000$$
$$1 \times 10^{-1} = .1$$
$$1 \times 10^{-2} = .01$$
$$1 \times 10^{-3} = .001$$
$$1 \times 10^{-4} = .0001$$
$$1 \times 10^{-5} = .00001$$
$$1 \times 10^{-6} = .000001$$

Follow these examples and you will learn to use the scientific notation method.

To express a number:

$$47,000 = 47 \times 10^3$$

$$.000100 = 100 \times 10^{-6}$$

$$.0025 = 25 \times 10^{-4}$$

$$3,500,000 = 3.5 \times 10^6$$

To multiply, add the exponents of ten.

$$47 \times 10^3 \times 25 \times 10^{-4}$$
$$= 47 \times 25 \times 10^{-1}$$

$$100 \times 10^{-6} \times 3.5 \times 10^6$$
$$= 100 \times 3.5$$

To divide, subtract exponents.

$$\frac{3.5 \times 10^6}{25 \times 10^{-4}} = \frac{3.5 \times 10^{10}}{25}$$

$$\frac{100 \times 10^{-6}}{25 \times 10^{-4}} = \frac{100 \times 10^{-2}}{25}$$

$$6 - (-4) = +10$$

$$-6 - (-4) = -2$$

To square, multiply exponent by 2.

$$(25 \times 10^2)^2 = 25^2 \times 10^4$$

$$(9 \times 10^{-3})^2 = 81 \times 10^{-6}$$

To extract square root, divide exponent by 2.

$$\sqrt{81 \times 10^4} = 9 \times 10^2$$

$$\sqrt{225 \times 10^{-8}} = 15 \times 10^{-4}$$

See Appendix 3 for using Powers of Ten in Conversions.

Appendix 2
COLOR CODES

RESISTORS: For your convenience, a standard Color Code has been adopted to determine the values of resistors and capacitors. The following table has been supplied through the courtesy of the Radio-Television Manufacturer's Association. (RETMA)

HOW TO READ THE CODE

Secure several resistors from stock. Hold a resistor so that the color bands are to the left as illustrated in Fig. A2-1. Let's assume the band colors are (from left to right) brown, black, green, silver.

The brown band is 1. The black band is 0. The green band is the multipler, x 10^5.

The silver band tells us that it is within the limits of ± 10 percent of the color code value.

Follow these examples for further experience:

Yellow, Violet, Brown, Silver –	470 Ω ± 10%
Brown, Black, Red, Gold –	1000 Ω ± 5%
Orange, Orange, Red, None –	3300 Ω ± 20%
Green, Blue, Red, None –	5600 Ω ± 20%
Red, Red, Green, Silver –	2,200,000 Ω ± 10%

CAPACITORS: Recent practices include the numerical marking of capacitors. Their values may be read directly. However, there are many coded capacitors in use. The current standard JAN and RETMA code will only be described. There are many special marking systems and a detailed manual describing all codes should be consulted when necessary.

COLOR	NUMERICAL FIGURE	DECIMAL MULTIPLIER		VALUE TOLERANCE
		Power of 10	Multiplier Value	
Black	0	10^0	1	
Brown	1	10^1	10	
Red	2	10^2	100	
Orange	3	10^3	1,000	
Yellow	4	10^4	10,000	
Green	5	10^5	100,000	
Blue	6	10^6	1 Million	
Violet	7	10^7	10 Million	
Gray	8	10^{-2} (alternate)	0.01 *	
White	9	10^{-1} (alternate)	0.1 *	
Silver	–	10^{-2} (preferred)	0.01	±10%
Gold	–	10^{-1} (preferred)	0.1	±5%
None	–	—	—	±20%

Color code for resistors and capacitors.

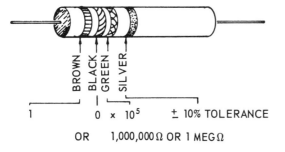

BROWN BLACK GREEN SILVER

1   0 x 10^5 ± 10% TOLERANCE

OR   1,000,000 Ω OR 1 MEG Ω

Fig. A2-1. When reading color code, resistor should be held with bands at left.

In Fig. A2-2, let us assume that the upper left hand dot is WHITE. WE WILL USE THE RETMA CODE. The next dot is RED. We will read 2. The third dot is GREEN. We will read 5. Our first two significant figures are 25. Going to the multiplier we see that it is BROWN which means x 10^1. Our value then is 250 pF. The tolerance value and the class may also be found but their meanings will not be discussed in this basic text.

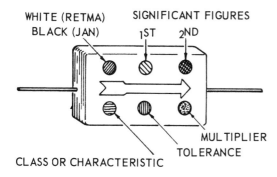

WHITE (RETMA)
BLACK (JAN)

SIGNIFICANT FIGURES
1ST   2ND

MULTIPLIER

TOLERANCE

CLASS OR CHARACTERISTIC

VALUES IN PF

Fig. A2-2. Capacitor codes.

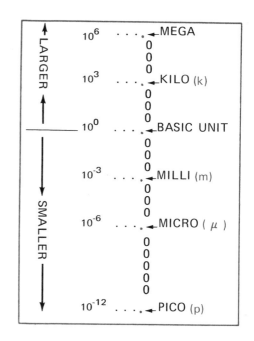

Appendix 3
CONVERSIONS

When using formulas to determine unknown values in a circuit, you will find the formula may be given in basic units, but the quantities you wish to use are given in larger or smaller units. A conversion must be made. For example: Ohm's Law states that:

$$I \text{ (in amperes)} = \frac{E \text{ (in volts)}}{R \text{ (in ohms)}}$$

If R were given in megohms such as 2.2 megohms, it would be necessary to change it to 2,200,000 ohms. If E were given in microvolts, such as 500 μV, it would be necessary to change it to .0005 volts.

Powers of Ten offer a simple solution to making conversions, but first, let us look at the common prefixes used in electronics.

mega = 1,000,000 times

kilo = 1000 times

milli = $\frac{1}{1000}$ of

micro = $\frac{1}{1,000,000}$ of

pico = $\frac{1}{1,000,000,000,000}$ of

The prefix "PICO" has now been adopted for "micromicro." Compare the meaning of these prefixes to powers of ten in Appendix 1 and you will discover that:

10,000 ohms = $10,000 \times 10^{-3}$ kilohms

47 kilohms = 47×10^3 ohms

950 kilohertz = 950×10^3 hertz or

950×10^{-3} megahertz

100 milliamperes = 100×10^{-3} amperes or

100×10^3 microamperes

.01 microfarads = $.01 \times 10^{-6}$ farads or

$.01 \times 10^6$ pF

250 pF = 250×10^{-6} μF or

250×10^{-12} F

8 μF = 8×10^{-6} F or

8×10^6 pF

75 milliwatts = 75×10^{-3} watts

A simple rule will also help you:

If the exponent of ten is negative, the decimal point is moved to the left in the answer.

$$447 \times 10^{-3} = .447$$

$$250 \times 10^{-6} = .00025$$

If the exponent is positive, the decimal point is moved to the right.

$$447 \times 10^{3} = 447,000$$

$$250 \times 10^{6} = 250,000,000$$

Here is a problem: What is the reactance of a 2.5 mH choke at 1000 KHz?

$$X_L = 2 \pi fL$$

$$= 2 \pi f \text{ (in Hz)} \times L \text{ (in henrys)}$$

Two conversions must be made:

$$1000 \text{ KHz} = 1000 \times 10^{3} \text{ Hz}$$

$$2.5 \text{ mH} = 2.5 \times 10^{-3} \text{ henrys}$$

Then:

$$X_L = 2 \times 3.14 \times 1000 \times 10^{3}$$

$$\times 2.5 \times 10^{-3}$$

$$= 6.28 \times 2.5 \times 10^{3}$$

$$(1000 \text{ changed to } 10^{3})$$

$$= 15.6 \times 10^{3} = 15,600 \text{ ohms}$$

Appendix 4
TRIGONOMETRY

Skill in using trigonometry is a part of the "kit of tools" of the electronic technician. It simplifies the solution of alternating current problems. It finds many applications in the designing and understanding of electronic circuits. Basically, trigonometry is the relationship between the angles and sides of a triangle. These relationships are called functions and represent numerically the ratio between two sides of the right triangle.

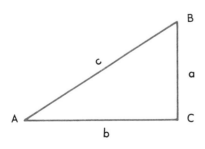

$$\text{Sine of angle A} = \frac{a \text{ opposite side}}{c \text{ hypotenuse}}$$

$$\text{Cosine of angle A} = \frac{b \text{ adjacent side}}{c \text{ hypotenuse}}$$

$$\text{Tangent of angle A} = \frac{a \text{ opposite side}}{b \text{ adjacent side}}$$

There are other functions, but these three are widely used in solving problems in electronics. By the above equations, if two values are known, the third may be found.

EXAMPLE: In a triangle, side a = 6 and b = 10. Find angle A. The tangent equation is used.

$$\text{Tan A} = \frac{a}{b} = \frac{6}{10} = .6$$

NATURAL TRIGONOMETRIC FUNCTIONS

Angle	Sine	Cosine	Tangent	Angle	Sine	Cosine	Tangent
1°	.0175	.9998	.0175	46°	.7193	.6947	1.0355
2°	.0349	.9994	.0349	47°	.7314	.6820	1.0724
3°	.0523	.9986	.0524	48°	.7431	.6691	1.1106
4°	.0698	.9976	.0699	49°	.7547	.6561	1.1504
5°	.0872	.9962	.0875	50°	.7660	.6428	1.1918
6°	.1045	.9945	.1051	51°	.7771	.6293	1.2349
7°	.1219	.9925	.1228	52°	.7880	.6157	1.2799
8°	.1392	.9903	.1405	53°	.7986	.6018	1.3270
9°	.1564	.9877	.1584	54°	.8090	.5878	1.3764
10°	.1736	.9848	.1763	55°	.8192	.5736	1.4281
11°	.1908	.9816	.1944	56°	.8290	.5592	1.4826
12°	.2079	.9781	.2126	57°	.8387	.5446	1.5399
13°	.2250	.9744	.2309	58°	.8480	.5299	1.6003
14°	.2419	.9703	.2493	59°	.8572	.5150	1.6643
15°	.2588	.9659	.2679	60°	.8660	.5000	1.7321
16°	.2756	.9613	.2867	61°	.8746	.4848	1.8040
17°	.2924	.9563	.3057	62°	.8829	.4695	1.8807
18°	.3090	.9511	.3249	63°	.8910	.4540	1.9626
19°	.3256	.9455	.3443	64°	.8988	.4384	2.0503
20°	.3420	.9397	.3640	65°	.9063	.4226	2.1445
21°	.3584	.9336	.3839	66°	.9135	.4067	2.2460
22°	.3746	.9272	.4040	67°	.9205	.3907	2.3559
23°	.3907	.9205	.4245	68°	.9272	.3746	2.4751
24°	.4067	.9135	.4452	69°	.9336	.3584	2.6051
25°	.4226	.9063	.4663	70°	.9397	.3420	2.7475
26°	.4384	.8988	.4877	71°	.9455	.3256	2.9042
27°	.4540	.8910	.5095	72°	.9511	.3090	3.0777
28°	.4695	.8829	.5317	73°	.9563	.2924	3.2709
29°	.4848	.8746	.5543	74°	.9613	.2756	3.4874
30°	.5000	.8660	.5774	75°	.9659	.2588	3.7321
31°	.5150	.8572	.6009	76°	.9703	.2419	4.0108
32°	.5299	.8480	.6249	77°	.9744	.2250	4.3315
33°	.5446	.8387	.6494	78°	.9781	.2079	4.7046
34°	.5592	.8290	.6745	79°	.9816	.1908	5.1446
35°	.5736	.8192	.7002	80°	.9848	.1736	5.6713
36°	.5878	.8090	.7265	81°	.9877	.1564	6.3138
37°	.6018	.7986	.7536	82°	.9903	.1392	7.1154
38°	.6157	.7880	.7813	83°	.9925	.1219	8.1443
39°	.6293	.7771	.8098	84°	.9945	.1045	9.5144
40°	.6428	.7660	.8391	85°	.9962	.0872	11.4301
41°	.6561	.7547	.8693	86°	.9976	.0698	14.3006
42°	.6691	.7431	.9004	87°	.9986	.0523	19.0811
43°	.6820	.7314	.9325	88°	.9994	.0349	28.6363
44°	.6947	.7193	.9657	89°	.9998	.0175	57.2900
45°	.7071	.7071	1.0000	90°	1.0000	.0000	

Now look at the table of Natural Functions and find the angle whose tangent is .6. It is 30° (answer).

EXAMPLE: Angle A = 45°, a = 6. What is the value of side c? Use sine equation.

$$\text{Sine } 45° = \frac{6}{c}$$

Look up the sine of 45°. It is .707. Then,

$$c = \frac{6}{.707} = 8.5 \text{ (approx.)}$$

In electronics, the right triangle used in the above example is labeled other ways, but the problems are worked out in exactly the same manner.

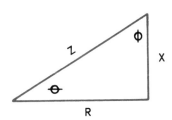

EXAMPLE: A circuit contains 40 Ω resistance (R) and 50 Ω reactance (X). What is the circuit impedance Z in ohms?

Find angle θ : Tan θ $= \dfrac{X}{R} = \dfrac{50}{40} = 1.25$

Look up in Tables: θ $= 51°$ approx.

Find Z: Sine θ $= \dfrac{X}{R}$

Look up in Tables: $.777 = \dfrac{50}{Z}$

$$Z = \dfrac{50}{.777} = 64.3 \ \Omega$$

These applications of trigonometry must be thoroughly understood. The best way to realize this is by practice. Problems have been included in the text for this purpose.

APPENDIX 5

STANDARD SYMBOLS
AND
ABBREVIATIONS

LETTER SYMBOLS

Symbols and abbreviations used in this text conform to Military Abbreviations and Contractions and IRE STANDARDS 54 IRE 21S1 and represent accepted practice in trade and industry.

β	beta
I	current
E	voltage
R	resistance
X	reactance
X_L	inductive reactance
X_c	capacitive reactance
Z	impedance

L	inductance
C	capacitor
M	mutual inductance
f	frequency
f_r	resonant frequency
λ	wave length
θ	phase displacement
Δ	a change in
Ω	ohms
Φ	phi-magnetic flux
G	conductance
Q	Q factor
r_p	plate resistance
g_m	transconductance
μ	amplification factor, permeability, micro.
A	gain
t	time
ϕ	phase angle
ω	angular velocity
V	vacuum tube or volts in transistor circuits
P	watts
p	pico

COMMON SCHEMATIC SYMBOLS

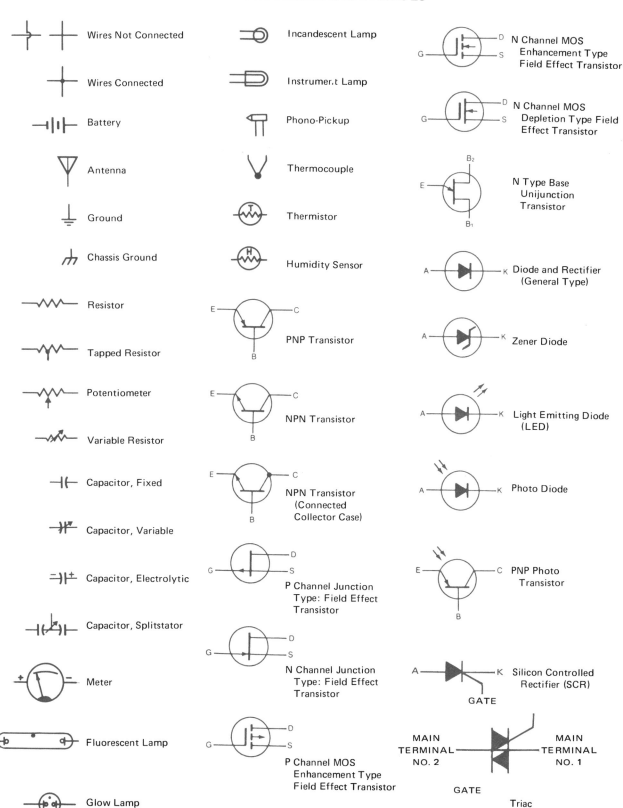

Wires Not Connected

Wires Connected

Battery

Antenna

Ground

Chassis Ground

Resistor

Tapped Resistor

Potentiometer

Variable Resistor

Capacitor, Fixed

Capacitor, Variable

Capacitor, Electrolytic

Capacitor, Splitstator

Meter

Fluorescent Lamp

Glow Lamp

Incandescent Lamp

Instrument Lamp

Phono-Pickup

Thermocouple

Thermistor

Humidity Sensor

PNP Transistor

NPN Transistor

NPN Transistor
(Connected
Collector Case)

P Channel Junction
Type: Field Effect
Transistor

N Channel Junction
Type: Field Effect
Transistor

P Channel MOS
Enhancement Type
Field Effect Transistor

N Channel MOS
Enhancement Type
Field Effect Transistor

N Channel MOS
Depletion Type Field
Effect Transistor

N Type Base
Unijunction
Transistor

Diode and Rectifier
(General Type)

Zener Diode

Light Emitting Diode
(LED)

Photo Diode

PNP Photo
Transistor

Silicon Controlled
Rectifier (SCR)

Triac

Electricity and Electronics

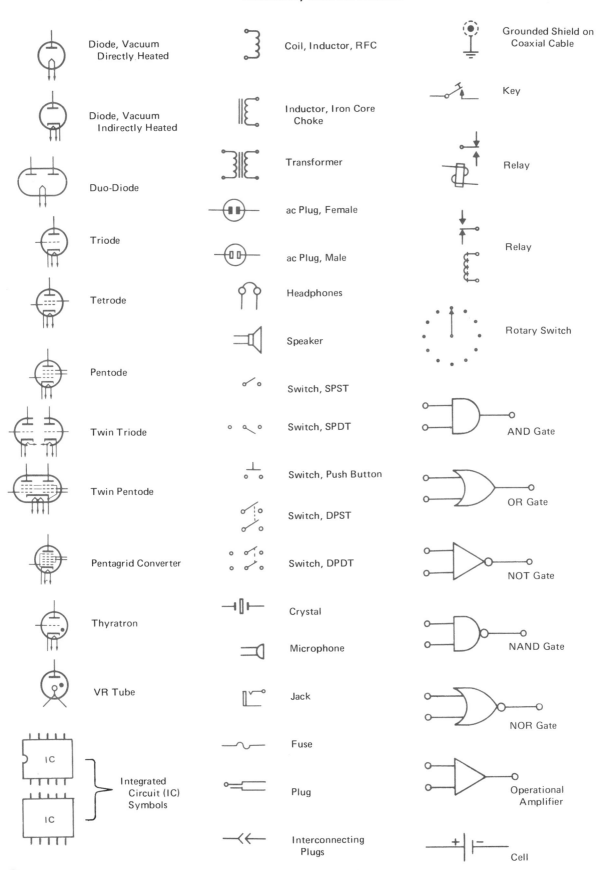

Diode, Vacuum Directly Heated

Diode, Vacuum Indirectly Heated

Duo-Diode

Triode

Tetrode

Pentode

Twin Triode

Twin Pentode

Pentagrid Converter

Thyratron

VR Tube

Integrated Circuit (IC) Symbols

Coil, Inductor, RFC

Inductor, Iron Core Choke

Transformer

ac Plug, Female

ac Plug, Male

Headphones

Speaker

Switch, SPST

Switch, SPDT

Switch, Push Button

Switch, DPST

Switch, DPDT

Crystal

Microphone

Jack

Fuse

Plug

Interconnecting Plugs

Grounded Shield on Coaxial Cable

Key

Relay

Relay

Rotary Switch

AND Gate

OR Gate

NOT Gate

NAND Gate

NOR Gate

Operational Amplifier

Cell

MATHEMATICAL SYMBOLS

$\times$ or $\cdot$ multiplied by	$\equiv$ identity	$\leqq$ less than or equal to
$\div$ or $:$ divided by	$\cong$ is similar to	$\therefore$ therefore
$+$ positive; add	$\neq$ does not equal	$\angle$ angle
$-$ negative; subtract	$>$ is greater than	Δ increment or decrement
$\pm$ plus or minus	$<$ is less than	$\perp$ perpendicular to
		$\parallel$ parallel to
$=$ or $::$ equals to	$\geqq$ greater than or equal to to	∞ infinity

---

STANDARD ABBREVIATIONS

ac	alternating current
AWG	American Wire Gage
amp	ampere
AM	Amplitude Modulation
af	audio frequency
afc	automatic frequency control
avc	automatic volume control
bfo	beat-frequency-oscillator
CRT	Cathode Ray Tube
cw	continuous wave
cps	cycles per second
db	decibel
dc	direct current
DPST	double pole, single throw
DPDT	double pole, double throw
emf	electromotive force
F	farad
FET	Field Effect Transistor
FM	Frequency Modulation
gnd	ground
H	henrys
Hz	hertz
hf	high frequency
hp	horsepower
if	intermediate frequency
JAN	Joint-Army-Navy
KHz	kilohertz
k	kilohm
kV	kilovolt
kwh	kilowatt hour
lf	low frequency
mmf	magnetomotive force
max	maximum
MHz	megahertz
meg	megohm
μ	micro
μA	microampere
μF	microfarad
μH	microhenry
$\mu\mu F$	micromicrofarad
mike	microphone
μV	microvolt
mH	millihenry
mA	milliampere
mV	millivolt
mW	milliwatt
min	minimum
osc	oscillator
pF	picofarad
pot	potentiometer
PF	Power Factor
rf	radio frequency
rpm	revolutions per minute
rms	root mean square
SPDT	single pole, double throw
SPST	single pole, single throw
spkr	speaker
sw	switch
TR	Transmit-Receive
uhf	ultra high frequency
VTVM	vacuum tube voltmeter
vhf	very high frequency
vf	video frequency
V	volts
W	watts

Appendix 6

ELECTRODEMONSTRATOR KIT

The Electrodemonstrator kit contains parts needed to build quickly and inexpensively, sixteen projects illustrated and described in preceding chapters of this text.

Projects which may be built from the parts include: 1. Permanent magnetic fields. 2. Electromagnetic field. 3. Electromagnet. 4. Relay. 5. Circuit breaker. 6. Door chime. 7. Buzzer. 8. Series dc motor. 9 Shunt dc motor. 10. ac motor. 11. Faraday's experiment. 12. Solenoid sucking coil. 13. Reactance dimmer. 14. Induction coil. 15. Series circuits. 16. Parallel circuits.

Each project is assembled on a peg-board base. After the project has been built and tested, and has served the purpose for which it was intended, it is then disassembled, and the parts reused to make other projects.

Items in the Electrodemonstrator kit are listed below. This includes items like sockets, leads, bolts and nuts, which are purchased; and brackets, coils, etc. Constructional details on student-made parts are given in the drawings, Fig. A6-1.

THE ELECTRODEMONSTRATOR PARTS LIST

1 – Base, 1/8 x 7 1/2 x 10 Masonite peg-board
4 – Base feet, rubber 3/8
4 – Feet bolts, 3/8 x 8–32 round heat with nuts
1 – Base support, wood (See Plan, Detail M)
2 – Coils, (Plan, Detail F)
4 – Coil brackets, brass 1/32" (Plan, Detail G)
2 – Coil cores, 1/2 x 2 1/2 mild steel (Plan, Detail E)
1 – Solenoid core, mild steel (Plan, Detail H)
2 – Magnets, 1/2 x 2 1/2 round (Miami Magnet Co.)
1 – Armature bracket brass 1/32" (Plan, Detail J)
1 – dc armature (Plan, Detail A)
1 – ac armature (Plan, Detail B)
1 – Brush support (Plan, Detail L)
1 – Relay armature (Plan, Detail N)
1 – Armature plate (Plan, Detail K)
1 – Circuit breaker stop (Plan, Detail P)
2 – Contact brackets (Plan, Detail R)
1 – Induction coil (Plan, Detail D)

1 – .01 mfd. paper capacitor, Sprague 4TM-S10, Mount on subbases.
1 – Push button switch, Hickok teaching systems, No. 152-12-678
1 – Toggle switch, Hickok teaching systems, No. 152-12-628
3 – Subbase, Hickok teaching systems, No. 6-626
6 – Jiffy leads, 8", Hickok teaching systems, No. 6-517
3 – Miniature Sockets with bulbs 6V
1 – Door bell transformer
1 – 50 Potentiometer with subbase for potentiometer (2w), Hickok teaching systems, No. 152-12-406
1 – Terminal base
3 – Battery holders D cell, Hickok teaching systems, No. 152-12-605
1 – Chime, hard steel tube or bar
1 – Galvanometer, 500-0-500 μa in meter case, Hickok teaching systems No. 3-913
12 – Bolts, 3/8 x 8-32 round head
12 – Nuts, Hex. 8-32
4 – Stove bolts, 3/16 x 2 1/2
8 – Nuts, 3/16"

Construction Hints:

1. The Masonite peg-board base with attached rubber feet at each corner is used as a base.
2. The two coils F are interchangeable in all projects. The coil brackets are bolted to the Masonite base. The saw kerf at each end of coil fits snugly into bracket.
3. The permanent magnets are interchangeable with the iron cores, when used for magnetic field demonstrations or in the dc motor.
4. Induction coil D is made to slide over coil F. Wind with more than 8 layers of No. 30 wire if higher voltages are desired. Capacitor is connected across breaker points.
5. If difficulty is experienced with the ac motor, try starting it by winding a string around the armature shaft and pulling on it. Its speed must be synchronized with the 60-hertz line voltage for operation.
6. All motors are constructed by placing armatures between plate R, bolted to base, and armature bracket as the top bearing.
7. The brush support L is constructed so that it may be locked in several positions to demonstrate a change in commutation angle.

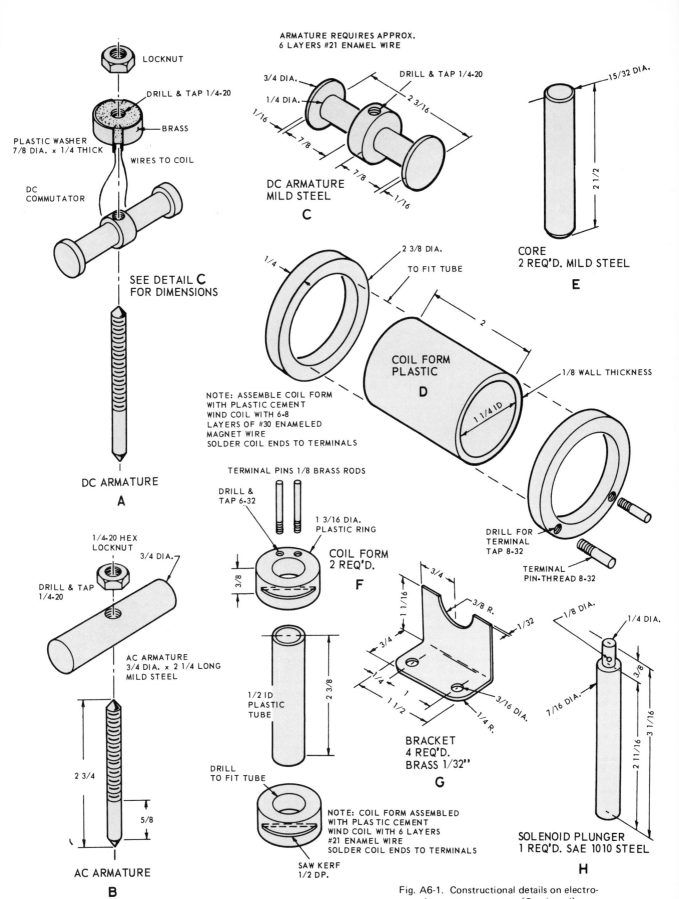

ARMATURE REQUIRES APPROX.
6 LAYERS #21 ENAMEL WIRE

LOCKNUT

DRILL & TAP 1/4-20

BRASS

PLASTIC WASHER
7/8 DIA. x 1/4 THICK

WIRES TO COIL

DC
COMMUTATOR

SEE DETAIL C
FOR DIMENSIONS

DC ARMATURE

A

3/4 DIA.
1/4 DIA.
1/16
7/8
7/8
1/16
2 3/16
DRILL & TAP 1/4-20

DC ARMATURE
MILD STEEL

C

15/32 DIA.
2 1/2

CORE
2 REQ'D. MILD STEEL

E

1/4
2 3/8 DIA.
TO FIT TUBE
2
COIL FORM
PLASTIC
D
1/8 WALL THICKNESS
1 1/4 ID

NOTE: ASSEMBLE COIL FORM
WITH PLASTIC CEMENT
WIND COIL WITH 6-8
LAYERS OF #30 ENAMELED
MAGNET WIRE
SOLDER COIL ENDS TO TERMINALS

DRILL FOR
TERMINAL
TAP 8-32

TERMINAL
PIN-THREAD 8-32

1/4-20 HEX
LOCKNUT
3/4 DIA.

DRILL & TAP
1/4-20

AC ARMATURE
3/4 DIA. x 2 1/4 LONG
MILD STEEL

2 3/4

5/8

AC ARMATURE

B

TERMINAL PINS 1/8 BRASS RODS

DRILL &
TAP 6-32

1 3/16 DIA.
PLASTIC RING

3/8

COIL FORM
2 REQ'D.

F

1/2 ID
PLASTIC
TUBE

2 3/8

DRILL
TO FIT TUBE

NOTE: COIL FORM ASSEMBLED
WITH PLASTIC CEMENT
WIND COIL WITH 6 LAYERS
#21 ENAMEL WIRE
SOLDER COIL ENDS TO TERMINALS

SAW KERF
1/2 DP.

3/4
1 11/16
3/8 R.
1/32
3/4
1/4
1
1 1/2
3/16 DIA.
1/4 R.

BRACKET
4 REQ'D.
BRASS 1/32"

G

1/8 DIA.
1/4 DIA.
3/8
7/16 DIA.
2 11/16
3 1/16

SOLENOID PLUNGER
1 REQ'D. SAE 1010 STEEL

H

Fig. A6-1. Constructional details on electrodemonstrator parts. (Continued)

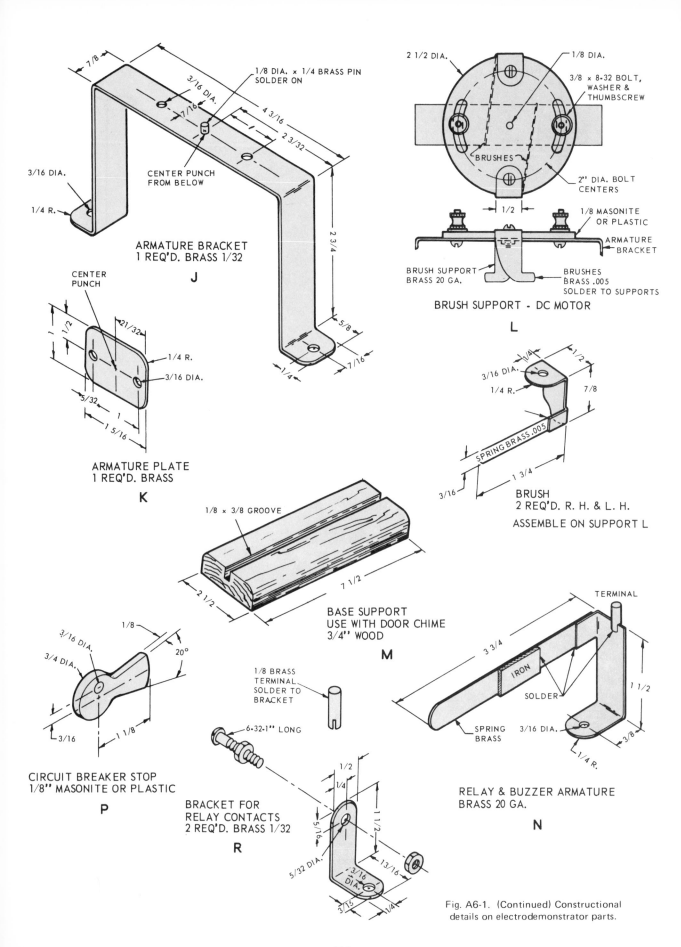

ARMATURE BRACKET
1 REQ'D. BRASS 1/32

J

1/8 DIA. x 1/4 BRASS PIN
SOLDER ON

CENTER PUNCH
FROM BELOW

7/8
3/16 DIA.
7/16
4 3/16
2 3/32
3/16 DIA.
1/4 R.
2 3/4
5/8
1/4
7/16

BRUSH SUPPORT - DC MOTOR

L

2 1/2 DIA.
1/8 DIA.
3/8 x 8-32 BOLT,
WASHER &
THUMBSCREW
BRUSHES
2" DIA. BOLT
CENTERS
1/2
1/8 MASONITE
OR PLASTIC
ARMATURE
BRACKET
BRUSH SUPPORT
BRASS 20 GA.
BRUSHES
BRASS .005
SOLDER TO SUPPORTS

ARMATURE PLATE
1 REQ'D. BRASS

K

CENTER
PUNCH
1/2
1
21/32
1/4 R.
3/16 DIA.
5/32
1
1 5/16

BRUSH
2 REQ'D. R. H. & L. H.
ASSEMBLE ON SUPPORT L

3/16 DIA.
1/4
1/2
1/4 R.
7/8
SPRING BRASS .005
1 3/4
3/16

BASE SUPPORT
USE WITH DOOR CHIME
3/4'' WOOD

M

1/8 x 3/8 GROOVE
2 1/2
7 1/2

CIRCUIT BREAKER STOP
1/8'' MASONITE OR PLASTIC

P

3/16 DIA.
1/8
20°
3/4 DIA.
3/16
1 1/8

BRACKET FOR
RELAY CONTACTS
2 REQ'D. BRASS 1/32

R

1/8 BRASS
TERMINAL
SOLDER TO
BRACKET

6-32-1'' LONG

1/2
1/4
5/16
1 1/2
5/32 DIA.
13/16
3/16
DIA.
3/16
1/4

RELAY & BUZZER ARMATURE
BRASS 20 GA.

N

TERMINAL
3 3/4
IRON
SOLDER
1 1/2
SPRING
BRASS
3/16 DIA.
3/8
1/4 R.

Fig. A6-1. (Continued) Constructional
details on electrodemonstrator parts.

316

DICTIONARY OF TERMS

A-BATTERY: Battery used to supply heater voltage for electron tubes.

AC: Alternating current.

ACCEPTOR CIRCUIT: Series tuned circuit at resonance. Accepts signal at resonant frequency.

ACCEPTOR IMPURITY: Impurity added to semiconductor material which creates holes for current carriers.

AC GENERATOR: Generator using slip rings and brushes to connect armature to external circuit. Output is alternating current.

AC PLATE RESISTANCE (symbol r_p): Variational characteristic of vacuum tube representing ratio of change of plate voltage to change in plate current, while grid voltage is constant.

AERIAL: An antenna.

AGC: Abbreviation for AUTOMATIC GAIN CONTROL. Circuit employed to vary gain or amplifier in proportion to input signal strength so output remains at constant level.

AGC: Automatic gain control.

AIR-CORE INDUCTOR: Inductor wound on insulated form without metallic core. Self-supporting coil without core.

ALNICO: Special alloy used to make small permanent magnets.

ALPHA: Greek letter $\propto$, representing current gain of a transistor. It is equal to change in collector current caused by change in emitter current for constant collector voltage.

ALPHA CUT-OFF FREQUENCY: Frequency at which current gain drops to .707 of its maximum gain.

ALTERNATOR: An ac generator.

ALTERNATING CURRENT (ac): Current of electrons that moves first in one direction and then in the other.

ALMALGAM: Compound or mixture containing mercury.

ALMALGAMATION: Process of adding small quantity of mercury to zinc during manufacture.

AM MODULATION, Types of: GRID. Audio signal is applied in series with the grid of the power amplifier. PLATE. Audio signal is injected into the plate circuit of the modulated stage.

AMMETER: Meter used to measure current.

AMPERE (Symbol I): Electron or current flow representing flow of one coulomb per second past given point in circuit.

AMPERE-HOUR: Capacity rating measurement of batteries. A 100 ampere-hour battery will produce, theoretically, 100 amperes for one hour.

AMPERE TURN (IN): Unit of measurement of magnetomotive force. Represents product of amperes times number of turns in coil of electromagnet. $F = 1.257$ IN.

AMPLIFICATION: Ability to control a relatively large force by a small force. In a vacuum tube, relatively small variation in grid input signal is accompanied by relatively large variation in output signal.

AMPLIFICATION FACTOR: Expressed as μ (mu). Characteristic of vacuum tube to amplify a voltage. Mu is equal to change in plate voltage as result of change in grid voltage while plate current is constant.

AMPLIFIERS: POWER. Electron tube used to increase power output. Sometimes called a current amplifier. VOLTAGE. An electron tube used to amplify a voltage.

Class A. An amplifier, so biased that plate current flows during entire cycle of input signal.

Class B. An amplifier, so biased that plate current flows for approximately one-half the cycle of input signal.

Class C. An amplifier, so biased that plate current flows for appreciably less than half of each cycle of applied signal.

Class AB. A compromise between class A and class B.

Class AB_1. Same as class AB, only grid is never driven positive and no grid current flows.

Class AB_2. Same as AB except that signal drives grid positive and grid current does flow.

DC AMPLIFIER. Directly coupled amplifiers. Amplifies without loss of dc component.

AF AMPLIFIER. Used to amplify audio frequencies.

IF AMPLIFIER. Used to amplify intermediate frequencies.

RF AMPLIFIER. Used to amplify radio frequencies.

VIDEO AMPLIFIER. Used to amplify video frequencies.

PULSE AMPLIFIER. Used to amplify pulses.

AMPLITUDE: Extreme range of varying quantity. Size, height of.

AMPLITUDE MODULATION (AM): Modulating a transmitter by varying strength of rf carrier at audio rate.

AND GATE: A logic gate that is used to determine the presence of yes signals of 1's.

ANGLE OF COMPENSATION: Correcting angle applied to compass reading to compensate for local magnetic influence.

ANGLE OF DECLINATION: See DECLINATION.

ANGULAR PHASE: Position of rotating vector in respect to reference line.

ANGULAR VELOCITY (ω): Speed of rotating vector in radians per second. ω (omega) = $2 \pi F$.

ANODE: Positive terminal, such as plate in electron tube.

ANTENNA: Device for radiating or receiving radio waves.

APERTURE MASK: A thin sheet of perforated material placed directly behind the viewing screen in a three-gun color picture tube.

APPARENT POWER: Power apparently used in circuit as product of current times voltage.

ARCBACK: Current flowing opposite direction in a diode, when plate has a high negative voltage.

ARMATURE: Revolving part in generator or motor. Vibrating or moving part of relay or buzzer.

ARMATURE REACTION: Effect on main field of generator by armature acting as electromagnet.

ARRL: American Radio Relay League. The association of radio amateurs.

ARTIFICIAL MAGNETS: Manufactured magnets.

ASPECT RATIO: Ratio of width to height of TV picture. This is standarized as 4:3.

A-SUPPLY: Voltages supplied for heater circuits of electron tubes.

AT-CUT CRYSTAL: Crystal cut at approximately a 35 deg. angle with Z axis.

ATOM: Smallest particle that makes up a type of material, called an element.

ATOMIC NUMBER: Number of protons in nucleus of a given atom.

ATOMIC WEIGHT: Mass of nucleus of atom in reference to oxygen, which has a weight of 16.

ATTENUATION: Decrease in amplitude or intensity.

ATTENUATOR: Networks of resistance used to reduce voltage, power or current to a load.

AUTOTRANSFORMER: Transformer with common primary and secondary winding. Step-up or step-down action is accomplished by taps on common winding.

AVC: Automatic volume control.

AVERAGE VALUE: Value of alternating current or voltage of sine wave form that is found by dividing area under one alternation by distance along X axis between 0 and 180 deg. E_{avg} = .637 E_{max}

AWG: American Wire Gage — used in sizing wire by numbers.

BAND: Group of adjacent frequencies in frequency spectrum.

BAND PASS FILTER: Filter circuit designed to pass currents of frequencies within continuous band and reject or attenuate frequencies above or below the band.

BAND REJECT FILTER: Filter circuit designed to reject currents in continuous band of frequencies, but pass frequencies above or below the band.

BAND SWITCHING: Receiver employing switch to change frequency range of reception.

BANDWIDTH: Band of frequencies allowed for transmitting modulated signal.

BARRIER REGION: Potential difference across a PN junction due to diffusion of electrons and holes across junction.

BASE: Semiconductor between emitter and collector of transistor.

BASS: Low frequency sounds in audio range.

BATTERY: Several voltaic cells connected in series or parallel. Usually contained in one case.

B-BATTERY: Group of series cells in one container producing high voltage for plate circuits of electronic devices. Popular batteries include 22 1/2, 45 and 90 volts.

BACK EMF: See COUNTER EMF.

BEAM POWER TUBE: Tube so constructed that electrons flow in concentrated beams from cathode through grids to plate.

BEAT FREQUENCY: The resultant frequency obtained by combining two frequencies.

BEAT FREQUENCY OSCILLATOR: Oscillator whose output is beat with continuous wave to produce beat frequency in audio range. Used in cw reception.

BEL: Unit of measurement of gain equivalent to 10 to 1 ratio of power gain.

BETA: Greek letter β , represents current gain of common-emitter connected transistor. It is equal to ratio of change in collector current to change in base current, while collector voltage is constant.

BIAS: CATHODE SELF-BIAS. Bias created by voltage drop across cathode resistor.

FIXED. Voltage supplied by fixed source.

GRID LEAK. Bias created by charging capacitor in grid circuit. Bias level is maintained by leak resistor.

BIAS, FORWARD: Connection of potential to produce current across PN junction. Source potential connected so it opposes potential hill and reduces it.

BIAS, REVERSE: Connection of potential so little or no current will flow across PN junction. Source potential has same polarity as potential hill and adds to it.

BINARY: Number system having base of 2, using only the symbols 0 and 1.

BLACK BOX: Box containing unknown and possibly complicated circuit.

BLANKING PULSE: Pulses transmitted by TV transmitter, which are used in receiver to cut off scanning beam during retrace time.

BLEEDER: Resistor connected across power supply to discharge filter capacitors.

BRIDGE CIRCUIT: Circuit with series-parallel groups of components that are connected by common bridge. Bridge frequently is meter in measuring devices.

BRIDGE RECTIFIER: Full-wave rectifier circuit employing four rectifiers in bridge configuration.

BRIGHTNESS: In television. Overall intensity of illumination of picture.

BRUSH: Sliding contact, usually carbon, between commutator and external circuit in dc generator.

B-SUPPLY: Voltages supplied for plate circuits of electron tubes.

BUFFER: Type of amplifier placed between oscillator and power amplifier to isolate from load.

BYPASS CAPACITOR: Fixed capacitor which bypasses unwanted ac to ground.

BX CABLE: Cable sheathed in metallic armor.

CABLE: May be stranded conductor or group of single conductors insulated from each other.

CAPACITANCE: Inherent property of electric circuit that opposes change in voltage. Property of circuit whereby energy may be stored in electrostatic field.

CAPACITIVE COUPLING: Coupling resulting from capacitive effect between components or elements in electron tube.

CAPACITIVE REACTANCE (X_c): Opposition to ac as a result of capacitance.

CAPACITOR: Device which possesses capacitance. Simple capacitor consists of two metal plates separated by insulator.

CAPACITOR INPUT FILTER: Filter employing capacitor as its input.

CAPACITOR MOTOR: Modified version of split-phase motor, employing capacitor in series with its starting winding, to produce phase displacement for starting.

CAPACITY: Ability of battery to produce current over a given length of time. Capacity of a battery is measured in ampere-hours.

CARRIER: Usually radio frequency continuous wave to which modulation is applied. Frequency of transmitting station.

CARRIER (in a semiconductor): Conducting hole of electron.

CASCADE: Arrangement of amplifiers where output of one stage becomes input of next, throughout series of stages.

CASCODE: Electron tubes connected so second tube acts as plate load for first. Used to obtain higher input resistance and retain low noise factor.

CATHODE: Emitter in electron tube.

CATHODE FOLLOWER: Single-stage Class A amplifier, output of which is taken from across unbypassed cathode resistor.

CATHODE RAY TUBE: Vacuum tube in which electrons emitted from cathode are shaped into narrow beam and accelerated to high velocity before striking phosphor coated viewing screen.

CAT WHISKER: Fine wire used to contact a crystal.

C-BATTERY: Battery used to supply grid bias voltages.

CENTER FREQUENCY: Frequency of transmitted carrier wave in FM when no modulation is applied.

CENTER TAP: Connection made to center of coil.

CHARACTERISTIC CURVE: Graphical representation of characteristics of component, circuit or device.

CHOKE INPUT FILTER: Filter employing choke as its input.

CIRCUIT BREAKER: Safety device which automatically opens circuit if overloaded.

CIRCULAR MIL: Cross-sectional area of conductor one mil in diameter.

CIRCULAR-MIL-FOOT: Unit conductor one foot long with cross-sectional area of one circular mil.

CIRCULATING CURRENT: Inductive and capacitive currents flowing around parallel circuit.

COAXIAL LINE: Concentric transmission line in which inner conductor is insulated from tubular outer conductor.

COEFFICIENT OF COUPLING (K): Percentage of coupling between coils, expressed as a decimal.

COLLECTOR: The electrode that collects electrons in a tube after they have performed their function.

COLPITTS OSCILLATOR: A basic type of oscillator that is characterized by tapped capacitors in the tank circuit.

COMMON BASE: Transistor circuit, in which base is common to input and output circuits.

COMMON COLLECTOR: Transistor circuit in which collector is common to input and output circuits.

COMMON EMITTER: Transistor circuit in which emitter is common to input and output circuits.

COMMUTATION: Process of reversing current in armature coils and conducting direct circuit to external circuit by means of commutator segments and brushes.

COMMUTATOR: Group of bars providing connections between armature coils and brushes. Mechanical switch to maintain current in one direction in external circuit.

COMPOUND GENERATORS, Degree of: FLAT. When no-load and full-load voltages have same value.

OVER. Full-load voltage is higher than no-load voltage.

UNDER. Full-load voltage is less than no-load voltage.

CONDENSER: Older name for a capacitor.

CONDUCTANCE (symbol G): Ability of circuit to conduct current. It is equal to amperes per volt and is measured in mhos. $G = \dfrac{1}{R}$.

CONDUCTION BAND: Outermost energy level of atom.

CONDUCTIVITY, N type: Conduction by electrons in N-type crystal.

CONDUCTIVITY, P type: Conduction by holes in a P-type crystal.

CONDUCTOR: Material which permits free motion of large number of electrons.

CONTINUOUS WAVE (cw): Uninterrupted sinusoidal rf wave radiated into space, with all wave peaks equal in amplitude and evenly spaced along time axis.

CONTRAST: In television. Relative difference in intensity between blacks and white in reproduced picture.

CONTROL GRID: Grid in vacuum tube closest to cathode. Grid to which input signal is fed to tube.

CONVERTER: Electromechanical system for changing alternating current to direct current.

CO-ORDINATES: Horizontal and vertical distances to locate point on graph.

COPPER LOSSES: Heat losses in motors, generators and transformers as result of resistance of wire. Sometimes called the I^2R loss.

COULOMB: Quantity of electrons representing approximately 6.28×10^{18} electrons.

COUNTER EMF (cemf): Voltage induced in conductor moving through magnetic field which opposes source voltage.

COUPLING: Percentage of mutual inductance between coils. LINKAGE.

COVALENT BOND: Atoms joined together, sharing each other's electrons to form stable molecule.

CROSSOVER FREQUENCY: Frequency in crossover network at which equal amount of energy is delivered to each of two loudspeakers.

CROSSOVER NETWORK: Network designed to divide audio frequencies into bands for distribution to loudspeakers.

CROSS NEUTRALIZATION: Method of neutralization used with push-pull amplifiers, where part of output from each tube is fed back to grid circuit of each opposite tube through capacitor.

CROSS TALK: Leakage from one audio line to another which produces objectional background noise.

CRYSTAL DIODE: Diode formed by small semiconductor crystal and cat whisker.

CRYSTAL LATTICE: Structure of material when outer electrons are joined in covalent bond.

C-SUPPLY: Voltages supplied for grid bias of electron tubes, usually negative voltage.

CURRENT (symbol I): Transfer of electrical energy in conductor by means of electrons moving constantly and changing positions in vibrating manner.

CUTOFF BIAS: Value of negative voltage applied to grid of tube which will cut off current flow through tube.

CW: Abbreviation for continuous wave.

CYCLE: Set of events occurring in sequence. One complete reversal of an alternating current from positive to negative and back to starting point.

DAMPER: Tube used in television set as half-wave rectifier to prevent oscillations in horizontal output transformer.

DAMPING: Gradual decrease in amplitude of oscillations in tuned circuit, due to energy dissipated in resistance.

D'ARSONVAL METER: Stationary-magnet moving coil meter.

DBM: Loss or gain in reference to arbitrary power level of one milliwatt.

DC: Direct current.

DC COMPONENT: A dc value of ac wave which has axis other than zero.

DC GENERATOR: Generator with connections to armature through a commutator. Output is direct current.

DECAY: Term used to express gradual decrease in values of current and voltage.

DECIBEL: One-tenth of a Bel.

DECLINATION: Angle between true north and magnetic north.

DECODER: Part of the communications block diagram that changes the coded message into an uncoded message.

DEFLECTION: Deviation from zero of needle in meter. Movement or bending of an electron beam.

DEFLECTION ANGLE: Maximum angle of deflection of electron beam in TV picture tube.

DEGENERATIVE FEEDBACK: Feedback 180 deg. out of phase with input signal so it subtracts from input.

DELTA-TYPE ELECTRON GUN: The electron gun arrangement where the guns are placed 120 deg. apart in a color picture tube.

DEMODULATION: Process of removing modulating signal intelligence from carrier wave in radio receiver.

DEPLETION LAYER (in a semiconductor): Region in which mobile carrier charge density is insufficient to neutralize net fixed charge of donors and acceptors. (IRE).

DEPOLARIZER: Chemical agent, rich in oxygen, introduced into cell to minimize polarization.

DETECTION: See DEMODULATION.

DIAPHRAGM: Thin disk, used in an earphone for producing sound.

DIELECTRIC: Insulating material between plates and capacitor.

DIELECTRIC CONSTANT: Numerical figure representing ability of dielectric or insulator to support electric flux. Dry air is assigned number 1.

DIELECTRIC FIELD OF FORCE: See ELECTRO-STATIC FIELD.

DIFFUSION: Movement of carriers across semiconductor junction in absence of external force.

DIGITAL METER: An electronic meter where the output is displayed in numerical readouts rather than on a meter.

DIGITAL INTEGRATED CIRCUIT: A switching type (on or off) integrated circuit.

DIODE: Two-element tube containing cathode and plate.

DIODE DETECTOR: Detector circuit utilizing unilateral conduction characteristics of diode.

DIRECT CURRENT (dc): Flow of electrons in one direction.

DISCRIMINATOR: A type of FM detector.

DISTORTION: The deviations in amplitude, phase and frequency between input and output signals of amplifier or system.

 AMPLITUDE. Distortion resulting from non-linear operation of electron tube when peaks of input signals are reduced or cut off by either excessive input signal or incorrect bias.

 FREQUENCY. Distortion resulting from signals of some frequencies being amplified more than others or when some frequencies are excluded.

 PHASE. Distortion resulting from shift of phase of some signal frequencies.

DOMAIN THEORY: Theory concerning magnetism, assuming that atomic magnets produced by movement of planetary electrons around nucleus have strong tendency to line up together in groups. These groups are called "domains."

DONOR IMPURITY: Impurity added to semiconductor material which causes negative electron carriers.

DOPING: Adding impurity to semiconductor material.

DRY CELL: Non-liquid cell, which is composed of zinc case, carbon positive electrode and paste of ground carbon, manganese dioxide and ammonium chloride as electrolyte.

DYNAMIC CHARACTERISTICS: Characteristics of tube describing the actual control of grid voltage over plate current when tube is operating as an amplifier.

DYNAMIC PLATE RESISTANCE: See AC PLATE RESISTANCE.

DYNAMIC SPEAKER: Loudspeaker which produces sound as result of reaction between fixed magnetic field and fluctuating field of voice coil.

DYNAMOMETER: Measuring instrument based on opposing torque developed between two sets of current carrying coils.

DYNAMOTOR: Motor-generator combination using two windings on single armature. Used to convert ac to dc.

EDISION CELL: Cell using positive electrodes of nickel oxide and negative electrodes of powdered iron. The electrolyte is dilute solution of sodium hydroxide.

EDISION EFFECT: Effect, first noticed by Thomas Edison, that emitted electrons were attracted to positive plate in vacuum tube.

EDDY CURRENT LOSS: Heat loss resulting from eddy current flowing through resistance of core.

EDDY CURRENTS: Induced current flowing in rotating core.

EFFECTIVE RESISTANCE: Ratio between true power absorbed in circuit to square of effective current flowing in circuit.

EFFECTIVE VALUE: That value of alternating current of sine wave form that has equivalent heating effect of a direct current. $(.707 \times E_{peak})$

EFFICIENCY: Ratio between output power and input power.

ELECTRIC FIELD OF FORCE: See ELECTRO-STATIC FIELD.

ELECTRODE: Elements in a cell.

ELECTRODYNAMIC SPEAKER: Dynamic speaker that uses electromagnetic fixed field.

ELECTROLYTE: Acid solution in a cell.

ELECTROLYTIC CAPACITOR: Capacitor with positive plate of aluminum and dry paste or liquid forms negative plate. Dielectric is thin coat of oxide on aluminum plate.

ELECTROMAGNET: Coil wound on soft iron core. When current runs through coil, core becomes magnetized.

ELECTROMOTIVE FORCE (emf): Force that causes free electrons to move in conductor. Unit of measurement is the volt.

ELECTRON: Negatively charged particle.

ELECTROSTRICTION: Piezoelectric property of some elements of undergoing changes in shape and size when voltage is applied and conversely, of producing voltage when subjected to pressure or stress.

ELECTRON TUBE: Highly evacuated metal or glass shell which encloses several elements.

ELECTROSTATIC FIELD: Space around charged body in which its influence is felt.

ELEMENT: One of the distinct kinds of substances which either singly or in combination with other elements, makes up all matter in the universe.

EMISSION: Escape of electrons from a surface.

EMISSION, COLD CATHODE: Phenomena of electrons leaving surface of a material caused by high potential field.

EMISSION, TYPES OF: A0 – Continuous wave, no modulation.

A1 – Continuous wave, keyed.

A2 – Telegraphy by keying modulating audio frequency.

A3 – Telephony.

A4 – Facsimile.

A5 – Televison.

F0 – Continuous wave, no FM.

F1 – Telegraphy by frequency shift keying.

F2 – Telegraphy by keying modulating audio frequency.

F3 – Telephone – FM.

F4 – Facsimile.

F5 – Television.

EMISSION, Types of: PHOTOELECTRIC. Emission of electrons as result of light striking surface of certain materials.

SECONDARY. Emission caused by impact of other electrons striking surface.

THERMIONIC. Process where heat produces energy for release of electrons from surface of emitter.

EMITTER: In transistor, semiconductor section, either P or N, which emits minority carriers.

EMITTER: Element in a vacuum tube from which electrons are emitted. The CATHODE.

ENCODER: Part of the communications block diagram that changes the information source into coded form.

ENERGY: That which is capable of producing work.

ENVELOPE: Enclosed wave form made by outlining peaks of modulated rf waves.

EXCITER: Small dc generator used to excite or energize field windings of large alternator.

FARAD: Unit of measurement of capacitance. A capacitor has a capacitance of one farad when charged of one coulomb raises its potential one volt.

$$C = \frac{Q}{E}$$

FEEDBACK: Transferring voltage from output of circuit back to its input.

FIELD MAGNETS: Electromagnets which make field of motor or generator.

FILAMENT: Heating element in vacuum tube coated with emitting material so it acts also as cathode.

FILTER: Circuit used to attenuate specific band or bands of frequencies.

FLUORESCENT: Property of a phosphor which indicates that radiated light will be extinguished when electron bombardment ceases.

FLUX DENSITY (symbol B): Number of lines of flux per cross-sectional area of magnetic circuit.

FREQUENCY: Number of complete cycles per second measured in hertz (Hz).

FREQUENCY BANDS: Abbreviations and ranges as follows.

vlf – Very low frequencies 10-30 KHz.

lf – Low frequencies 30-300 KHz.

mf – Medium frequencies 300-3000 KHz.

hf – High frequencies 3-30 MHz.

vhf – Very high frequencies 30-300 MHz.

uhf – Ultra high frequencies 300-3000 MHz.

shf – Super high frequencies 3000-30,000 MHz.

ehf – Extremely high frequencies 30,000-300,000 MHz.

FREQUENCY DEPARTURE: Instantaneous change from center frequency in FM as result of modulation.

FREQUENCY DOUBLER: Amplifier stage in which plate circuit is tuned to twice the frequency of grid tank circuit.

FREQUENCY METER: Meter used to measure frequency of an ac source.

FREQUENCY MODULATING (FM): Modulating transmitter by varying frequency of rf carrier wave at an audio rate.

FREQUENCY RESPONSE: Rating of device indicating its ability to operate over specified range of frequencies.

FREQUENCY TRIPLER: Amplifier stage in which plate circuit is tuned to three times the frequency (second harmonic) of grid circuit.

FULL-WAVE RECTIFIER: Rectifier circuit which produces a dc pulse output for each half-cycle of applied alternating current.

FUNDAMENTAL: A sine wave that has same frequency as complex periodic wave. Component tone of lowest pitch in complex tone. Reciprocal of period of wave.

FUSE: Safety protective device which opens an electric circuit if overloaded. Current above rating of fuse will melt fusible link and open circuit.

GAIN: Ratio of output ac voltage to input ac voltage.

GALVANOMETER: Meter which indicates very small amounts of current and voltage.

GAS FILLED TUBE: Tubes designed to contain specific gas in place of air, usually nitrogen, neon, argon or mercury vapor.

GAUSS: Measurement of flux density in lines per square centimetre.

GENERATOR: Rotating electric machine which provides a source of electrical energy. A generator converts mechanical energy to electric energy.

GENERATORS, Types of: COMPOUND. Uses both series and shunt windings.

INDEPENDENTLY EXCITED. Field windings are excited by separate dc source.

SERIES. Field windings are connected in series with armature and load.

SHUNT. Field windings are connected across armature in shunt with load.

GHOST: In TV. Duplicate image of reproduced picture, caused by multipath reception of reflected signals.

GILBERT: Unit of measurement of magnetomotive force. Represents force required to establish one maxwell in circuit with one Rel of reluctance.

GRID: Grid of fine wire placed between cathode and plate of an electron tube.

GRID BIAS: Voltage between the grid and cathode, usually negative.

GRID CURRENT: Current flowing in grid circuit of electron tube, when grid is driven positive.

GRID DIP METER: A test instrument for measuring resonant frequencies, detecting harmonics and checking relative field strength of signals.

GRID LEAK DETECTOR: Triode amplifier connected so it functions like a diode detector and an amplifier. Detection takes place in grid circuit.

GRID MODULATION: Modulation circuit where modulating signal is fed to grid of modulated stage.

GRID VOLTAGE: Bias or C voltage applied to grid of a vacuum tube.

GROUND: The common return circuit in electronic equipment whose potential is zero; a connection to earth by means of plates or rods.

HALF-WAVE RECTIFIER: Rectifier which permits one-half of an alternating current cycle to pass and rejects reverse current of remaining half-cycle. Its output is pulsating dc.

HARMONIC FREQUENCY: Frequency which is multiple of fundamental frequency. Example: If fundamental frequency is 1000 KHz, then second harmonic is 2 x 1000 KHz or 2000 KHz; third harmonic is 3 x 1000 KHz or 3000 KHz and so on.

HARTLEY OSCILLATOR: A basic type of oscillator that has a tapped oscillator coil.

HEATER: Resistance heating element used to heat cathode in vacuum tube.

HEAT SINK: Mass of metal used to carry heat away from component.

HENRY (H): Unit of measurement of inductance. A coil has one henry of inductance if an emf of one volt is induced when current through inductor is changing at rate of one ampere per second.

HERTZ (Hz): Basic unit for frequency. One hertz equals one cycle per second.

HETERODYNE: Process of combining two signals of different frequencies to obtain different frequency.

HOLE: Positive charge. A space left by removed electron.

HOLE INJECTION: Creation of holes in semiconductor material by removal of electrons by strong electric field around point contact.

HORIZONTAL POLARIZATION: An antenna positioned horizontally, so its electric field is parallel to earth's surface.

HORSEPOWER: 33,000 ft. lb. of work per minute or 550 ft. lb. of work per second equals one horsepower. Also 746 watts = 1 HP.

HUM: Form of distortion introduced in an amplifier as result of coupling to stray electromagnetic and electrostatic fields or insufficient filtering.

HYDROMETER: Bulb-type instrument used to measure specific gravity of a liquid.

HYSTERESIS: Property of magnetic substance that causes magnetization to lag behind force that produced it.

HYSTERESIS LOOP: Graph showing density of magnetic field as magnetizing force is varied uniformly through one cycle of alternating current.

HYSTERESIS LOSS: Energy loss in substance as molecule or domains move through cycle of magnetization. Loss due to molecular friction.

IMPEDANCE (Z): Total resistance to flow of an alternating current as a result of resistance and reactance.

INDIRECTLY HEATED: Electron tube employing separate heater for its cathode.

INDUCED CURRENT: Current that flows as result of

induced electromotive force.

INDUCED EMF: Voltage induced in conductor as it moves through magnetic field.

INDUCTANCE: Inherent property of electric circuit that opposes a change in current. Property of circuit whereby energy may be stored in magnetic field.

INDUCTION MOTOR: An ac motor operating on principle of rotating magnetic field produced by out of phase currents. Rotor has no electrical connections, but receives energy by transformer action from field windings. Motor torque is developed by interaction of rotor current and rotating field.

INDUCTIVE CIRCUIT: Circuit in which an appreciable emf is induced while current is changing.

INDUCTIVE REACTANCE (X_L): Opposition to an ac current as a result of inductance.

INSTANTANEOUS VALUE: Any value between zero and maximum depending upon instant selected.

INSULATION RESISTANCE: Resistance to current leakage through and over surface of insulating material.

INSULATORS: Substances containing very few free electrons and requiring large amounts of energy to break electrons loose from influence of nucleus.

INTEGRATED CIRCUIT: A packaged electronic circuit containing resistors, transistors, diodes, and capacitors with their interconnected leads. These are usually processed from a chip of silicon.

INTENSITY: Magnetizing force per unit length of magnetic circuit.

INTERELECTRODE CAPACITANCE: Capacitance between metal elements in an electron tube.

INTERLACE SCANNING: Process in television of scanning all odd lines and then all even lines to reproduce picture. Used in United States.

INTERNAL RESISTANCE: Refers to internal resistance of source of voltage or emf. A battery or generator has internal resistance which may be represented as a resistor in series with source.

INTERPOLES: Auxillary poles located midway between main poles of generator to establish flux or satisfactory commutation.

INTERRUPTED CONTINUOUS WAVE (icw): Continuous wave radiated by keying transmitter into long and short pulses of energy, (dashes and dots), conforming to code such as Morse Code.

INTERSTAGE: Existing between stages, such as an interstage transformer between two stages of amplifiers.

INTRINSIC SEMICONDUCTOR: Semiconductor with electrical characterisics similar to a pure crystal.

IONIZATION: An atom is said to be ionized when it has lost or gained one or more electrons.

IONIZATION POTENTIAL: Voltage applied to a gas-filled tube at which ionization occurs.

IONOSPHERE: Atmospheric layer from 40 to 350 miles above the earth, containing a high number of positive and negative ions.

IR DROP: See VOLTAGE DROP.

IRON VANE METER: Meter based on principle of repulsion between two concentric vanes placed inside a solenoid.

ISOLATION: Electrical separation between two locations.

JOULE: Unit of energy equal to one watt-second.

JUNCTION IDODE: PN junction, having unidirectional current characteristics.

JUNCTION TRANSISTOR: Transistor consisting of thin layer of N or P type crystal between P or N type crystals. Designated as NPN or PNP.

KEY: Manually-operated switch used to interrupt rf radiation of transmitter.

KEY CLICK FILTER: Filter in keying circuit of a transmitter to prevent surges of current and prevent sparking at key contacts.

KEYING: Process of causing cw transmitter to radiate an rf signal when key contacts are closed.

KEYING, SYSTEMS OF: CATHODE. Key is inserted in grid and cathode circuits of keyed stage.

GRID-BLOCK. Keying stage by applying high negative voltage on grid of tube.

PLATE. Key is inserted in plate circuit of stage to be keyed.

KILO: Prefix meaning one thousand times.

KILOGAUSS: One thousand gausses.

KILOWATT-HOUR (kWh): Means 1000 watts per hour. Common unit of measurement of electrical energy for home and industrial use. Power is priced by the kWh.

KIRCHHOFF'S CURRENT LAW: At any junction of conductors in a circuit, algebraic sum of currents is zero.

KIRCHHOFF'S LAW OF VOLTAGES: In simple circuit, algebraic sum of voltages around circuit is equal to zero.

LAGGING ANGLE: Angle current lags voltage in inductive circuit.

LAMBDA: Greek letter λ. Symbol for wavelength.

LAMINATIONS: Thin sheets of steel used in cores of transformers, motors and generators.

LAWS OF MAGNETISM: Like poles repel; unlike poles attract.

L/C RATIO: Ratio of inductance to capacitance.

LEAD ACID CELL: Secondary cell which uses lead peroxide and sponge lead for plates, and sulfuric acid and water for electrolyte.

LEADING ANGLE: Angle current leads voltage in capacitive circuit.

LECLANCHE CELL: Scientific name for common dry cell.

LEFT-HAND RULE: A method, using your left hand, to determine polarity of an electromagnetic field or direction of electron flow.

LENZ'S LAW: Induced emf in any circuit is always in such a direction as to oppose effect that produces it.

LIMITER: A stage or circuit that limits all signals at the same maximum amplitude.

LINEAR AMPLIFIER: An amplifier whose output is in exact proportion to its input.

LINEAR DETECTOR: Detector using linear portions of characteristic curve on both sides of knee. Output is proportional to input signal.

LINEAR DEVICE: Electronic device or component whose current-voltage relation is a straight line.

LINEAR INTEGRATED CIRCUITS: An amplifying (variable output) integrated circuit.

LINEARITY: Velocity of the scanning beam. It must be uniform for good linearity.

LINES OF FORCE: Graphic representation of electrostatic and magnetic fields showing direction and intensity.

LIQUID CRYSTAL DISPLAYS: A digital or alphanumeric display unit that can be used in visual outputs for information.

LOAD: Resistance connected across circuit which determines current flow and energy used.

LOADING A CIRCUIT: Effect of connecting voltmeter across circuit. Meter will draw current and effective resistance of circuit is lowered.

LOAD LINE: Line drawn on characteristic family of curves of electron tube, when used with specified load resistor, representing plate current at zero and maximum plate voltage.

LOCAL ACTION: Defect in voltaic cells caused by impurities in zinc, such as carbon, iron and lead. Impurities form many small internal cells which contribute nothing to external circuit. Zinc is wasted away, even when cell is not in use.

LOCAL OSCILLATOR: Oscillator in superheterodyne receiver, output of which is mixed with incoming signal to produce intermediate frequency.

LODESTONE: Natural magnet, so called a "leading stone" or lodestone because early navigators used it to determine directions.

LOUDSPEAKER: Device to convert electrical energy into sound energy.

L-SECTION FILTER: Filter consisting of capacitor and an inductor connected in an inverted L configuration.

MAGNET: Substance that has the property of magnetism.

MAGNETIC AMPLIFIER: Transformer type device employing a dc control winding. Control current produces more or less magnetic core saturation, thus varying output voltage of amplifier.

MAGNETIC CIRCUIT: Complete path through which magnetic lines of force may be established under influence of magnetizing force.

MAGNETIC FIELD: Imaginary lines along which magnetic force acts. These lines emanate from N pole and enter S pole, forming closed loops.

MAGNETIC FLUX (symbol Φ phi): Entire quantity of magnetic lines surrounding a magnet.

MAGNETIC LINE OF FORCE: Magnetic line along which compass needle aligns itself.

MAGNETIC MATERIALS: Materials such as iron, steel, nickel and cobalt which are attracted to magnet.

MAGNETIC PICKUP: Phono cartridge which produces an electrical output from armature in magnetic field. Armature is mechanically connected to reproducing stylus.

MAGNETIC SATURATION: This condition exists in magnetic material when further increase in magnetizing force produces very little increase in flux density. Saturation point has been reached.

MAGNETOMOTIVE FORCE (F) (mmf): Force that produces flux in magnetic circuit.

MAGNETIZATION: Graph produced by plotting intensity of magnetizing force on X axis and relative magnetism on Y axis.

MAGNETIZING CURRENT: Current used in transformer to produce transformer core flux.

MAGNET POLES: Point of maximum attraction on a magnet; designated as North and South poles.

MAJOR CARRIER: Conduction through semiconductor as result of majority of electrons or holes.

MATTER: Physical substance of common experience. Everything about us is made up of matter.

MAXIMUM POWER TRANSFER: This condition exists when resistance of load equals internal resistance of source.

MAXIMUM VALUE: Peak value of sine wave either in positive or negative direction.

MAXWELL: One single line of magnetic flux.

MEGA: Prefix meaning one million times.

MERCURY VAPOR RECTIFIER: Hot cathode diode tube which uses mercury vapor instead of high vacuum.

METALLIC RECTIFIER: Rectifier made of copper oxide, based on principle that electrons flow from copper to copper oxide but not from copper oxide to copper. It is unidirectional conductor.

MHO: Unit of measurement of conductance.

MICA CAPACITOR: Capacitor made of metal foil plates separated by sheets of mica.

MICRO: Prefix meaning one millionth of.

MICROFARAD (μF): One millionth of a farad.

MICROHENRY (μH): One millionth of a henry.

MICROMHO: One millionth of a mho.

MICROMICRO: Prefix for one millionth of one millionth of.

MICROMICROFARAD ($\mu\mu$F): One millionth of one millionth of a farad. Same as picofarad.

MICROPHONE: Energy converter that changes sound energy into corresponding electrical energy.

MICROSECOND: One millionth of a second.

MIL: One thousandth of an inch (.001 inches).

MIL-FOOT: A wire which is one mil in diameter and one foot long.

MILLI: Prefix meaning one thousandth of.

MILLIAMMETER: Meter which measures in milliammeter range of currents.

MILLIHENRY (mH): One thousandth of a henry.

MINOR CARRIER: Conduction through semiconductor opposite to major carrier. Example: If electrons are major carrier, then holes are minor carrier.

MINUS (symbol $-$): Negative terminal or junction of circuit.

MISMATCH: Incorrect matching of load to source.

MIXER: Multi-grid tube used to combine several input signals.

MODULATED CONTINUOUS WAVE (mcw): Carrier wave amplitude modulated by tone signal of constant frequency.

MODULATION: Process by which amplitude or frequency of sine wave voltage is made to vary according to variations of another voltage or current called modulation signal.

MOLECULE: Smallest division of matter. If further subdivision is made, matter will lose its identity.

MOTOR: Device which converts electrical energy into mechanical energy.

MOTOR REACTION: Opposing force to rotation developed in generator, created by load current.

MOTORS, Types of dc: COMPOUND. Uses both series and parallel field coils.

SERIES. Field coils are conencted in series with armature circuit.

SHUNT. Field coils are connected in parallel with armature circuit.

MU (μ): Greek letter representing the amplifcation factor of a tube.

MULTIELEMENT TUBE: Electron tube with more elements than cathode, plate and grid.

MULTIGRID TUBE: Special tube with 4, 5 or 6 grids.

MULTIMETER: Combination volt, ampere and ohm meter.

MULTIPLER: Resistance connected in series with meter movement to increase its voltage range.

MULTIUNIT TUBES: HEXODE. Six elements with four grids.

HEPTODE. Seven elements with five grids.

OCTODE. Eight elements with six grids.

TWIN DIODE. Two diodes in one envelope.

TWIN DIODE-TRIODE. Diode and triode in one envelope.

TWIN DIODE TETRODE. Diode and tetrode in one envelope.

TWIN PENTODE. Two pentodes in one envelope.

MULTIVIBRATORS, Type of: ASTABLE. A free-running multivibrator.

BISTABLE. A single trigger pulse switches conduction from one tube to the other.

CATHODE COUPLED. Both tubes have a common cathode resistor.

FREE RUNNING. Frequency of oscillation depending upon value of circuit components. Continuous oscillation.

MONOSTABLE. One trigger pulse is required to complete one cycle of operation.

ONE SHOT. Same as MONOSTABLE.

PLATE COUPLED. The plates of the tubes and grids are connected by RC networks.

MUTUAL INDUCTANCE (M): When two coils are located so that magnetic flux of one coil can link with turns of other coil, there is mutual inductance. The change in flux of one coil will cause an emf in other.

NAND GATE: A negative AND logic gate.

NATURAL MAGNET: Magnets found in natural state in form of mineral called magnetite.

NEGATIVE ION: Atom which has gained electrons and is negatively charged.

NETWORK: Two or more components connected in either series or parallel.

NEUTRALIZATION: Process of feeding back voltage from plate of amplifier to grid, 180 deg. out of phase, to prevent self-oscillation.

NEUTRON: Particle which is electrically neutral.

NICKEL CADMIUM CELL: Alkaline cell with paste electrolyte hermetically sealed. Used in aircraft.

NOISE: Any desired interference to a signal.

NO LOAD VOLTAGE: Terminal voltage of battery or supply when no current is flowing in external circuit.

NONLINEAR DEVICE: Electronic device or component whose current-voltage relation is not a straight line.

NOR GATE: A negative OR logic gate.

NOT GATE: An inverter that changes the polarity of

an incoming signal in the output.

NUCLEUS: Core of the atom.

OERSTED: Unit of magnetic intensity equal to one gilbert per centimetre.

OHM (symbol Ω): Unit of measurement of resistance.

OHMMETER: Meter used to measure resistance in ohms.

OHM'S LAW: Mathematical relationship between current, voltage and resistance discovered by George Simon Ohm.

$$I = \frac{E}{R} \qquad E = IR \qquad R = \frac{E}{I}$$

OHMS PER VOLT: Unit of measurement of sensitivity of a meter.

OPEN CIRCUIT: Circuit broken or load removed. Load resistance equals infinity.

OPERATIONAL AMPLIFIER: A type of a linear integrated circuit.

OR GATE: A logic gate that will provide an output signal if there is a signal on either of its inputs.

OSCILLATOR: An electron tube generator of alternating current voltages.

OSCILLATORS: Types of: ARMSTRONG. An oscillator using tickler coil for feedback.

COLPITTS. An oscillator using split tank capacitor as feedback circuit.

CRYSTAL-CONTROLLED. Oscillator controlled by piezoelectric effect.

ELECTRON COUPLED OSCILLATOR (ECO). Combination oscillator and power amplifier utilizing electron stream as coupling medium between grid and plate tank circuits.

HARTLEY. Oscillator using inductive coupling of tapped tank coil for feedback.

PUSH-PULL. Push-pull circuit utilizing interelectrode capacitance of each tube to feed back energy to grid circuit to sustain oscillations.

RC OSCILLATORS. Ocsillators depending upon charge and discharge of capacitor in series with resistance.

TRANSITRON. Oscillator utilizing negative transconductance.

TUNED-GRID TUNED-PLATE. Oscillator utilizing tuned circuits in both grid and plate circuits.

ULTRAUDION. Oscillator, similar to Colpitts, but employing grid-to-cathode and plate-to-cathode interelectrode capacitance for feedback.

OSCILLOSCOPE: Test instrument, using cathode ray tube, permitting observation of signal.

OVERMODULATION: Condition when modulating wave exceeds amplitude of continuous carrier wave, resulting in distortion.

PARALLEL CIRCUIT: Circuit which contains two or more paths for electrons supplied by common voltage source.

PARALLEL RESONANCE: Parallel circuit of an inductor and capacitor at frequency when inductive and capacitive reactances are equal. Current in capacitive branch is 180 deg. out of phase with inductive current and their vector sum is zero.

PARASITIC OSCILLATION: Oscillations in circuit resulting from circuit components or conditions, occurring at frequencies other than that desired.

PEAK: Maximum value of sine wave.

PEAK INVERSE VOLTAGE: Value of voltage applied in reverse direction across diode.

PEAK INVERSE VOLTAGE RATING: Inverse voltage diode will withstand without arcback.

PEAK REVERSE VOLTAGE: Same as peak inverse voltage.

PEAK TO PEAK: Measured value of sine wave from peak in positive direction to peak in negative direction.

PEAK VALUE: Maximum value of an alternating current or voltage.

PENTAGRID CONVERTER: Tube with five grids.

PENTAVALENT: Semiconductor impurity having five valence electrons. Donor impurities.

PENTODE: Electron tube with five elements including cathode, plate, control grid, screen grid and suppressor grid.

PERCENTAGE OF MODULATION: Maximum deviation from normal carrier value as result of modulation expressed as a percentage.

PERCENTAGE OF RIPPLE: Ratio of rms value of ripple voltage to average value of output voltage expressed as a percentage.

PERIOD: Time for one complete cycle.

PERMEANCE (P): Ability of a material to carry magnetic lines of force. The reciprocal of reluctance.

$$P = \frac{1}{R}$$

PERMEABILITY (symbol μ): Relative ability of substance to conduct magnetic lines of force as compared with air.

PERMANENT MAGNET: Bars of steel and other substances which have been permanently magnetized.

PHASE: Relationship between two vectors in respect to angular displacement.

PHASE INVERTER: Device or circuit that changes phase of a signal 180 deg.

PHASE SPLITTER: Amplifier which produces two waves that have exactly opposite polarities from single input wave form.

PHOTOSENSITIVE: Term used to describe the characteristic of a material which emits electrons from its surface when energized by light.

PHOTOTUBE: Vacuum tube employing photo sensitive material as its emitter or cathode.

PICOFARAD (pF): Same as micromicrofarad.

PICTURE ELEMENT: Small areas or dots of varying intensity from black to white which contain visual image of scene.

PIERCE OSCILLATOR: Crystal oscillator circuit in which crystal is placed between plate and grid circuit of tube.

PIEZOELECTRIC EFFECT: Property of certain crystalline substances of changing shape when an emf is impressed upon crystal. Action is also reversible.

PI-SECTION FILTER: Filter consisting of two capacitors and an inductor connected in a π configuration.

PITCH: Property of musical tone determined by its frequency.

PLATE: Anode of vacuum tube. Element in tube which attracts electrons.

PLATE DETECTOR: An rf signal is amplified and detected in plate circuit. Tube is biased to approximately cut-off by cathode resistor.

PLATE EFFICIENCY: Ratio between useful output power to dc input power to plate of electron tube.

PLATE MODULATION: Modulation circuit where modulating signal is fed to plate circuit of modulated stage.

PLUS (symbol +): Positive terminal or junction of circuit.

PM SPEAKER: Loudspeaker employing permanent magnet as its field.

PN JUNCTION: Piece of N-type and a piece of P-type semiconductor material joined together.

POINT CONTACT DIODE: Diode consisting of point and a semiconductor crystal.

POLARITY: Property of device or circuit to have poles such as North and South or positive and negative.

POLARIZATION: Defect in cell caused by hydrogen bubbles surrounding positive electrode and effectively insulating it from chemical reaction.

POLARIZATION: Producing magnetic poles or polarity.

POLES: Number of poles in motor or generator field.

POLYPHASE: Consisting of currents having two or more phases.

POSITIVE ION: Atom which has lost electrons and is positively charged.

POWER: Rate of doing work. In dc circuits P = I x E.

POWER AMPLIFICATION: Ratio of output power to input grid driving power.

POWER DETECTOR: Detector designed to handle signal voltages having amplitudes greater than one volt.

POWER FACTOR: Relationship between true power and apparent power of circuit.

POWER SUPPLY: Electronic circuit designed to provide various ac and dc voltages for equipment operation. Circuit may include transformers, rectifiers, filters and regulators.

PREAMPLIFIER: Sensitive low-level amplifier with sufficient output to drive standard amplifier.

PREEMPHASIS: Process of increasing strength of signals or higher frequencies in FM at transmitter to produce greater frequency swing.

PRIMARY CELL: Cell that cannot be recharged.

PRIMARY WINDING: Coil of transformer which receives energy from ac source.

PROTON: Positively charged particle.

PULSE: Sudden rise and fall of a voltage or current.

PUSH-PULL AMPLIFIER: Two tubes used to amplify signal in such a manner that each tube amplifies one half cycle of signal. Tubes operate 180 deg. out of phase.

Q: Letter representation for quantity of electricity (coulomb).

Q: Quality, figure of merit; ratio between energy stored in inductor during time magnetic field is being established to losses during same time. $Q = \dfrac{X_L}{R}$

QUANTA: Definite amount of energy required to move an electron to higher energy level.

QUIESCENT: At rest. Inactive.

RADIO FREQUENCY CHOKE (RFC): Coil which has high impedance to rf currents.

RADIO SPECTRUM: Division of electromagnetic spectrum used for radio.

RASTER: Area of light produced on screen of TV picture tube by electron beam. Contains no picture information.

RATIO DETECTOR: Type of FM detector.

REACTANCE (X): Opposition to alternating current as result of inductance or capacitance.

REACTIVE POWER: Power apparently used by reactive component of circuit.

RECIPROCAL: Reciprocal of number is one divided by the number.

RECTIFIER: Component or device used to convert ac into a pulsating dc.

REGENERATIVE FEEDBACK: Feedback in phase with input signal so it adds to input.

REGULATION: Voltage change that takes place in output of generator or power supply when load is changed.

REGULATION, PERCENTAGE OF: Percentage of change in voltage from no-load to full-load in respect to full-load voltage. Expressed as:

$$\frac{E_{no\ load} - E_{full\ load}}{E_{full\ load}} \times 100 = \%$$

REJECT CIRCUIT: Parallel tuned circuit at resonance. Rejects signals at resonant frequency.

REL: Unit of measurement of reluctance.

RELATIVE CONDUCTANCE: Relative conductance of material in reference to silver which is considered as 100 percent.

RELATIVE RESISTANCE: Numerical comparison of resistance of a material compared to silver which is assigned value 1.0.

RELAY: Magnetic switch.

RELAXATION OSCILLATOR: Non-sinusoidal oscillator whose frequency depends upon time required to charge or discharge capacitor through resistor.

RELUCTANCE: Resistance to flow of magnetic lines of force.

REMOTE CUT-OFF TUBE: Tube which gradually approaches its cut-off point at remote bias point, due to special grid construction.

REPULSION-START MOTOR: Motor which develops starting torque by interaction of rotor currents and single-phase stator field.

RESIDUAL MAGNETISM: Magnetism remaining in material after magnetizing force is removed.

RESISTANCE: Quality of electric circuit that opposes flow of current through it.

RESONANT FREQUENCY: Frequency at which tuned circuit oscillates. See TUNED CIRCUIT.

RETENTIVITY: Ability of material to retain magnetism after magnetizing force is removed.

RETMA: Radio Electronics Television Manufacturer's Association.

RETRACE: Process of returning scanning beam to starting point after one line is scanned.

REVERSE CURRENT CUTOUT: Relay which permits current to flow only in one direction.

RIPPLE VOLTAGE: An ac component of dc output of power supply due to insufficient filtering.

RMS VALUE: ROOT-MEAN-SQUARE value. The same as effectivie value. ($.707 \times E_{peak}$)

ROLL-OFF: Gradual attenuation with increase or decrease in frequency of signal.

ROTOR: Rotating part of an ac generator.

ROWLAND'S LAW: Low for magnetic circuits which states that number of lines of magnetic flux is in direct proportion to magnetomotive force and inversely proportional to reluctance of circuit.

$$\Phi = \frac{F}{R}$$

RUBEN CELL (RM cell): Mercury cell employing mercuric oxide and zinc. Electrolyte is potassium hydroxide.

RUMBLE: Low-frequency mechanical vibration of a turntable which is transmitted to recorded sound.

SATURATION CURRENT: Current through electron tube when saturation voltage is applied to plate.

SATURATION VOLTAGE: Voltage applied to plate of vacuum tube so all emitted electrons are attracted to plate.

SAWTOOTH GENERATOR: Electron tube oscillator producing sawtooth wave form.

SAWTOOTH WAVE: Wave shaped like teeth of saw.

SCAN: Process of sweeping electron beam across each element of picture in successive order, to reproduce total picture in television.

SCHEMATIC: Diagram of electronic circuit showing electrical connections and identification of various components.

SCREEN GRID: Second grid in electron tube between control grid and plate, to reduce interelectrode capacitance.

SECONDARY CELL: Cell that can be recharged by reversing chemical action with electric current.

SECONDARY EMISSION: Emission of electrons as result of electrons striking plate of electron tube.

SECONDARY WINDING: Coil which receives energy from primary winding by mutual induction and delivers energy to load.

SECOND HARMONIC DISTORTION: Distortion of wave by addition of its second harmonic.

SELECTIVITY: Relative ability of receiver to select desired signal while rejecting all others.

SELF-INDUCTANCE: Emf is self-induced when it is induced in conductor carrying current.

SEMICONDUCTOR: Conductor with resistivity somewhere in range between conductors and insulators.

SEMICONDUCTOR, N type: Semiconductor which uses electrons as major carrier.

SEMICONDUCTOR, P type: Semiconductor which uses holes as major carrier.

SENSITIVITY: Ability of circuit to respond to small signal voltages.

SENSITIVITY OF METER: Indication of loading effect of meter. Resistance of moving coil and multiplier divided by voltage for full scale deflection. Sensitivity equals one divided by current required for full scale deflection. Example: A 100 μ A meter movement has sensitivity of $\frac{1}{.0001}$ or 10,000 ohms/volt.

SERIES CIRCUIT: Circuit which contains only one possible path for electrons through circuit.

SERIES PARALLEL: Groups of series cells with output terminals connected in parallel.

SERIES RESONANCE: Series circuit of inductor, a capacitor and resistor at a frequency when inductive and capacitive reactances are equal and cancelling. Circuit appears as pure resistance and has minimum impedance.

SHADED POLE MOTOR: Motor in which each of its field poles is split to accomodate a short-circuit copper strap called a shading coil. This coil produces a sweeping movement of field across pole face for starting.

SHADOW MASK: Same as aperture mask.

SHIELD: Partition or enclosure around components in circuit to minimize effects of stray magnetic and radio frequency fields.

SHORT CIRCUIT: Direct connection across source which provides zero resistance path for current.

SHOT EFFECT: Noise produced in lectron tube as a result of variation in rate of electron emission from the cathode.

SHUNT: To connect across or parallel with circuit or component.

SHUNT: Parallel resistor to conduct excess current around meter moving coil. Shunts are used to increase range of meter.

SIDEBANDS: Frequencies above and below carrier frequency as result of modulation.

LOWER. Frequencies equal to difference between carrier and modulating frequencies.

UPPER. Frequencies equal to carrier plus modulating frequencies.

SIDE CARRIER FREQUENCIES: Waves of frequencies equal to sum and difference between carrier wave frequency and modulating wave frequency.

SILICON CONTROLLED RECTIFIER (SCR): A three junction device (anode, gate, and cathode) that is usually open until a signal on the gate switches it on.

SINE WAVE: A graphical representation of a wave whose strength is proportional to the sine of an angle that is a linear function of time or distance.

SINGLE-PHASE MOTOR: Motor which operates on single-phase alternating current.

SINUSOIDAL: Wave varying in proportion to sine of angle.

SKY WAVE: Waves emanating toward sky from radio antenna.

SLIP RINGS: Metal rings connected to rotating armature windings in generator. Brushes sliding on these rings provide connections for external circuit.

SOCKET: Device for holding lamp or electron tube.

SOFT TUBE: Gaseous tube.

SOLENOID: Coil of wire carrying electric current possessing characteristics of magnet.

SOURCE OF SUPPLY: The device attached to input of circuit which produces electromotive force. May be generator, battery or other device.

SPACE CHARGE: Cloud of electrons around cathode of an electron tube.

SPECIFIC GRAVITY: Weight of liquid in reference to water which is assigned value 1.0.

SPLIT-PHASE MOTOR: Single-phase induction motor, which develops starting torque by phase displacement between field windings.

SQUARE LAW DETECTOR: Detector whose output voltage is proportional to square of effective input voltage.

SQUIRREL CAGE ROTOR: Rotor used in an induction motor made of bars placed in slots of rotor core and all joined together at ends.

STAGE: Section of an electronic circuit, usually containing one lectron tube and associated components.

STANDING WAVE: Wave in which ratio of instantaneous value at one point to that at another point does not vary with time. Waves appearing on transmission line as a result of reflections from termination of line.

STANDING WAVE RATIO: Ratio of effective voltage at loop of standing wave to effective voltage at node. It may also be expressed effective current. Also ratio of characteristic impedance to load impedance.

STATIC CHARACTERISTICS: Characteristics of tube taken with constant plate voltage.

STATIC CHARGE: Charge on body either negatively or positively.

STATIC ELECTRICITY: Electricity at rest as opposed to electric current.

STATOR: Stationary coils of an ac generator.

STEADY STATE: Fixed nonvarying condition.

STORAGE BATTERY: Common name for lead-acid battery used in automotive equipment.

STRATOSPHERE: Atmosphere above troposphere in which temperature is constant and there is no cloud formation.

STYLUS: Phonograph needle or jewel, which follows grooves in a record.

SUBHARMONIC: Frequency below harmonic, usually fractional part of fundamental frequency.

SULFATION: Undesirable condition of lead-acid battery caused by leaving it in discharged condition or by improper care. Sulfates forming on plates make battery partially inactive.

SUPERSONIC: Frequencies above audio frequency range.

SURFACE ALLOY TRANSISTOR: Silicon junction transistor, in which aluminum electrode are deposited in shallow pits etched on both sides of thin silicon crystal, forming P regions.

SUPERHETERODYNE: Radio receiver in which incoming signal is converted to fixed intermediate frequency before detecting audio signal component.

SUPPRESSOR GRID: Third grid in electron tube, between screen grid and plate, to repel or suppress secondary electrons from plate.

SWEEP CIRCUIT: Periodic varying voltage applied to deflection circuits of cathode ray tube to move electron beam at linear rate.

SWITCH: Device for directing or controlling current flow in circuit.

SYNCHRO: Electromechanical device used to transmit angular position of shaft from one position to another without mechanical linkage.

SYNC PULSE: Abbreviation for synchronization pulse, used for triggering an oscillator or circuit.

SYNCHRONOUS: Having same period or frequency.

SYNCHRONOUS MOTOR: Type of ac motor which uses a separate dc source of power for its field. It runs at synchronous speed under varying load conditions.

SYNCHRONOUS VIBRATOR: Vibrator with additional contact points to switch output circuit so current is maintained in one direction through load.

TANK CIRCUIT: Parallel resonant circuit.

TAP: Connection made to coil at point other than its terminals.

TELEVISION: Method of transmitting and receiving visual scene by radio broadcasting.

TELEVISION CHANNEL: Allocation in frequency spectrum of 6 MHz assigned to each television station for transmission of picture and sound information.

TETRODE: Electron tube with four elements including cathode, plate, control grid and screen grid.

THERMAL RUNAWAY: In transistor, regenerative increase in collector current and junction temperature.

THERMISTOR: Semiconductor device which changes resistivity with change in temperature.

THERMOCOUPLE METER: Meter based on principle that if two dissimilar metals are welded together and junction is heated, a dc voltage will develop across open ends. Used for measuring radio frequency currents.

THETA (θ): Angle of rotation of vector representing selected instants at which sine wave is plotted. Angular displacement between two vectors.

THORIATED TUNGSTEN: Tungsten emitter coated with thin layer of thorium.

THREE-PHASE ALTERNATING CURRENT: Combination of three alternating currents have their voltages displaced by 120 deg. or one-third cycle.

THRESHOLD OF SOUND: Minimum sound at particular frequency which can be heard.

THYRATHRON: Gas-filled tube in which grid is used to control firing potential.

TICKLER: Coil used to feed back energy from output to input circuit.

TIME CONSTANT (RC): Time period required for the voltage of a capacitor in an RC circuit to increase to 63.2 percent of maximum value or decrease to 36.7 percent of maximum value.

TONE CONTROL: Adjustable filter network to emphasize either high or low frequencies in output of audio amplifer.

TRANSCONDUCTANCE (Symbol g_m): Grid plate transconductance of vacuum tube expressed as ratio of small change in plate current to small change in grid voltage while plate voltage is held constant. Measured in mhos.

TRANSDUCER: Device by which one form of energy may be converted to another form, such as electrical, mechanical or acoustical.

TRANSFER CHARACTERISTIC: Relation between input and output characteristics of device.

TRANSFORMER: Device which transfer energy from one circuit to another by electromagnetic induction.

TRANSFORMERS, Types. ISOLATION. Transformer with one-to-one turns ratio.

STEP-DOWN. Transformer with turns ratio greater than one. The output voltage is less than input voltage.

STEP-UP. Transformer with turns ratio of less than one. Output voltage is greater than input voltage.

TRANSIENT RESPONSE: Response to momentary signal or force.

TRANSISTOR: Semiconductor device derived from two words, transfer and resistor.

TRANSMISSION LINE: Wire or wires used to conduct or guide electrical energy.

TRANSMITTER: Device for converting intelligence into electrical impulses for transmission through lines or through space from radiating antenna.

TRF: Abbreviation for tuned radio frequency.

TRIAC: A full wave silicon switch.

TRIODE: Three-element vacuum tube, consisting of cathode, grid and plate.

TRIVALENT: Semiconductor impurity having three valence electrons. Acceptor impurity.

TROPOSPHERE: Lower part of atmosphere where clouds form and temperature decreases with altitude.

TRUE POWER: Actual power absorbed in circuit.

TUNED AMPLIFIER: Amplifier employing tuned circuits for input and/or output coupling.

TUNED CIRCUIT: Circuit containing capacitance, inductance and resistance in series or parallel, which when energized at specific frequency known as its resonant frequency, an interchange of energy occurs between coil and capacitor.

TURNS RATIO: Ratio of number of turns of primary winding of transformer to number of turns of secondary winding.

ULTRA HIGH FREQUENCY (UHF): Television frequencies that cover channels 14-83.

UNIJUCTION TRANSISTOR (UJT): A three terminal transistor that has an emitter and two bases.

UNITY COUPLING: If two coils are positioned so all lines of magnetic flux of one cell will cut across all turns of second coil, it is called UNITY COUPLING.

UNIVERSAL MOTOR: Series ac motor which operates also on dc. Fractional horsepower ac-dc motor.

UNIVERSAL TIME CONSTANT CHART: Graph with curves representing growth and decay of voltages and currents in RC and RL circuits.

VALVE: British name for vacuum tube.

VARIABLE MU TUBE: Tube with increased range of amplification due to its remote cutoff characteristics.

VECTOR: Straight line drawn to scale, showing direction and magnitude of a force.

VECTOR DIAGRAM: Diagram showing direction and magnitude of several forces, such as voltage and current, resistance, reactance and impedance.

VELOCITY FACTOR: Speed of propagation of signal along transmission line compared to speed of light.

VERTICAL POLARIZATION: Antenna positioned vertically so its electric field is perpendicular to earth's surface.

VIBRATOR: Magnetically operated interrupter, similar to buzzer, to change steady state dc to pulsating ac.

VIDEO SIGNAL: Electrical signal from studio camera used to modulate TV transmitter.

VOICE COIL: Small coil attached to speaker cone, to which signal is applied. Reaction between field of voice coil and fixed magnetic field causes mechanical movement of cone.

VOLT (symbol E in electricity, symbol V in semiconductor circuits): Unit of measurement of electromotive force or potential difference.

VOLTAGE DIVIDER: Tapped resistor or series resistors across source voltage to multiply voltages.

VOLTAGE DOUBLER: Rectifier circuit which produces double the input voltage.

VOLTAGE DROP: Voltage measured across resistor. Voltage drop is equal to product of current times resistance in ohms. $E = IR$.

VOLTAGE MULTIPLIER: Rectifier circuits which produce output voltage at multiple greater than input voltage, usually doubling, tripling or quadrupling.

VOLTAGE REGULATOR TUBE: Cold cathode gas-filled tube which maintains constant voltage drop independent of current, over its operating range.

VOLTAIC CELL: Cell produced by suspending two dissimilar elements in acid solution. Potential difference is developed by chemical action.

VOLTMETER: Meter used to measure voltage.

VOLT-OHM-METER: A portable multimeter.

VR TUBE: Gas-filled, cold cathode tube used for voltage regulation.

VTVM: Vacuum Tube Volt Meter.

VU: Number numerically equal to number of decibels above or below reference volume level. Zero vu represents power level of one milliwatt dissipated in 600 ohm load or voltage of .7746 volts.

WATT: Unit of measurement of power.

WATT-HOUR: Unit of energy measurement, equal to one watt per hour.

WATT-HOUR METER: Meter that shows instantaneous rate of power consumption of device or circuit.

WATTLESS POWER: Power not consumed in an ac circuit due to reactance.

WATTMETER: Meter used to measure power in watts.

WAVE LENGTH: Distance between point on loop of wave to corresponding point on adjacent wave.

WAVE METER: Meter to measure frequency of wave.

WAVE TRAP: Type of band reject filter.

WEAK-SIGNAL DETECTOR: Unit that detects signal voltages having amplitudes of less than one volt.

WHEATSTONE BRIDGE: Bridge circuit used for precision measurement of resistors.

WORK: When a force moves through a distance, work is done. Work in foot-pounds = Force x Distance.

WORKING VOLTAGE: Maximum voltage that can be steadily applied to capacitor without arc-over.

WOW: Low-frequency flutter resulting from variation in turntable speeds.

WWV and WWVH: National Bureau of Standards Radio Stations in Washington, DC and Hawaii respectively.

X AXIS: Horizontal axis of a graph.

X AXIS: Axis through corners of hexagonal crystal.

X-CUT CRYSTAL: Cut perpendicular to X axis.

YAGI ANTENNA: Dipole with two or more director elements.

Y AXIS: Vertical axis of graph.

Y AXIS: Axis drawn perpendicular to faces of hexagonal crystal.

Y-CUT CRYSTAL: Cut perpendicular to Y axis.

YOKE: Coils places around neck of TV picture tube for magnetic deflection of beam.

Z AXIS: Optical axis of crystal.

ZENER DIODE: Silicon diode which makes use of the breakdown properties of a PN junction. If a reverse voltage across the diode is progressively increased, a point will be reached when the current will greatly increase beyond its normal cutoff value. This voltage point is called the Zener voltage.

ZERO REFERENCE LEVEL: Power level selected as reference for computing gain of amplifier or system.

INDEX

A

Abbreviations, 313
AC meters, 152
AC-DC supply, 175
Activities and projects.
 See list on page 6.
AF amplifiers, television, 262
AM radio, 223
AM radio receiver, 235, 236
AM transmitter, 232, 233, 234
Acceptor circuit, 112, 113, 114
Alignment,
 radio receivers, 247, 248
Alternating current, 72, 73, 74
 generator, 74
Alternator, 74, 75, 76
American Wire Gage System, 28, 29
Ammeter, 144, 145, 146
Ampere, 13
Amplification factor, 180
Amplifier,
 paging, 207, 208
 telephone, 205, 206
 transistor, 198, 199
Amplifiers,
 cascaded, 199
 Class A, 192, 193
 Class B, 193, 194
 electron, 179
 electron tube, 179
 magnetic, 278, 279, 280
 push-pull, 194, 195
 semiconductor, 196
Amplitude modulation, 223, 224
 radio receiver, 235, 236
AND gates, 294
Antenna coil, 236, 237
Apparent power, 91, 105, 106
Applause meter, 154, 155
Armstrong oscillator, 210, 211
Atom, 7, 8
Atomic
 characteristics, 160
 structure, 8
Audio generator, 290
Automatic control for garage lights,
 107, 108, 109

Automatic gain control,
 television, 263
Automobile battery trickle charger,
 176, 177

B

Base, transistor, 196
Basic meter movement, 143, 144
Basic photo control circuits, 273
Basic television camera, 256, 257
Battery, 15, 16, 19, 20
 capacity, 22, 23
Beam power tube, 184
Binary numbering system, 292, 293
Black and white
 picture tube, 259, 260, 261
 television camera, 256, 257
 TV receiver, 261, 262, 263
Bleeder, load resistor, 171
Blocking capacitor, 202
Bridge rectifier, 168, 169
Broadcast band, 223
Bypassing, 121, 122

C

Camera, television, 256
Capacitance, 96, 97
 interelectrode, 182, 183
Capacitance and resistance in ac
 circuit, 106, 107
Capacitance in ac circuits, 104, 105
Capacitance in electrical circuits, 96
Capacitive reactance, 104, 105
Capacitor, 96, 97, 98, 99, 100
 input filter, 170
 start induction motor, 140, 141
 transient response, 100, 101
 types, 98, 99, 100
Carbon microphone, 222
Career
 clusters, 300, 301
 decisions, 304
 opportunities, 300
Careers, 300
 education, 303
 educational requirements, 304
 in electronic servicing, 301, 302
Cascaded amplifiers, 199
Cathode, 158

bias, 191
bias, degeneration, 192
Center tap transformer, 166
Ceramic disc capacitors, 99, 100
Charges, Law of, 9
Choke-input filter, 170
Circuit breaker, 58
Circuit configuration, transistor,
 197, 198
Circuit loading, 153
Circuits and power, 28
Citizens band radio, 226, 227
Class A amplifiers, 192, 193
Class B amplifiers, 193, 194
Code practice oscillator, 214, 215
Coil, induction, 87
Collector, transistor, 196
Color codes, 306
Color television, 263, 264, 265,
 266, 267, 268
Colpitts oscillator, 212
Combination circuits,
 series and parallel, 43, 44
Common base, transistor, 197
Common emitter, 198
Commutation and interpoles, 131, 132
Composite video signal, 258, 259
Compound dc motors, 133
Compound generators, 69
Conductance, 41
Conduction of electricity, 160, 161
Conductor, 12, 13
Constant speed motor, 132
Construction of a generator, 66, 67
Control grid, 179
Controls, photoelectric, 271, 272
Conversions, 307, 308
Cooling, thermoelectric, 281
Copper losses, 68, 86
Copper wire table, 30
Coulomb, 13
Counter emf, 131
Coupling, 82
 direct, 203, 204
 RC, 200, 201, 202, 203
CRT, black and white picture tube,
 259, 260, 261

Crystal
  diode, 163
  microphones, 222
  oscillators, 212, 213
  radio, 251, 252
Cumulative compound motor, 133
Current,
  alternating, 72, 73, 74
  electric, 12, 13
  flow, direction, 13
Cycle, 72, 209

D

DC restoration, television, 263
Damper, television, 262
Decibel, 205
Defects in primary cells, 16, 17
Degeneration by cathode bias, 192
Demodulation, 236
Detection, waves, 237
Dectector, diode, 237, 238, 239
Dictionary of terms, 317
Dielectric
  constant, 97
  field, 12
  heating, 278
Differential compound motor, 133
Digital integrated circuits, 292
Digital meters, 285
Diode, 159
  characteristics and ratings, 164
  crystal, 163
  detector, 237, 238, 239
  semiconductor, 162
  zener, 173
Direct coupling, 203, 204
Direction of current flow, 13
Discrimination, radio receivers, 249
Door bell and buzzer, 58
Dry cell, 17
Dynamic
  characteristics, electron tube, 180
  microphone, 222

E

Eddy-current losses, 86
Eddy currents, 68
Electric
  current, 12, 13
  current and magnetism, 52, 53
  motors, 127
  stoves, 45, 46
  shocker, 87
Electrical energy from
  heat, 25
  light, 23, 24
  mechanical energy, 64
  mechanical pressure, 25
Electricity
  from magnetism, 26
  sources of, 15
  static, 9, 10, 11
Electrodemonstrator, 46, 47, 314,
  315, 316
Electroluminescent
  night light, 216, 217

Electrolyte, 16
Electrolytics,
  can type, 99
  tubular, 99
Electromagnets, 54, 55
Electromotive force, 13
Electron amplifiers, 179
Electron tube
  amplifers, 179
  characteristics, 180, 181, 182
  how amplifies, 187, 188, 189
Electronic
  heating, 276
  oscillators, 209
  servicing, careers in, 301, 302
Electronics in industry, 271
Electronics, science of, 7
Electrons, 8, 160, 161
Electroscope, 9, 10
Electrostatic fields, 12
Emitter, transistor, 196

F

FET-VOM, 284
FM
  audio detector, television, 262
  detection, 248, 249
  radio, 226
  receiver, 248
Feedback circuit, 213
Field, dielectric, 12
  electrostatic, 12
Field windings, 127, 128, 129
Filter, 119, 122, 123, 169
  network, 169, 170
Filtering circuits, 119, 120, 121
Fixed bias, 190
Fixed paper capacitors, 98, 99
Floating ground, 176
Foster-Seeley discriminator, 249
Four transistor radio, 252, 253
Frequency, 72, 209
  distortion, 205
  meters, 289
  modulation, 226, 227, 228
  spectrum, 218
Full-wave rectifier, 167, 168

G

Garage lights, automatic control,
  107, 108, 109
Gates, logic, 294, 295, 296
Generators, 64
  construction of, 66, 67
  losses, 68
  types, 68, 96
  voltage and current regulation,
  70, 71
Germanium, 160
Grid bias voltage, 190, 191
Grid leak, bias voltage, 190
Ground, floating, 176
Ground waves, 231, 232

H

Half-wave rectification, 166, 167
Harmonic distortion, 205

Hartley oscillator, 211, 212
Headphones, 243, 246
Heat, electrical energy from, 25
Heating,
  electronic, 276
  dielectric, 278
  induction, 276, 277
High-pass filter, 122
High voltage regulator,
  television, 262
Horizontal AFC, television, 262
Horizontal oscillator,
  television, 262
How electron tube amplifies, 187,
  188, 189
How to use a meter, 153, 154
Hydrometer, 21
Hysteresis loss, 68, 86

I

Ignition system, 88
Impedance, 92
Independently excited field
  generator, 68
Induced current and voltage, 89, 90
Inductance, 78, 79
  and RL circuits, 78
  in ac circuits, 89, 91, 92, 93
  series and parallel, 89
Induction
  coil, 87
  heating, 276, 277
  motors, 138
Inductive reactance, 89
Industry, electronics, 271
Instruments and measurements, 143
Insulators, 13, 14
Integrated circuits,
  digital and linear, 292
Interelectrode capacitance, 182, 183
Interpoles, 131, 132
Ionization, 8, 9
Iron vane meter movement, 149, 150
Isolation, 86

J

Junction transistor, 196, 197

K

Kirchhoff's Laws, 39

L

L section filter, 170
Law of Charges, 9
Laws of Magnetism, 50
Lead acid cell, 20, 21
Left Hand Rule, 52, 53
Length of wire and resistance, 30
Lenz's Law, 65, 78
Lesson in safety, 7, 12, 16, 21, 22,
  28, 29, 30, 31, 34, 50, 51, 52,
  58, 87, 100, 166, 175, 260
Letter symbols, 310
Light, electrical energy from, 23, 24
Linear integrated circuits, 292, 297
Liquid crystal display, 154
Load resistor, 171, 172, 173
Loading a circuit, 153

Index

Loading the tank circuit, 119
Lodestones, 49
Logic gates, 294
Loudspeakers, 243, 244, 245
Low amperage power supply, 177, 178
Low-pass filter, 122

M

Magnetic
  amplifiers and reactors, 278, 279
  circuits, 49, 53, 54
  fields, 51
  flux, 52
  putting green, 60, 61, 62
  shields, 59
Magnetism, 49
  and electric current, 52, 53
  electricity from, 26
Mathematical symbols, 313
Matter, nature of, 7
Measurements and instruments, 143
Mechanical pressure,
    electrical energy from, 25
Mercury cell, 17
Meter, how to use, 153, 154
Meter movement, 143
Meters, digital, 285
Mica capacitor, 100
Microphones, 221, 222
Mixer, television, 262
Modulation, 222, 223, 224
  amplitude, 223, 224
  frequency, 226, 227, 228
  index, 229
  patterns, 224, 225
  percent of, 229
Molecule and the atom, 7, 8
Motor, 127
  controls, thyristor, 136, 137
  operation, 127, 128, 129, 130
  starting circuits, 134, 135
Multi-element tubes, 186
Multimeter, 148, 149
Mutual conductance, 181, 182
Mutual inductance, 82

N

NAND gates, 296
Narrow band FM, 229
Nature of matter, 7
Neutrons, 8
Nickel-cadmium rechargeable cell, 18
Noise limiting, radio receivers, 250
Nomograph, 124, 125
NOR gates, 296, 297
NOT gates, 295, 296

O

Ohmmeters, 148
Ohm's law, 31, 32, 33, 34, 35
  for ac circuits, 93, 94
  and the Power Law, 34, 35
Operational amplifier, 297, 299
OR gates, 295
Oscillator,
  Armstrong, 210, 211
  code practice, 214, 215

Hartley, 211, 212
Oscillators,
  Colpitts, 212
  crystal, 212, 213
  electronic, 209
  Pierce, 213
Oscilloscope, 285, 286
  familiarization, 286, 287

P

Paging amplifier, 207, 208
Parallel,
  cells, 19, 20
  circuits, 40, 41, 42
  and series circuits, 103
  and series combination
    circuits, 43, 44
Paraphase amplifier, 195, 196
Peak amplitude, 72
Pentagrid, 186
Pentode, 184
Period, 72, 209
Permeability, 54
Phase
  displacement, 74
  inverter, 195, 196
  relationship in transformer, 89
  splitter, 194, 195
Photo control circuits, 273
Photodiodes, 272
Photoelectric controls,
    24, 25, 271, 272
Photoemissive cells, 271
Phototubes, 271
Photovoltaic cells, 23, 272
Picture tube, black and white,
    259, 260, 261
Pierce oscillator, 213
Piezoelectric effect, 25, 213
Pi-section filter, 170
Plate resistance, 181
Polyphase motor, 139, 140
Potentiometer, 36, 37
Power, 33, 34
  and circuits, 28
  in a series circuit, 39
  Law, 34, 35
  oscillators and converters, 214
  reactive, 91, 105, 106
  supplies, 165, 177, 178
  transformers, 165
Precautions in use of transistors, 205
Primary cells, defects in, 16, 17
Primary winding, 82
Principles of motor operation,
    127, 128, 129, 130
Projects and activities.
    See list on page 6.
Protons, 8
Push button starter, 135
Push-pull amplifiers, 194, 195
Pyrometer, 25
Pythagorean theorem, 92

Q

Q of tuned circuits, 117, 118, 119

R

RC coupling, 200, 201, 202, 203
RC time constant, 101, 102
RCL networks, 111
Radio
  control, 273, 274, 275
  crystal, 251, 252
  frequency range, 223
  four transistor, 252, 253
  receivers, 235
  six transistor superheterodyne,
    253, 254
  transmitters, 218, 220
  wave, 229, 230
  wave propagation, 231
Ratio detector, radio receivers, 250
Reactance, 89
Reactive power, 91, 105, 106
Reactors, magnetic, 278, 279
Reasons for using transformer, 87
Receivers, radio, 235
Rectification, half-wave, 166, 167
Rectifier
  arrangements,
    series and parallel, 164
  bridge, 168, 169
  full-wave, 167, 168
Rectifiers, silicon, 163
Rectangular oil filled capacitors, 99
Reed relay, 57, 58
Reference Section, 305
RF
  amplification, 237
  amplifier, television, 261, 262
  generator, 290
Reject circuit, 115, 116, 117
Relay, 56
  reed, 57, 58
Reluctance, 53
Repulsion induction motor, 141
Resistance, 28
  in ac circuit, 91, 92, 93
  and capacitance in
    ac circuit, 106, 107
Resistor, load, 171, 172, 173
Resistors, 36, 37
Resonance, 111
Retentivity, 55
Ripple, 169
Rotor, 138
Rowland's Law, 53
RPM meter, 76

S

SCR, 136, 137
Safety, 7, 12, 16, 21, 22, 28, 29,
    30, 31, 34, 50, 51, 52, 58, 87,
    100, 166, 175, 260
Scanning, television, 257, 258
Schematic
  diagram, 28
  symbols, 311, 312
Science of electronics, 7
Scientific notation, 305
Screen grid, 183, 184

Secondary
  cells, 20
  winding, 82
Semiconductor
  amplifiers, 196
  diodes, 162
Semiconductors, 157, 159, 161
Series,
  cells, 19, 20
  circuit, power, 39
  circuits, 38, 39
  dc motor, 133
  generator, 69
  motor, 130
  parallel, cells, 20
Series and parallel
  combination circuits, 43, 44
  inductance, 89
  rectifier arrangements, 164
Shaded pole motor, 141, 142
Shunt
  dc motor, 132
  generator, 68, 69
  motor, 130
Sideband
  frequencies, 223, 224
  power, 225, 226
Signal generator, 287
Silicon
  controlled rectifier, 136, 137
  rectifiers, 163
Sine wave, 72, 73
Single-phase induction motor, 140
Six transistor superheterodyne
  radio, 253, 254
Sky waves, 232
Solar cells, 272
Solenoid, 53
  sucking coil, 55
Sources of electricity, 15
Special purpose circuits,
  television, 263
Speed regulation, 132
Squawker horn, 215, 216
Squirrel cage, 139
Static characteristics,
  electron tube, 180
Static electricity, 9, 10, 11
Stator, 138
Step-down transformer, 84
Step-up transformer, 84
Stranded wires, 31
Superheterodyne
  radio, 253, 254
  receiver, 236, 239,
    240, 241, 242, 243
Suppressor grid, 184
Sweep generator, 289
Switch, 28
Symbols, 310, 311, 312, 313

Sync
  amplifier, television, 262
  separator, television, 262

T

Tachometer, 76
Tank circuit, 114, 115, 119
Telephone amplifier, 205, 206
Television, 256
  camera, 256, 257
  channel, 268, 269
  color, 263, 264, 265, 266, 267, 268
  composite video signal, 258, 259
  of the future, 269
  receiver, black and white,
    261, 262, 263
  scanning, 257, 258
  special purpose circuits, 263
Temperature and resistance, 30, 31
Test instruments, 282
Tetrode, 183
Thermionic
  emission, 157, 271
  emitters, 157
Thermistor, 37, 38
Thermocouple, 25
Thermoelectric cooling, 281
Third Law of Magnetism, 52
Three-phase induction motor, 139
Thyristor motor controls, 136, 137
Time control circuits, 273
Tone control,
  frequency distortion, 205
  radio receivers, 246
Transconductance, 181
Transducers, 243, 244, 245, 246
Transformer, 82
  coupling, 199, 200
  losses, 86
  phase relationship, 89
  power, 165
  reasons for using, 87
Transient responses, 79, 80, 81,
  100, 101
Transistor, 159
  amplifier, 198, 199
  circuit configuration, 197, 198
  junction, 196, 197
  precautions, 205
  radio, 252, 253
  testers, 288
Transistorized
  telephone amplifier, 205, 206
  transmitters, 226
Transmitter, AM, 232, 233, 234
Transmitters,
  radio, 218, 220
  transistorized, 226
Triac motor control, 136
Trickle charger,
  automobile battery, 176, 177

Trigonometry, 308, 309, 310
Trimmer capacitor, 100
Triode, 179, 180
True power, 91, 105
Truth tables, 295
Tube,
  beam power, 184
  checkers, 288
  pentode, 184
  multi-element, 186
  tetrode, 183
Tune-radio-frequency receiver, 236
Tuned circuit filters, 123
Tuned circuits,
  RCL networks, 111
Tuning circuit, 236
Turns ratio, 83, 84, 85
Tweeters, 244
12 volt automobile battery trickle
  charger, 176, 177

U

Ultrasonic cleaning, 280
Unity coupling, 82
Universal motor, 137

V

Vacuum tube voltmeter, 282
Vacuum tubes, 157
Variable capacitors, 98
Velocity microphone, 222
Vertical
  oscillator, television, 263
  output, television, 263
Vestigial sideband filter, 224
Video
  amplifier, television, 262
  detector, television, 262
Voltage, 13
  and current regulation, 70, 71
  divider, 172
  doublers, 174, 175
  drop, 38, 39
  regulation, 171
  regulators, 174
Voltaic cell, 16
Voltmeter, 146, 147
  sensitivity, 147, 148
Volt-Ohm-Meter, 283, 284
VOM, 281, 284
VTVM, 280, 283

W

Walkie-talkie, 226
Watt-hour meter, 151, 152
Wattless power, 105
Wattmeter, 151
Wave length, 230
Wheatstone bridge, 149
Wise Owl project, 102, 103
Woofers, 244

Z

Zener diode, 173